T0211204

Lecture Notes in Computer Science 13885

Founding Editors

Gerhard Goos
Juris Hartmanis

Editorial Board Members

The series Lecture Notes in Computer Science (LNCS), including its subseries Lecture Notes in Artificial Intelligence (LNAI) and Lecture Notes in Bioinformatics (LNBI), has established itself as a medium for the publication of new developments in computer science and information technology research, teaching, and education.

LNCS enjoys close cooperation with the computer science R & D community, the series counts many renowned academics among its volume editors and paper authors, and collaborates with prestigious societies. Its mission is to serve this international community by providing an invaluable service, mainly focused on the publication of conference and workshop proceedings and postproceedings. LNCS commenced publication in 1973.

Rikke Gade · Michael Felsberg ·
Joni-Kristian Kämäräinen
Editors

Image Analysis

22nd Scandinavian Conference, SCIA 2023
Sirkka, Finland, April 18–21, 2023
Proceedings, Part I

Springer

Editors
Rikke Gade 🆔
Aalborg University
Aalborg, Denmark

Michael Felsberg 🆔
Linköping University
Linköping, Sweden

Joni-Kristian Kämäräinen 🆔
Tampere University
Tampere, Finland

ISSN 0302-9743   ISSN 1611-3349 (electronic)
Lecture Notes in Computer Science
ISBN 978-3-031-31434-6   ISBN 978-3-031-31435-3 (eBook)
https://doi.org/10.1007/978-3-031-31435-3

This Springer imprint is published by the registered company Springer Nature Switzerland AG
The registered company address is: Gewerbestrasse 11, 6330 Cham, Switzerland

Preface

This book constitutes the refereed proceedings of the 22nd Scandinavian Conference on Image Analysis, SCIA 2023, held in Levi Ski Resort, Kittilä (Lapland, Finland), April 18–21, 2023. SCIA 2023 was scheduled to be held in 2021, but postponed due to the COVID-19 outbreak.

SCIA 2023 received 31 submissions during the first round and 77 during the second (108 submissions in total). Nineteen first-round submissions were accepted (61% acceptance rate), and 48 submissions from the second round (62%). The first-round submissions were reviewed by two, and the second round by two to three reviewers. The program chairs made the final decision on each paper based on the review grades and comments. Revisions were requested for a small number of the first round submissions, and final grades were decided after the second round. The number of submissions increased 71% from the previous SCIA in 2019, and the final SCIA 2023 acceptance rate was 62%.

The second-round submissions were assigned to three reviewers. One submission was desk rejected, 54 submissions received three reviews, and 22 received two reviews. A small number of emergency reviewers was used to have at least two reviews for every paper. The Microsoft CMT3 system was used to manage the reviewing process, and the reviews were double-blind.

The most popular topics of the accepted papers are the following:

- Machine learning and deep learning (16 papers),
- Image and video processing, analysis, and understanding (9),
- Detection, recognition, and localization in 2D and/or 3D (9),
- Datasets and evaluation (8),
- 3D Vision from multiview and other sensors (7),
- Vision for robotics and autonomous vehicles (5),
- Segmentation, grouping, and shape (4),
- Biometrics, faces, body gestures, and pose (4),
- Vision applications and systems (3), and
- Action and behavior recognition (2).

97 of the submitted papers were from European universities, companies, and research institutions, and the rest from North America and Asia. All papers were physically presented in the conference, except for a small number of papers whose authors had insuperable problems to obtain a visa or travel to Finland.

The scientific program consisted of the accepted papers and three invited talks. The invited presentations were given by Timo Aila from Nvidia, Martin Danelljan from ETH Zurich, and Ondrej Dusek from Charles University. In addition, Niki Loppi and the Nvidia team provided a full-day tutorial about GPU programming, which was a success.

The two main sponsors of SCIA 2023 were Huawei and Nvidia, and the organizers would like to thank them for their generous support.

March 2023

Rikke Gade
Michael Felsberg
Joni-Kristian Kämäräinen

The original version of the book was revised: the book cover and front matter were revised, the conference edition number has been corrected as 22. The correction to the book is available at https://doi.org/10.1007/978-3-031-31435-3_29

Organization

General Chair

Jiri Matas
    Czech Technical University in Prague, Czech Republic

Program Chairs

Rikke Gade
    Aalborg University, Denmark
Michael Felsberg
    Linköping University, Sweden[1]
Joni-Kristian Kämäräinen
    Tampere University, Finland

Publicity Chair

Guoying Zhao
    University of Oulu, Finland

Local Organizing Committee

Esa Rahtu
    Tampere University, Finland
Nataliya Strokina
    Tampere University, Finland
Juho Kannala
    Aalto University, Finland

Program Committee

Aasa Feragen
    Technical University of Denmark, Denmark
Aleksei Tiulpin
    University of Oulu, Finland
Anders Skaarup Johansen
    Aalborg University, Denmark
Andreas Aakerberg
    Aalborg University, Denmark
Andreas Møgelmose
    Aalborg University, Denmark
Andreas Robinson
    Linköping University, Sweden
Axel Carlier
    IRIT, University of Toulouse, France
Axel Flinth
    Umeå University, Sweden
Bastian Wandt
    Linköping University, Sweden

---

[1] Also honorary professor at University of KwaZulu-Natal, South Africa.

Carl Olsson	Lund University, Sweden
Christopher Zach	Chalmers University of Technology, Sweden
Constantino Álvarez Casado	University of Oulu, Finland
Dimitri Bulatov	Fraunhofer IOSB, Germany
Dirk Kraft	University of Southern Denmark, Denmark
Domenico Bloisi	University of Basilicata, Italy
Fredrik Kahl	Chalmers University of Technology, Sweden
Gunilla Borgefors	Uppsala University, Sweden
Heikki Kälviäinen	LUT University, Finland
Helene Schulerud	SINTEF, Norway
Ida-Maria Sintorn	Vironova AB & Uppsala University, Sweden
Jan Flusser	UTIA, Czech Academy of Sciences, Czech Republic
Janne Heikkila	University of Oulu, Finland
Janne Mustaniemi	University of Oulu, Finland
Joakim Haurum	Aalborg University, Denmark
Johan Edstedt	Linköping University, Sweden
Jon Sporring	University of Copenhagen, Denmark
Jorma Laaksonen	Aalto University, Finland
Juan Lagos	Tampere University, Finland
Jukka Peltomäki	Tampere University, Finland
Jussi Tohka	University of Eastern Finland, Finland
Karl Åström	Lund University, Sweden
Ke Chen	South China University of Technology, China
Kjersti Engan	University of Stavanger, Norway
Lasse Lensu	LUT University, Finland
Lauri Suomela	Tampere University, Finland
Lazaros Nalpantidis	Technical University of Denmark, Denmark
Le Nguyen	University of Oulu, Finland
Maciej Plocharski	Aalborg University, Denmark
Magnus Oskarsson	Lund University, Sweden
Malte Pedersen	Aalborg University, Denmark
Marike van den Broek	Aalborg University, Denmark
Marius Pedersen	NTNU Gjøvik, Norway
Mark Philip Philipsen	Aalborg University, Denmark
Mårten Wadenbäck	Linköping University, Sweden
Mia Siemon	Milestone Systems & Aalborg University, Denmark
Miguel Bordallo Lopez	University of Oulu, Finland
Nataliya Strokina	Tampere University, Finland
Neelu Madan	Aalborg University, Denmark

Contents – Part I

Detection, Recognition, Classification, and Localization in 2D and/or 3D

Contents – Part II

Segmentation, Grouping, and Shape

Vision for Robotics and Autonomous Vehicles

Biometrics, Faces, Body Gestures and Pose

3D Vision from Multiview and Other Sensors

Vision Applications and Systems

Datasets and Evaluation

LiDAR Place Recognition Evaluation with the Oxford Radar RobotCar Dataset Revised

Jukka Peltomäki[1]([⊠])(iD), Farid Alijani[1](iD), Jussi Puura[2](iD), Heikki Huttunen[3](iD), Esa Rahtu[1](iD), and Joni-Kristian Kämäräinen[1](iD)

[1] Tampere University, Tampere, Finland
`jukka.peltomaki@tuni.fi`
[2] Sandvik Ltd, Stockholm, Sweden
[3] Visy Ltd, Tampere, Finland
`http://research.tuni.fi/vision`

Abstract. The Oxford Radar RobotCar dataset has recently become popular in evaluating LiDAR-based methods for place recognition. The Radar dataset is preferred over the original Oxford RobotCar dataset since it has better LiDAR sensors and location ground truth is available for all sequences. However, it turns out that the Radar dataset has serious issues with its ground truth and therefore experimental findings with this dataset can be misleading. We demonstrate how easily this can happen, by varying only the gallery sequence and keeping the training and test sequences fixed. Results of this experiment strongly indicate that the gallery selection is an important consideration for place recognition. However, the finding is a mistake and the difference between galleries can be explained by systematic errors in the ground truth. In this work, we propose a revised benchmark for LiDAR-based place recognition with the Oxford Radar RobotCar dataset. The benchmark includes fixed gallery, training and test sequences, corrected ground truth, and a strong baseline method. All data and code will be made publicly available to facilitate fair method comparison and development.

Keywords: place recognition · ground truth position accuracy · Oxford Radar RobotCar dataset · evaluation protocol

1 Introduction

Autonomous navigation in a known environment is a core skill of autonomous mobile robots and vehicles. The navigation skill requires that the robot is able to recognize different places on the map when it revisits them repeatedly over time. In the simplest form the map is stored as images or depth maps associated with location tags. Observed images, *queries*, are matched to the map, *gallery*, and the location of the best match is retrieved as the location estimate. This vision problem is known as *visual place recognition*. Mainstream methods and approaches can be found from the recent surveys of Lowry et al. [23] and Zhang et al. [46], focusing on the conventional and deep approaches, respectively. In robotics, the

R. Gade et al. (Eds.): SCIA 2023, LNCS 13885, pp. 3–16, 2023.
https://doi.org/10.1007/978-3-031-31435-3_1

depth sensors, such as stereo, ToF, LiDAR and RADAR, are popular as they are more robust to visual changes such as illumination variation (time-of-day) and weather conditions. In this work we focus on LiDAR-based place recognition.

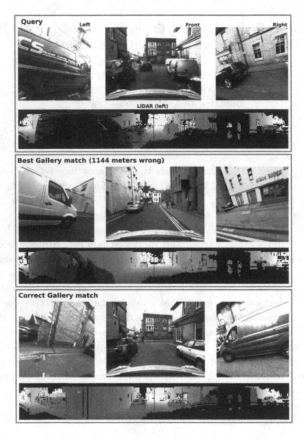

Fig. 1. Example problem that inspired our original work: dynamic objects clutter gallery LiDAR depth maps and therefore gallery sequence selection can affect place recognition performance. For example, the white van in the best gallery match matches the LiDAR shape of the black van in the query. Our original hypothesis was verified by the experimental results with the Oxford Radar RobotCar dataset, but this finding was found to be false. This paper explains why and how to fix the problem.

LiDAR-based place recognition does not strongly differ from visual place recognition since the observed depth maps can be used similar to grayscale images. The training and testing procedures of state-of-the-art deep methods, such as NetVLAD [2] and GeM [30], are also the same for depth maps [27]. The main bottleneck for LiDAR-based methods is the lack of suitable training and test data. The Oxford RobotCar dataset [24] is a large scale dataset suitable for place recognition and it has been widely used in visual place recognition.

The dataset challenges vision as the same 9 km route in downtown Oxford is travelled in various weather conditions and various times of day over the time period of one year. The early LiDAR place recognition works also used the older Oxford RobotCar dataset [19,21,41], but it has the following shortcomings for LiDAR place recognition: **i)** its LiDAR data is of poor quality due to outdated sensors, and **ii)** location ground truth is inaccurate or completely lacking in many sequences. To fix these issues the Oxford research group recently collected and published a smaller but more LiDAR-friendly *Oxford Radar Robot-Car* dataset [4]. The Oxford Radar RobotCar dataset has recently become more popular in LiDAR place recognition works [20,27,38,44].

It is noteworthy that the works using the Oxford Radar RobotCar dataset in LiDAR place recognition evaluation use only a small number of varying sequences [20,27,38,44]. If the gallery, training or test set sequences are different, then the numbers in different papers are not anymore comparable. The researchers would benefit from a unified evaluation protocol. In addition, it turns out that the Oxford Radar RobotCar dataset location ground truth needs to be carefully curated as the errors may lead to misleading findings. Our work provides the following contributions that help to run LiDAR-based place recognition evaluations on the Oxford Radar RobotCar dataset: **1)** we perform a well justified experiment that however provides a misleading finding that the gallery sequence has strong impact on the place recognition performance; **2)** we demonstrate that the misleading finding is due to errors in the Radar dataset location ground truth; and **3)** we propose a fix to the ground truth. As the main result **4)** we propose a unified benchmark and evaluation protocol for LiDAR-based place recognition with the Oxford Radar RobotCar dataset. The benchmark includes fixed gallery, training and test sequences, correct and curated ground truth locations, and a strong baseline using the state-of-the-art metric learning (GeM) of Radenovic et al. [30]. All data and code will be made publicly available to facilitate fair comparisons and method development.

2 Related Work

Visual place recognition has traditionally been done by extracting handcrafted local features from images and by combining them into global feature vectors. Local features such as SIFT [22], SURF [6], and HOG [11] have been used. Global vectors have been based on the visual bag-of-features model, for example DBoW [13] and FAB-MAP [10]. Conventional place recognition methods have been surveyed by Lowry in 2016 [23].

More recently, deep learning has been incorporated for learned features, which outperform the conventional features. A convolutional deep learning backbone network is often ImageNet [33] pre-trained, and can be fine-tuned for place recognition. Local features such as DELF [25] are usually more accurate but slower to process than global features such as GeM [30] and NetVLAD [2]. Systems such as DELG [8] combine the local and global features for better performance. Deep place recognition methods have been surveyed by Zhang et al. [46].

Place recognition is typically the first step in visual localization, which combines place recognition with 3D pose estimation. Hierarchical localization with Super-Point [34] and SuperGlue [35] is currently one of the best performing methods for visual localization [28].

LiDAR place recognition methods have been studied in a number of works based on both the conventional and deep features and the principles of these methods are similar to visual place recognition [15,17,27,32,37,41,43,45]. However, LiDAR-based place recognition needs LiDAR datasets for evaluation and recently autonomous driving datasets have become popular. MulRan [18] is an urban outdoors dataset with high quality LiDAR data, for which the recommended place recognition evaluation is done with a precision-recall curve at a threshold of 5 m. The used baseline method is their own egocentric spatial descriptor [17]. Kim et al. [18] also assessed the suitability of other similar urban outdoor datasets for LiDAR place recognition. Mapillary Street-Level Sequences [42] is an RGB only dataset for lifelong place recognition, but they have an involved evaluation methodology. They evaluate their baseline with GeM and NetVLAD methods using recall@1/5/10. They have typical image to image place recognition as well as several multiview setups. Some datasets offer high quality LiDAR and ground truth position data, but are not place recognition specific and thus offer no place recognition evaluation or methodology: The Newer College Dataset [31], University of Michigan North Campus Long-Term (NCLR) [9], nuScenes [7], Kitti [14], Complex Urban [16], Ford Campus [26], and Ford Multi-AV Seasonal [1].

In this work, we focus on the Oxford Radar RobotCar dataset [4] that is popular in the recent LiDAR place recognition works. Moreover, we propose a strong baseline based on the GeM method [30] which has been successfully used in LiDAR place recognition [27].

3   Oxford Radar RobotCar Place Recognition Benchmark

Here we propose our protocol for evaluating LiDAR-based place recognition methods with the Oxford Radar RobotCar dataset [4]. A well-defined protocol facilitates fair comparison between methods.

3.1   Overview

The Oxford Radar RobotCar dataset [4] consists of 32 traversals of the same approximately 9.0 km route in the downtown Oxford, UK. The sequences were collected during the period of nine consequent days from January 10 to January 18 in 2019. The time of day varies from 11:46:21 GMT (11am46) to 15:20:12 (3pm20) being bright daytime during which the weather varied from cloudy to partly sunny. Therefore, the data presents typical short-term variations of outdoor conditions for autonomous cars, including small illumination changes due to time of day and weather, and random partial and full occlusions due to other traffic on the streets and sidewalks. In this sense, the Oxford Radar

RobotCar dataset is ideal for studying how the selection of gallery and training sets affects to the visual place recognition performance.

3.2 Sensors

In our work we focus on LiDAR data as it provides 3D information for navigation and control, and depth is more robust to illumination changes than RGB. The Oxford Radar RobotCar vehicle has three radars installed on top of the car. In the middle is a Navtech FMCW radar that provides 400 measurements per 4 Hz and on its both side 20 Hz Velodyne LiDARs of 41.3° vertical FoV sensors. Suitable sensors for our studies are the two side-view LiDARs that provide 360-degree panoramic point clouds. See Fig. 1 for LiDAR examples and note the fixture on the right-hand-side of all depth maps; this is the RobotCar radar on top of the vehicle.

3.3 Gallery, Training and Test Sequences

For the selection of gallery, training and query (test) sequences we used the following assumptions that are based on practical workflow: i) data should represent a real use case where gallery and training data are recorded during one day and testing is performed after their collection; ii) the test sequences should span uniformly over the available time period (January 10-18 2019). The velocity of the car is moderate and thus the distance between two measurements is rarely more than 0.5 m. We selected the first (11:46) and last (15:19) traversals of the first collection day (Jan 10) as the two gallery sets. For training the two sequences between the training sequences were selected, at 14:02 and 14:50. For testing, three sequences from each day were selected thus forming the total of 18 query sequences. Note that sequences from Jan 12 to 13 were missing from the original data. Details of the place recognition dataset are in Table 1.

3.4 Evaluation Metrics

Place recognition evaluation metrics often revolve around different ways of evaluating recall and with differing test threshold distances based on the situation at hand. For example, recall@N with 25 m test thresholding is used [2,3,39,40] as well as recall@N and recall@1 with 5 m thresholding [5,18].

Our evaluation focus was on varying the threshold to highlight the achievable spatial accuracy while only using the very best match found by the system.

We use recall@1, which is the percentage of correct top-1 matches in the query set. The correctness of the match is determined by the distance threshold variable τ. If the query image I_q ground truth position $p(I_q)$ is within the threshold τ from the first match gallery image ground truth position $p(I_g)$, the images are considered a match:

$$match(I_q, I_g) = \begin{cases} 1, & \text{if } dist(p(I_q), p(I_g)) \leq \tau \\ 0, & \text{otherwise} \end{cases} \tag{1}$$

Table 1. Selected place recognition gallery, training and test sequences. All sequences are during daylight as the starting times are between 11:46 and 15:20, single traversal takes about 30 min and in January the sun sets approximately at 16:50 (4pm50) in Oxford.

Name	Data	Time (GMT 24h)	Weather[†]
Gallery 1	Day00 (Jan 10)	Noon (11:46)	Low clouds
Gallery 2	Day00	Afternoon (15:19)	Low clouds
Train 1	Day00	Afternoon (14:02)	Low clouds
Train 2	Day00	Afternoon (14:50)	Low clouds
Query 01	Day01 (Jan 11)	Noon (12:26)	Broken clouds
Query 02	Day01	Afternoon (14:37)	Broken clouds
Query 03	Day01	Afternoon (13:24)	Broken clouds
Query 04	Day04 (Jan 14)	Noon (12:05)	Broken clouds
Query 05	Day04	Afternoon (14:48)	Broken clouds
Query 06	Day04	Afternoon (13:38)	Broken clouds
Query 07	Day05 (Jan 15)	Noon (12:01)	Partly sunny
Query 08	Day05	Afternoon (14:24)	Partly sunny
Query 09	Day05	Afternoon (13:06)	Partly sunny
Query 10	Day06 (Jan 16)	Noon (11:53)	Light train, partly sunny
Query 11	Day06	Afternoon (14:15)	Light train, partly sunny
Query 12	Day06	Afternoon (13:09)	Light train, partly sunny
Query 13	Day07 (Jan 17)	Noon (11:46)	Passing clouds
Query 14	Day07	Afternoon (14:03)	Passing clouds
Query 15	Day07	Noon (12:48)	Passing clouds
Query 16	Day08 (Jan 18)	Noon (12:42)	Partly sunny
Query 17	Day08	Afternoon (15:20)	Partly sunny
Query 18	Day08	Afternoon (14:14)	Partly sunny

[†]weather conditions obtained from https://www.timeanddate.com

The top-N recall metric determines whether or not there is a correct match within the best N matches found for the query. If we have all the found gallery matches ordered from best to worst for the query image I_q, we can check if the ordered gallery has any correct matches in the first N places:

$$topN(I_q, N) = \begin{cases} 1, & \text{if } \sum_{k=1}^{N} match(I_q, I_g(k)) \geq 1 \\ 0, & \text{otherwise} \end{cases} \tag{2}$$

Recall@N for all the query images is the fraction of per image top-N sum over the whole query set:

$$recall@N = \frac{\sum_{j=1}^{Q} topN(I_q(j), N)}{Q} \tag{3}$$

We report recall@1 performance for several different test distance thresholds: $\tau = 2, 5, 10,$ or 25 m.

4   Baseline Method

We selected the state-of-the-art deep learning based image retrieval method by Radenović et al. [29,30] as our baseline method. This method performed the best in our previous works [27]. The method uses metric learning optimized global feature vectors and it thus offers a good balance between accuracy and computational complexity. It is also well known, flexible, and freely available from its original authors.

The main components of their CNN Retrieval is generalized mean pooling (GeM) on top of a fully convolutional backbone network. The training is done with contrastive loss set up in a siamese configuration. The final feature vector is whitened and dimension reduced. The image retrieval is done by exhaustively comparing the euclidean distances of feature vectors between all the query and gallery set images. The top match for any query image is the gallery set image with the shortest Euclidean distance to the query image.

The pipeline starts with a convolutional backbone network. The 3D tensor output χ of size $W * H * K$ from the last layer is fed into a generalized mean pooling layer,

$$
\mathbf{f} = [f_1...f_k...f_K]^\top, f_k = \left(\frac{1}{|\chi_k|} \sum_{x \subset \chi_k} x^{p_k} \right)^{\frac{1}{p_k}},
\tag{4}
$$

which transforms χ into a feature vector $\mathbf{f}$. χ_k is the set of activations for the feature map $k \in \{1...K\}$. The pooling parameter p_k is learned. After pooling, the vector is l_2-normalized for its final form.

For training, the data is fed as pairs of images (i, j) and their corresponding labels $Y(i, j) \in \{0, 1\}$, such that 1 is a match, and 0 is a non-match. The feature space Euclidean distance is decreased for the matches and increased for the non-matches by the contrastive loss function

We extended the CNN Retrieval system to accommodate for the Radar RobotCar dataset training and testing. A training batch is created by choosing any image in the training set, then one positive match for it, and a few negative matches for it. A match is defined by a distance threshold for training. We used 5.0 m, so a positive match was any image within five meters and a negative match any image over five meters away.

All the LiDAR scans are exported as flat 8-bit greyscale pixel images, and represent the depth map of the environment.

For the experiments, we use an ImageNet [12,33] pretrained VGG-16 [36] backbone network.

5   Preliminary Experiments

In order to demonstrate problems with the original Oxford Radar RobotCar ground truth we conducted an experiment where we tested the strong baseline (Sect. 4) with two different gallery sequences from the same day (Gallery 1 and 2 in Table 1). This experiment is meaningful as it experimentally evaluates whether and how much gallery selection affects to the place recognition performance. An additional research question is that if there is performance gap between different galleries then to what degree it can be mitigated with additional training sequences.

Table 2. Average top-1 recall rates (Rec@1) over all 18 query sequences (Query 01 - 18) for the two available galleries (Gallery 1 and 2) with and without training data (Train 1). Here we use the original Oxford RobotCar GPS ground truth.

Method	Rec@1-25m	Rec@1-10m	Rec@1-5m	Rec@1-2m
Gallery 1	0.870	0.802	0.637	0.237
Gallery 2	**0.876**	**0.840**	**0.731**	**0.330**
Gallery 1 + train	0.936	0.877	0.723	0.265
Gallery 2 + train	**0.941**	**0.920**	**0.838**	**0.394**

The sensor used in this experiment is the left-side Velodyne LiDAR that is pointed to the pedestrian walkway and buildings on the left hand side of the vehicle. The choice is fair as the left-hand side LiDAR is not blocked by oncoming vehicles of the left-hand traffic in the UK. In addition, the route is always travelled to the same direction.

The main difference between the two gallery sequences is the time-of-day, noon (Gallery 1) vs. afternoon (Gallery 2), which slightly affects to illumination but which should not affect LiDAR. The main difference should be the amounts of pedestrians, cyclist and other traffic that can vary between the lunch time and normal office hours. One training sequence, Train 1, was selected between the two gallery sequences. The route specific training data further fine-tune features for this dataset and suppress differences between the two galleries. Intuitively, the place recognition performance of the two galleries should be nearly identical.

The average performance over all 18 test sequences and for various distance thresholds using the original GPS ground truth are in Table 2. For all distance thresholds Gallery 2 provides better results. The performance difference is magnified when the accuracy threshold gets smaller. For example, on the tightest threshold at < 2.0 m the Gallery 2 recall is 40% better without training data (0.237 vs. 0.330) and 49% better (0.265 vs. 0.394) with the training data. For the both galleries the recall rate systematically improves with training data that fine-tunes the deep features more optimal for the Oxford route, as expected.

Findings. Based on these results, the selection of gallery sequence has significant impact on the place recognition performance. However, this finding is misleading and it is not possible to find any meaningful difference between the LiDAR or RGB images of the two galleries. The examples in Fig. 1 are berry picked and thus artificial. The actual reason for the finding is in the GPS ground truth that has systematic errors in Gallery 1.

6 Revised Location Ground Truth

Essential for successful place recognition is that the location tags, ground truth (GT), associated with gallery images are accurate. Various ground truth values are included to the data files of the Oxford Radar RobotCar dataset. The following location sources are available: 1) GPS, 2) GPS/INS, 3) Visual Odometry and 4) Radar Odometry. GPS and GPS/INS readings are raw data obtained from the sensors installed to the Oxford RobotCar vehicle. The location sensor is NovAtel SPAN-CPT ALIGN that integrates GPS and inertial navigation (INS) systems. Noteworthy, the location accuracy of GPS and GPS/INS measurements are the same, but due to higher sampling frequency of the INS sensor the GPS/INS data has ten times more points and it additionally contains the vehicle orientation information.

The recommended ground truth of the Oxford Radar RobotCar dataset is Radar Odometry [4]. Radar Odometry data is produced off-line by using the FAB-MAP library [10] constrained by the GPS/INS estimates of each point cloud. It is noteworthy that this ground truth which was produced by jointly optimizing all 32 original sequences provides correct paths, but it seems that there are translation and rotation shifts in some of the sequences due to undefined starting point. This is illustrated in Fig. 2.

In Fig. 2 GPS/INS looks better than the recommended ground truth Radar Odometry. This, however, is misinterpretation as the Radar Odometry plot is made using the closest timestamped GPS/INS starting pose which are not accurate enough for aligning the whole 9 km sequence.

Radar Odometry (Fixed). To get the best possible ground truth for tasks relying on aligned sequences, we manually tuned the radar odometry start poses for all the sequences.

To align the poses, we start with the closest time-stamped GPS/INS providing northing, easting, and yaw angle. This is the position point and compass angle of the car at the start of the odometry sequence. Then tuning is started by plotting a few sequences at time on a high resolution figure. Then we added or subtracted position coordinates (in meters) to northing and easting and angles of degrees to the yaw. After a few iterations is possible to see that the adjustments are as close as possible within the given resolution for that sequence.

In the end of the process, we have a list of absolute starting pose x_i, y_i, and α_i values for each radar odometry sequence i, and which align the sequences to each other and the GPS coordinates. The alignment allows us to experiment with any task depending on sequence alignment with higher accuracy than what is possible with the GPS/INS positioning. The post-processed radar odometry has significantly less drift than the GPS/INS positioning.

The odometry values are very good, but not perfect, and definitely not identical, so there are still some deviations. Few sequences had huge deviations making them, or parts of them, practically impossible to align with just an accurate starting pose. For these sequences we also made some slight modifications along the route to make them align well. The modifications were made in a smooth manner along a range of about one to a few dozen odometry transform points. The modifications per odometry transform point were either a small increase or decrease to the angle of rotation or the distance travelled. The final aligned routes are shown in Fig. 2(c).

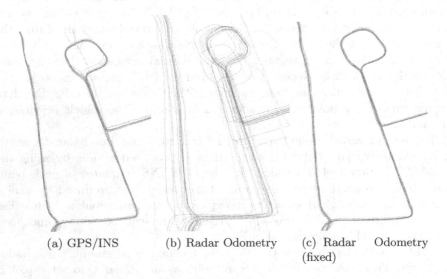

(a) GPS/INS   (b) Radar Odometry   (c) Radar   Odometry
                                                     (fixed)

Fig. 2. Various Oxford Radar RobotCar ground truth sources visualized for all 22 sequences used in our experiments (2 gallery, 2 train and 18 query sequences). "GPS" is omitted as it is practically GPS/INS with less samples and without orientation information and provides almost identical results for place recognition. In Radar Odometry (fixed) each sequence start pose is manually tuned (translation and rotation).

7   Experiments

The results from the preliminary experiment in Sect. 5 are repeated here but now with the manually corrected ground truth. In addition to the original research question, i.e., *how the gallery selection affects to the LiDAR place recognition performance*, we added another side study: if additional sequences of the same route are available, should they be used as training data to fine-tune map specific deep features or additional gallery images that would reduce effects of dynamic changes such as other traffic.

The new results are in Table 3. Now the findings are very different from the preliminary experiments in Sect. 5 and Table 2. With accurate and error-free ground truth, "Radar Odometry (fixed) gt", there is no significant difference between the two galleries from the same day (Day 0). Interestingly, using the additional sequence, Train 1, for training is more beneficial than adding it as additional gallery data, but with wrong ground truth even this finding would be the opposite (see the GPS/INS gt results for Gallery 1).

Table 3. Average top-1 recall rates for the same data as in Table 2 but now using various ground truth sources (Sect. 6). Best results for all cases are bold and if difference below 0.005 multiple marked.

Method	Rec@1-25m	Rec@1-10m	Rec@1-5m	Rec@1-2m
GPS gt				
Gallery 1	0.857	0.785	0.599	0.167
Gallery 1 + tr. (Train 1)	0.922	0.860	0.680	0.189
Gallery 1 + ext. (Train 1)	0.895	0.845	0.681	0.242
Gallery 2	0.864	0.824	0.709	0.278
Gallery 2 + tr. (Train 1)	**0.927**	**0.903**	**0.815**	**0.328**
Gallery 2 + ext. (Train 1)	0.894	0.859	0.741	0.312
GPS/INS gt				
Gallery 1	0.852	0.781	0.599	0.167
Gallery 1 + tr. (Train 1)	0.917	0.856	0.681	0.189
Gallery 1 + ext. (Train 1)	0.889	0.840	0.680	0.244
Gallery 2	0.857	0.818	0.706	0.282
Gallery 2 + tr. (Train 1)	**0.921**	**0.898**	**0.810**	**0.334**
Gallery 2 + ext. (Train 1)	0.888	0.853	0.737	0.317
Radar Odometry gt				
Gallery 1	0.615	0.453	0.249	0.051
Gallery 1 + tr. (Train 1)	0.666	**0.504**	**0.286**	0.063
Gallery 1 + ext. (Train 1)	0.644	0.484	0.264	**0.073**
Gallery 2	0.627	0.451	0.202	0.041
Gallery 2 + tr. (Train 1)	**0.675**	0.498	0.234	0.045
Gallery 2 + ext. (Train 1)	0.648	0.482	0.235	0.064
Radar Odometry (fixed) gt				
Gallery 1	0.871	0.837	0.753	0.471
Gallery 1 + tr. (Train 1)	0.938	0.918	0.857	**0.560**
Gallery 1 + ext. (Train 1)	0.910	0.882	0.801	0.495
Gallery 2	0.878	0.843	0.758	0.443
Gallery 2 + tr. (Train 1)	**0.943**	**0.922**	**0.861**	0.535
Gallery 2 + ext. (Train 1)	0.909	0.878	0.795	0.468

The best accuracies are obtained using the corrected ground truth (Radar Odometry (fixed) gt) for which the differences between Gallery 1 and Gallery 2 are insignificant for all distance thresholds. Moreover, the improvement using training data is systematic and substantial as compared to gallery only and using the training data as additional gallery data.

Findings. The results with correct ground truth reveal that gallery sets collected during similar conditions and same day do not significantly differ in their place recognition accuracy. Other traffic, that is the main cause of errors in this case (see Fig. 1), is not an important performance factor. For the very same reason, additional sequences should be used as training data to tune the features rather than as additional gallery data.

8   Conclusion

In this work we studied the impact of gallery data to the performance of LiDAR-based place recognition. In particular, we formed an evaluation protocol by selecting two gallery sequences, two training sequences and 18 query sequences from the recent Oxford Radar RobotCar dataset that has become popular for evaluating LiDAR-based place recognition. The protocol is based on a realistic scenario where gallery and training data are collected on a single day and then the system should autonomously navigate afterwards. Our main findings was that the original ground truth has problems despite that multiple modalities are available (GPS, GPS/INS, Visual Odometry and Radar Odometry). We proposed a fix to the Radar Odometry ground truth and re-run preliminary experiments on that. Strikingly, the results and the main findings are very different between the original (Sect. 5) and the fixed ground truth (Sect. 7). All data, including the new ground truth, and code will be published to facilitate fair method comparisons in the future.

References

1. Agarwal, S., Vora, A., Pandey, G., Williams, W., Kourous, H., McBride, J.: Ford multi-AV seasonal dataset. Int. J. Robot. Res. **39**(12), 1367–1376 (2020)
2. Arandjelovic, R., Gronat, P., Torii, A., Pajdla, T., Sivic, J.: NetVLAD: CNN architecture for weakly supervised place recognition. In: TPAMI (2018)
3. Arandjelović, R., Zisserman, A.: DisLocation: scalable descriptor distinctiveness for location recognition. In: Cremers, D., Reid, I., Saito, H., Yang, M.-H. (eds.) ACCV 2014. LNCS, vol. 9006, pp. 188–204. Springer, Cham (2015). https://doi. org/10.1007/978-3-319-16817-3_13
4. Barnes, D., Gadd, M., Murcutt, P., Newman, P., Posner, I.: The oxford radar robotcar dataset: a radar extension to the oxford RobotCar dataset. In: Proceedings of the IEEE International Conference on Robotics and Automation (ICRA). Paris (2020)
5. Barnes, D., Posner, I.: Under the radar: learning to predict robust keypoints for odometry estimation and metric localisation in radar. In: 2020 IEEE International Conference on Robotics and Automation (ICRA), pp. 9484–9490. IEEE (2020)
6. Bay, H., Ess, A., Tuytelaars, T., Van Gool, L.: Speeded-up robust features (surf). Comput. Vis. Image Underst. **110**(3), 346–359 (2008)
7. Caesar, H., et al.: nuScenes: a multimodal dataset for autonomous driving. In: Proceedings of the IEEE/CVF Conference on Computer Vision and Pattern Recognition, pp. 11621–11631 (2020)
8. Cao, B., Araujo, A., Sim, J.: Unifying deep local and global features for image search. In: Vedaldi, A., Bischof, H., Brox, T., Frahm, J.-M. (eds.) ECCV 2020. LNCS, vol. 12365, pp. 726–743. Springer, Cham (2020). https://doi.org/10.1007/978-3-030-58565-5_43
9. Carlevaris-Bianco, N., Ushani, A.K., Eustice, R.M.: University of Michigan north campus long-term vision and lidar dataset. Int. J. Robot. Res. **35**(9), 1023–1035 (2016)

10. Cummins, M., Newman, P.: FAB-MAP: probabilistic localization and mapping in the space of appearance. Int. J. Robot. Res. **27**(6), 647–665 (2008). https://doi.org/10.1177/0278364908090961
11. Dalal, N., Triggs, B.: Histograms of oriented gradients for human detection. In: 2005 IEEE Computer Society Conference on Computer Vision and Pattern Recognition (CVPR2005), vol. 1, pp. 886–893. IEEE (2005)
12. Deng, J., Dong, W., Socher, R., Li, L.J., Li, K., Fei-Fei, L.: ImageNet: a large-scale hierarchical image database. In: 2009 IEEE Conference on Computer Vision and Pattern Recognition, pp. 248–255. IEEE (2009)
13. Gálvez-López, D., Tardos, J.D.: Bags of binary words for fast place recognition in image sequences. IEEE Trans. Rob. **28**(5), 1188–1197 (2012)
14. Geiger, A., Lenz, P., Stiller, C., Urtasun, R.: Vision meets robotics: the KITTI dataset. Int. J. Robot. Res. **32**(11), 1231–1237 (2013)
15. He, L., Wang, X., Zhang, H.: M2DP: a novel 3d point cloud descriptor and its application in loop closure detection. In: 2016 IEEE/RSJ International Conference on Intelligent Robots and Systems (IROS), pp. 231–237. IEEE (2016)
16. Jeong, J., Cho, Y., Shin, Y.S., Roh, H., Kim, A.: Complex urban dataset with multi-level sensors from highly diverse urban environments. Int. J. Robot. Res. **38**(6), 642–657 (2019)
17. Kim, G., Kim, A.: Scan context: egocentric spatial descriptor for place recognition within 3d point cloud map. In: 2018 IEEE/RSJ International Conference on Intelligent Robots and Systems (IROS), pp. 4802–4809. IEEE (2018)
18. Kim, G., Park, Y.S., Cho, Y., Jeong, J., Kim, A.: MulRan: multimodal range dataset for urban place recognition. In: 2020 IEEE International Conference on Robotics and Automation (ICRA), pp. 6246–6253. IEEE (2020)
19. Komorowski, J.: MinkLoc3d: point cloud based large-scale place recognition. In: Proceedings of the IEEE/CVF Winter Conference on Applications of Computer Vision (WACV) (2021)
20. Komorowski, J., Wysoczanska, M., Trzcinski, T.: Large-scale topological radar localization using learned descriptors. In: Mantoro, T., Lee, M., Ayu, M.A., Wong, K.W., Hidayanto, A.N. (eds.) ICONIP 2021. LNCS, vol. 13109, pp. 451–462. Springer, Cham (2021). https://doi.org/10.1007/978-3-030-92270-2_39
21. Liu, Z., et al.: LPD-NET: 3D point cloud learning for large-scale place recognition and environment analysis. In: ICCV (2019)
22. Lowe, D.G.: Distinctive image features from scale-invariant keypoints. Int. J. Comput. Vision **60**(2), 91–110 (2004)
23. Lowry, S., et al.: Visual place recognition: a survey. IEEE Trans. Rob. **32**(1), 1–19 (2016). https://doi.org/10.1109/TRO.2015.2496823
24. Maddern, W., Pascoe, G., Linegar, C., Newman, P.: 1 Year, 1000 km: the Oxford RobotCar dataset. Int. J. Robot. Res. (IJRR) **36**(1), 3–15 (2017)
25. Noh, H., Araujo, A., Sim, J., Weyand, T., Han, B.: Large-scale image retrieval with attentive deep local features. In: Proceedings of the IEEE International Conference on Computer Vision, pp. 3456–3465 (2017)
26. Pandey, G., McBride, J.R., Eustice, R.M.: Ford campus vision and lidar data set. Int. J. Robot. Res. **30**(13), 1543–1552 (2011)
27. Peltomäki, J., Alijani, F., Puura, J., Huttunen, H., Rahtu, E., Kämäräinen, J.K.: Evaluation of long-term LiDAR place recognition. In: IEEE/RSJ International Conference on Intelligent Robots and Systems (IROS). Prague, Czech Rep. (2021)
28. Pion, N., Humenberger, M., Csurka, G., Cabon, Y., Sattler, T.: Benchmarking image retrieval for visual localization. In: International Conference on 3D Vision (3DV) (2020)

29. Radenović, F., Tolias, G., Chum, O.: CNN Image retrieval learns from BoW: unsupervised fine-tuning with hard examples. In: Leibe, B., Matas, J., Sebe, N., Welling, M. (eds.) ECCV 2016. LNCS, vol. 9905, pp. 3–20. Springer, Cham (2016). https://doi.org/10.1007/978-3-319-46448-0_1
30. Radenović, F., Tolias, G., Chum, O.: Fine-tuning CNN image retrieval with no human annotation. In: TPAMI (2018)
31. Ramezani, M., Wang, Y., Camurri, M., Wisth, D., Mattamala, M., Fallon, M.: The newer college dataset: handheld lidar, inertial and vision with ground truth. In: 2020 IEEE/RSJ International Conference on Intelligent Robots and Systems (IROS) (2020)
32. Rizzini, D.L., Galasso, F., Caselli, S.: Geometric relation distribution for place recognition. IEEE Robot. Autom. Lett. 4(2), 523–529 (2019)
33. Russakovsky, O., et al.: Imagenet large scale visual recognition challenge. Int. J. Comput. Vision 115(3), 211–252 (2015)
34. Sarlin, P.E., Cadena, C., Siegwart, R., Dymczyk, M.: From coarse to fine: robust hierarchical localization at large scale. In: CVPR (2019)
35. Sarlin, P.E., DeTone, D., Malisiewicz, T., Rabinovich, A.: SuperGlue: learning feature matching with graph neural networks. In: CVPR (2020)
36. Simonyan, K., Zisserman, A.: Very deep convolutional networks for large-scale image recognition. arXiv preprint arXiv:1409.1556 (2014)
37. Steder, B., Grisetti, G., Burgard, W.: Robust place recognition for 3D range data based on point features. In: ICRA (2010)
38. Tang, T., Martini, D.D., Newman, P.: Get to the point: learning lidar place recognition and metric localisation using overhead imagery. In: Robotics: Science and Systems (RSS) (2021)
39. Torii, A., Arandjelovic, R., Sivic, J., Okutomi, M., Pajdla, T.: 24/7 place recognition by view synthesis. In: Proceedings of the IEEE Conference on Computer Vision and Pattern Recognition, pp. 1808–1817 (2015)
40. Torii, A., Sivic, J., Pajdla, T., Okutomi, M.: Visual place recognition with repetitive structures. In: Proceedings of the IEEE Conference on Computer Vision and Pattern Recognition, pp. 883–890 (2013)
41. Uy, M.A., Lee, G.H.: PointNetVLAD: deep point cloud based retrieval for large-scale place recognition. In: Proceedings of the IEEE Conference on Computer Vision and Pattern Recognition, pp. 4470–4479 (2018)
42. Warburg, F., Hauberg, S., Lopex-Antequera, M., Gargallo, P., Kuang, Y., Civera, J.: Mapillary street-level sequences: a dataset for lifelong place recognition. In: IEEE Conference on Computer Vision and Pattern Recognition (CVPR) (2020)
43. Xie, S., Pan, C., Peng, Y., Liu, K., Ying, S.: Large-scale place recognition based on camera-lidar fused descriptor. Sensors 20(10), 2870 (2020)
44. Yin, H., Xu, X., Wang, Y., Xiong, R.: Radar-to-lidar: heterogeneous place recognition via joint learning. Front. Robot. AI 8, 661199 (2021)
45. Yin, P., et al.: Stabilize an unsupervised feature learning for lidar-based place recognition. In: 2018 IEEE/RSJ International Conference on Intelligent Robots and Systems (IROS), pp. 1162–1167. IEEE (2018)
46. Zhang, X., Wang, L., Su, Y.: Visual place recognition: a survey from deep learning perspective. Pattern Recogn. 113, 107760 (2020). https://doi.org/10.1016/j.patcog.2020.107760

BrackishMOT: The Brackish Multi-Object Tracking Dataset

Malte Pedersen[1,2(✉)] , Daniel Lehotský[1] , Ivan Nikolov[1,2] , and Thomas B. Moeslund[1,2]

[1] Visual Analysis and Perception Lab, Aalborg University, Aalborg, Denmark
mape@create.aau.dk
[2] Pioneer Center for AI, Aalborg, Denmark

Abstract. There exist no publicly available annotated underwater multi-object tracking (MOT) datasets captured in turbid environments. To remedy this we propose the BrackishMOT dataset with focus on tracking schools of small fish, which is a notoriously difficult MOT task. BrackishMOT consists of 98 sequences captured in the wild. Alongside the novel dataset, we present baseline results by training a state-of-the-art tracker. Additionally, we propose a framework for creating synthetic sequences in order to expand the dataset. The framework consists of animated fish models and realistic underwater environments. We analyse the effects of including synthetic data during training and show that a combination of real and synthetic underwater training data can enhance tracking performance. *Links to code and data can be found at* https://www.vap.aau.dk/brackishmot.

Keywords: Dataset · Multi-Object Tracking · Synthetic data · Underwater · Fish

1 Introduction

Humans have relied on the oceans as a steady food-source for millennia, but the marine ecosystems are now rapidly deteriorating due to human impact. This is especially a problem for coastal societies across the globe that rely on fish as their main food-source and for biodiversity in general. The severity is underlined by the fact that the United Nations included *Life Below Water* as the fourteenth Sustainable Development Goal (UN SDG #14) [44]. The increase in attention to monitoring the condition of the ocean has entailed pressure on marine researchers to gather data at an unprecedented pace. However, traditional marine data-gathering methods are often time-consuming, intrusive, and difficult to scale as they require organisms to be caught and measured manually. Therefore, it is critical that assistive solutions for optimizing and scaling data-gathering in marine environments are developed (Fig. 1).

During the past decade, computer vision solutions have increased dramatically in performance due to the utilization of strong graphics processing units (GPU) combined with the popularisation of deep learning algorithms. Simultaneously, underwater cameras have become significantly better and cheaper [22]. This calls for marine researchers to utilize both cameras and computer vision to scale data-gathering right away, however,

R. Gade et al. (Eds.): SCIA 2023, LNCS 13885, pp. 17–33, 2023.
https://doi.org/10.1007/978-3-031-31435-3_2

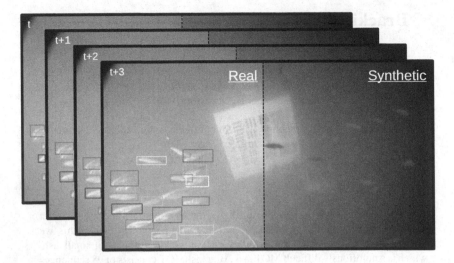

Fig. 1. We present and publish a bounding box annotated underwater multi-object tracking dataset captured in the wild named BrackishMOT, together with a synthetic framework for generating more data for which we publish both data and source code.

this is currently not feasible as there is a critical lack of marine datasets for training and evaluating marine computer vision models.

To remedy this gap we propose a novel multi-object tracking (MOT) dataset of marine organisms in the wild named BrackishMOT. In short, MOT describes the task of obtaining the trajectories of all objects in the scene. The trajectories are obtained by having a model detect objects spatially and associating the detections temporally. Tracking is a core component in marine research and can be used for multiple purposes such as counting or conducting behavioral analysis.

Manually annotated datasets are critical and necessary for evaluating the performance of trackers on data from the wild. However, they are not scalable and do not necessarily generalize well to environments that are not included in the dataset. Therefore, we investigate how synthetic underwater sequences can be used for training multi-object trackers. We develop a framework for creating synthetic sequences that resemble the BrackishMOT environment and analyse how key factors, namely turbidity, floating particles, and the background, affect tracker performance. This is a critical step toward the development of new high-quality underwater synthetic datasets. Our contributions are summarized below:

Contributions

- We present and publish BrackishMOT, a novel MOT dataset captured in brackish waters with a total of 98 sequences and six different classes.
- We propose a framework for creating synthetic underwater sequences based on phenomena observed in the wild and analyse their effect on tracker performance.
- We analyse different training strategies for a state-of-the-art tracker using both real and synthetic data and present baseline results for BrackishMOT.

2 Related Work

The current state-of-the-art MOT algorithms like Tracktor [1], CenterTrack [51], Fair-
MOT [50], and ByteTrack [49] have all been developed for tracking terrestrial objects
like pedestrians and vehicles. Common denominators for these types of objects are relatively predictable motion and typically strong visual cues. However, in the underwater domain, most objects like fish are prey animals which means that they may behave
erratically to avoid being tracked by predators. Furthermore, objects of the same species
often look very similar and provide weak visual features for re-identification. In other
words, the trackers need to be re-trained or fine-tuned to cope with the challenges of the
underwater domain. Modern trackers are generally based on deep learning and require
large amounts of training data. While there exist multiple terrestrial tracking datasets
like KITTI [9], MOTChallenge [6,17,26], and UAVDT [48], there are only a few publicly available datasets with underwater objects. In this section, we dive into the sparse
literature on underwater datasets and trackers.

2.1 Underwater MOT Datasets

Compared to its terrestrial counterpart, the underwater MOT domain has not witnessed
a noteworthy increase in novel algorithms during the past decade. One of the reasons
for this lack of algorithms dedicated to the underwater domain is the low number of
publicly available annotated underwater datasets suitable for training and evaluating
modern algorithms.

Underwater MOT datasets can generally be split into two categories: controlled
environments such as aquariums [31,38], and the more challenging uncontrolled natural underwater environments which we will focus on in this paper. One of the earliest
datasets used for tracking of fish captured in the wild was the Fish4Knowledge (F4K)
dataset [8,10]. The F4K dataset was captured more than a decade ago in mostly clear
tropical waters off the coast of Taiwan in very low resolution and low frame rate. More
recently, the two underwater object tracking datasets UOT32 [15] and UOT100 [29]
were published with annotated underwater sequences sourced from YouTube videos.
The UOT32 and UOT100 datasets provide sequences from diverse underwater environments but are focused on single object tracking. Lastly, a high-resolution underwater MOT dataset captured off the coast of Hawai'i island named FISHTRAC [24]
was recently proposed. However, at the time of writing only three training videos (671
frames in total) with few objects and little occlusion have been published.

The datasets captured in tropical waters only cover a tiny fraction of the diverse
underwater ecosystems. The conditions in many other areas are far less favorable with
less colorful fish and more turbid water. To advance the research in underwater MOT, it
is critical to developing new datasets captured in other and more challenging environments and we see the BrackishMOT dataset as an important contribution to this field.

2.2 Underwater Trackers

Relatively few multi-object trackers dedicated to the underwater domain exist with most
of them developed for tracking fish in controlled environments [28,31,38,45]. A common trait for trackers developed for controlled environments is the assumption of good

detections and strong visual cues for re-identification. This is generally not the case in uncontrolled environments, where the light may change, the water is turbid, the background varies, and algae may bloom on the lens [8, 33].

To tackle these problems the team behind the F4K dataset proposed a method for detecting fish using mixture models for background subtraction and handcrafted features based on motion and color for classifying fish from other objects [8]. Lastly, they modeled every track by feature-based covariance matrices based on representations from previous frames and associated new detections by minimizing the distance between the covariance matrices. Another group that also worked on the F4K tracking data experimented with AlexNet [16] and VGG-19 [39] as feature extractors for appearance-based association and used a directed acyclic graph in a two-step approach by first constructing strong local tracklets followed by a tracklet-association step for finalizing the tracks [13].

Recently, a few groups have proposed trackers evaluated on new annotated underwater datasets. Liu et al. proposed a multi-class tracker named RMFC [20] utilizing YOLOv4 [4] as the backbone for a detection and tracking branch running in parallel, which showed promising results. In the work by Martija et al. [25] they investigated the use of synthetic data to enhance tracker performance. They propose to use Faster R-CNN [35] for object detection and a deep hungarian network [46] for associating detections temporally using visual cues. Unfortunately, both groups evaluate their method on private datasets, and they have not shared their code.

The most recent work on fish tracking in the wild is the work done by Mandel et al. [24]. They propose an offline tracker utilizing a greedy approach that initializes a track from the strongest detection across all frames based on a confidence score. Detections in previous and future frames are associated with the track based on appearance and motion. When a track has been finalized, the next track is built in the same manner, and so forth. They evaluated their tracker with detections from YOLOv4 and RetinaNet [18].

Common for the aforementioned methods is a reliance on strong visual cues or predictable motion for associating detections. This works well for scenes with few objects, in clear tropical waters, or if the objects are visually distinct. This is a natural consequence caused by the limited datasets used in the development of the methods and it exposes the need for diverse datasets to represent the variety of underwater ecosystems.

2.3 Synthetic Underwater Data

A way to remedy the scarce amount of publicly available underwater datasets is to use synthetic data. A typical approach to produce synthetic underwater data (in 2D) is by pasting cutouts of the organisms onto some background. Mahmood et al. [23] used this method to place manually segmented parts of lobsters, like the body or the antennas, onto a diverse set of backgrounds sampled from the Benthoz15 [3] dataset to generate synthetic data suitable for lobster detection in heavily occluded scenes. Martija et al. [25] used weakly generated bounding boxes and masks to simulate the movement of fish across a background to create rough synthetic MOT data. And for developing an underwater litter detector Music et al. [27] pasted various 3D shapes into real-life underwater images to create training data.

An alternative to the 2D approach is data generation from animated 3D scenes. In 1987, Reynolds proposed the boid model for accurately simulating the behaviour of fish schools [36]. The boid behaviour model has since been extended multiple times [11, 34, 40], with [12] combining their variation of boids with a synthetic data generator to produce realistic annotated underwater data from animated sequences of fish schools. However, to the best of our knowledge, there has been no attempt to produce synthetic underwater MOT data based on boid behavior. In the next section, we will introduce and describe our new underwater MOT dataset followed by a description of our framework for creating synthetic underwater sequences.

| (a) Fish | (b) Small fish | (c) Crab |

Fig. 2. Image samples from the Brackish Dataset [30]. In a majority of the sequences containing the *small fish* class, there are multiple specimens forming a school of fish.

3   The BrackishMOT Dataset

In 2019 the Brackish Dataset [30] was published. Its purpose was to advance object detection in brackish waters. It has been popular in the community since it was the first underwater detection dataset captured in non-tropical waters. The recordings of the dataset were captured nine meters below the surface in brackish waters and consist of 89 sequences in total. The sequences contain manually annotated bounding boxes of six coarse classes: *fish, crab, shrimp, starfish, small fish,* and *jellyfish*. Examples from the original dataset can be seen in Fig. 2.

3.1   Dataset Overview

In this work, we propose to expand the Brackish Dataset to include a MOT task. Therefore, we provide a new set of ground truth annotations for every sequence, based on the MOTChallenge annotation style [6]. Additionally, we present 9 new sequences focused on the *small fish* class, which gives a total of 98 sequences for the MOT task of the Brackish Dataset which we name **BrackishMOT**. The *small fish* class is especially relevant for the MOT task as it contains species that exhibit social and schooling behavior as illustrated in Fig. 2b. The ground truth files are comma-separated and include annotations per object in the following structure:

```
<frame>, <id>, <left>, <top>, <width>, <height>,
<confidence>, <class>, <visibility>
```

where `left` and `top` are the x and y coordinates of the object's top-left corner of the bounding box. Together with the `width` and `height` they describe the object's bounding box in pixels. `confidence` and `visibility` are both set to 1 and an object keeps its `id` as long as it is within field of view. There are rare cases where objects gets fully occluded in-frame and it is ambiguous to decide the ID of the object as it re-appears; in these cases, the object acquires a new ID. The `class` is in the range 1–6 where: (1) *fish*, (2) *crab*, (3) *shrimp*, (4) *starfish*, (5) *small fish*, and (6) *jellyfish*.

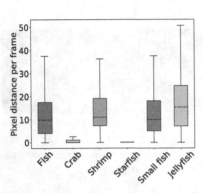

(a) Boxplot showing the distribution of the traveled distance between consecutive frames measured in pixels.

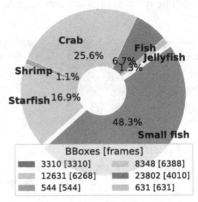

(b) Class distribution of the brackish-MOT dataset based on the number of bounding boxes.

Fig. 3. Plots describing the composition of the brackishMOT dataset with respect to motion and class distribution. For both plots, the data is from all the sequences.

In Fig. 3 we present two charts illustrating the motion and class distribution for the dataset. We see that the *crab* and *starfish* classes barely move compared to the rest. In addition to that, they are well-camouflaged and most often move along the seabed. This constitutes a specific task as they are both hard to detect visually and from motion cues. The class distribution presented in Fig. 3b shows that the dataset is imbalanced with few occurrences of the *shrimp*, *fish*, and *jellyfish* classes. Furthermore, as the number of bounding boxes and frame occurrences are equal we can decipher that these three classes occur in the sequences as single objects. As the *small fish* class is the only class that exhibits erratic motion and appears in groups it is deemed the most interesting class with respect to MOT.

3.2   BrackishMOT Splits

Creating balanced training and testing splits is important to ensure a fair evaluation of the tracker performance and to give an accurate depiction of the task. To a large degree, this is a problem that has been overlooked in the creation of most MOT datasets due to the lack of a suitable metric. This has changed with the recent introduction of the

MOTCOM framework [32]. MOTCOM is a metric that can estimate the complexity of MOT sequences based on the ground truth annotations and lay the foundation for creating more balanced data splits. The metric is a combination of three sub-metrics that describe the level of occlusion (OCOM), non-linear motion (MCOM), and visual similarity (VCOM) for every sequence. MOTCOM and the sub-metrics are all in the interval from 0 to 1 where a higher MOTCOM score means a more complex problem. We aim to create splits that are approximately evenly complex.

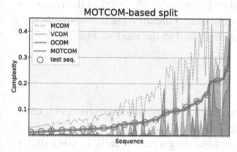
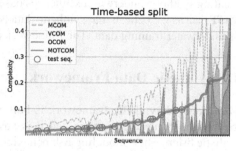

(a) Sorting the sequences based on MOTCOM and taking every fifth to be included in the test split. This is the approach we follow.

(b) A typical test split consisting of the 20 first recorded sequences. This approach clearly skews the splits with respect to complexity.

Fig. 4. These plots illustrate MOTCOM and the sub-metrics for all the BrackishMOT sequences. In both plots, the sequences are sorted based on their MOTCOM score. The circles mark the test sequences with respect to the split-scheme.

In Fig. 4a we present MOTCOM and the sub-metrics for each sequence of the BrackishMOT dataset. The metrics are calculated on basis of all six classes combined. We see that the motion varies a lot between the individual sequences. However, even though the motion is quite non-linear and complex for several sequences then both occlusion and visual similarity are very low. This is due to the generally low number of objects in the scenes. A single jellyfish or shrimp may move fast and non-linearly, but if they are alone or in a scene with just a few objects they are less likely to be occluded or confused with other individuals.

The sequences containing the *small fish* class are generally exceptions to the above as they tend to score higher values in all three sub-metrics compared to the other sequences. These sequences often include fish schools which means that they have a higher number of objects that moves more around and are more social compared to e.g., starfish and crabs on the seabed. Therefore, the objects are more likely to be occluded and they are easier to confuse with each other as they look visually similar.

We create the splits based on the following scheme: we sort the sequences according to their MOTCOM score, then we pick the sequence with the highest MOTCOM score to be in the train split, the second highest goes into the test split, and from then on every fifth sequence goes into the test split while the rest goes into the train split. This gives a total of 20 test sequences illustrated by the circles in Fig. 4a and 78 train sequences. If we, on the other hand, had chosen a typical scheme like picking the first

20 recorded sequences to be included in the test split and the rest in the train split, we would have had a significantly different dataset structure as illustrated in Fig. 4b. With such a composition it is likely that trackers would generally perform better when evaluated on the test split, but it would be on false terms as the train split is significantly more challenging compared to the test split. The opposite can of course also happen, but that is equally problematic. For this reason, we make an informed split based on the MOTCOM scores.

To extend the proposed BrackishMOT dataset, we have developed a framework for creating synthetic underwater sequences based on phenomena observed in the Brackish-MOT data as we believe that synthetic data is critical as a means to scale the availability of underwater training data. The framework is described in the next section.

4   Synthetic Data Framework

The proposed synthetic framework is built within the Unity game engine [41], using the built-in rendering pipeline and it is based on three main components: providing realistic fish meshes, modeling fish behavior, and building a realistically looking underwater environment. We provide options for each of the components in order to create a synthetic environment that resembles the BrackishMOT data, however, the proposed framework is easily extendable with other species, behavior models, and surroundings. We will describe each of the components in the following sections.

Fish Model. An illustration of the fish model used in our framework can be seen in Fig. 5. The model was taken from the fish database of images and photogrammetry 3D reconstructions [14] and was selected as it visually resembles a *stickleback*, which is the family of the *small fish* class that most often occurs in schools in the Brackish dataset sequences according to the authors [30]. The 3D input model, shown in Fig. 5a, was decimated to 11,000 vertices. In order to preserve finer details of the mesh, a normal map was created from the high-resolution texture. The down-sampled mesh can be seen in Fig. 5b. Lastly, the model was rigged using the bones system in Blender [5] to allow for smooth animations of the body and tail.

The number of spawned fish in each sequence is randomly selected within a range between 4 and 50 to resemble the diversity of the BrackishMOT sequences. The initial pose, scale, and appearance (texture albedo and glossiness) for each fish varies between the sequences. A table with all the randomized parameters and their respective ranges can be found in the supplementary material.

Behavior Model. To approximate realistic fish schooling behavior, we use a boid-based behavioral model inspired by the work of C.W. Reynolds [36] and C. Hartman and B. Benes [11]. Each fish considers the position and heading of all other fish in its neighborhood. For each fish the velocity and heading is dependent on four factors: separation s, cohesion k, alignment m, and leader l. Separation ensures avoidance of collisions with other members of the school. Cohesion is a force that drives the fish to seek the center of the neighborhood. Alignment is the drive of individual fish to match

| (a) High-resolution mesh. | (b) Low-resolution mesh with rig. |

Fig. 5. Illustration of the *stickleback* fish model used in our framework. (a) Initial high-resolution model and (b) decimated and rigged model for Unity.

the others' velocity. Leader is a direction towards where a given leader is heading and for each fish, the leader is the neighbor with a heading vector closest to the fish's own heading vector. The steering vector is given by

$$steer = Ss + Kk + Mm + Ll, \tag{1}$$

where S, K, M, and L are weights for the separation, cohesion, alignment, and leader forces, respectively. A more detailed description of the behavior model can be found in the supplementary material.

The Surrounding Environment. To investigate how changes in the environment impact tracker performance, we design the synthetic environment based on three variables: turbidity, background, and distractors.

Turbidity represents tiny floating particles in the water that engulfs the scene like a fog that intensifies as the distance between the object and the camera increases. The visibility varies to a large degree in the BrackishMOT sequences due to this phenomenon. We implement the turbidity effect using a custom-made Unity material with adjustable transparency and post-processing effects of depth of field and color grading. The color of the material spans between grey and green to resemble the turbidity observed in the BrackishMOT sequences. Both the color and intensity vary between the generated sequences.

We use videos from the Brackish dataset without fish as the background to make the scene more realistic. We include a range of background sequences and augment them by saturation, color, and blur to increase variation. When no background is present, we use a monotone color that matches the color of the turbidity.

Lastly, we introduce distractors [42,43], which represent floating particles. The BrackishMOT sequences have been captured in shallow water where the current is often strong. The combination of strong current and shallow water induces floating plant material and resuspended sediments often occur in front of the camera as unclear circular bodies. To simulate this phenomenon we implement distractors as spheres with varying scales, levels of transparency, and color. The color range spans between grey and green as with the turbidity and monotone background. The number of distractors vary between the sequences and each distractor is spawned in a random position and is randomly moved to a new position between each frame.

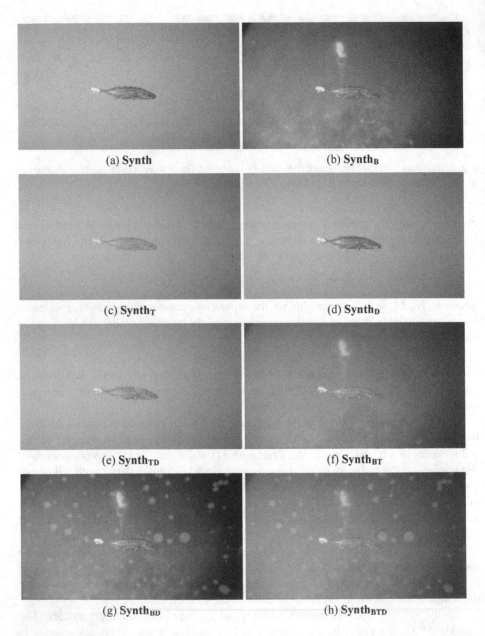

(a) **Synth**

(b) **Synth$_B$**

(c) **Synth$_T$**

(d) **Synth$_D$**

(e) **Synth$_{TD}$**

(f) **Synth$_{BT}$**

(g) **Synth$_{BD}$**

(h) **Synth$_{BTD}$**

Fig. 6. Visualisation of different conditions of our synthetic environment. (**a**) Plain background, no turbidity, no distractors (**Synth**). (**b**) Video background, no turbidity, no distractors (**Synth$_B$**). (**c**) Plain background, turbidity, no distractors (**Synth$_T$**). (**d**) Plain background, no turbidity, with distractors (**Synth$_D$**). (**e**) Plain background, with turbidityand distractors (**Synth$_{DT}$**). (**f**) Video background with turbidity, but without distractors (**Synth$_{BT}$**). (**g**) Video background with distractor, but no turbidity (**Synth$_{BD}$**). (**h**) Video background with turbidity and distractors (**Synth$_{BTD}$**). (Color figure online)

Each of the environmental variables adds a layer of complexity to the synthetic scene based on phenomena observed in the real sequences. Combinatorial variations of the variables give us eight synthetic environments, which can be seen in Fig. 6. Each generated video sequence is 10 s long and contains 150 frames animated with a frame rate of 15 FPS, which resembles the sequences of the BrackishMOT dataset. We include 50 sequences for each environment variation. The synthetic framework is general as all parameters can be adjusted to fit other underwater environments, e.g., using another video background would significantly alter the visuals of the sequences, or one could change the current or add new models to the scene. Source code and guides to using the framework can be found on the project page *(URL upon paper acceptance)*.

5   Experiments

It is notoriously difficult for humans to visually track fish of the same species in video sequences captured in the wild. This is especially true when the water is turbid, the camera resolution is low, and the objects swim close to each other as is the case in some of the BrackishMOT sequences. This indicates that visual cues for re-identifying the objects are not pronounced and likely not reliable for solving this specific problem. Therefore, we conduct experiments based on the state-of-the-art tracker Center-Track [51], which tracks objects as points and focuses on associating objects locally between consecutive frames with little emphasis on visual features.

As a basis for our experiments, we use two pre-trained models provided by the authors of CenterTrack and fine-tune on top of them to reduce training time and minimize the potential of overfitting. We name the base models CT-COCO and CT-ImNet, where CT-COCO has been pre-trained on the MS COCO dataset [19] and CT-ImNet has been pre-trained on the ImageNet dataset [7]. Both models have a similar architecture with a DLA-34 [47] backbone. We train all our models with a batch size of 12 and a learning rate of 1.25e-4 and we resize and pad the BrackishMOT images from 1920×1080 to 960×544 following the strategy proposed by the CenterTrack authors.

First, we evaluate how pre-training on the two large-scale datasets MS COCO and ImageNet affects CenterTrack's ability to learn from the BrackishMOT sequences. We then extend these results by introducing training strategies for including synthetic sequences, to investigate the potential benefits of using synthetic tracking data in underwater environments. We evaluate all our models based on conventional MOT performance metrics like the Multiple Object Tracking Accuracy metric (MOTA) from the CLEAR MOT metrics [2], the ID F1 score (IDF1) [37], and the recent Higher Order Tracking Accuracy metric (HOTA) [21].

Training on Real Data. We fine-tune the base models on the BrackishMOT train split, which consists of 78 sequences, for 30 epochs and name the new models CT-COCO-Brack and CT-ImNet-Brack. Evaluating these models on the 20 sequences of the BrackishMOT test split gives us an indication of the performance to be expected from fine-tuning a state-of-the-art tracker on manually annotated real data. The HOTA, MOTA, and IDF1 results are presented in Table 1 along with detections (*Dets*), ground truth detections (*GT dets*), IDs, ground truth IDs (*GT IDs*), and ID switches (*IDSW*).

Table 1. Performance of the CenterTrack models fine-tuned on the BrackishMOT train sequences and evaluated on the BrackishMOT test split.

Model	HOTA↑	MOTA↑	IDF1↑	Dets	GT dets	IDs	GT IDs	IDSW
CT-COCO-Brack	0.36	0.37	0.39	10270	14670	887	182	493
CT-ImNet-Brack	**0.38**	**0.43**	**0.44**	10056	14670	755	182	464

We see that both models deliver promising results although the model pre-trained on ImageNet outperforms the model pre-trained on MS COCO. This indicates that ImageNet is better suited as a foundation for detecting and tracking objects in this type of underwater environment. We will investigate whether this is also the case when including synthetic data in the following sections.

Training on Synthetic Data. Next, we investigate whether the base models can be taught to track fish in real sequences if they are fine-tuned strictly on synthetic data. We do this by studying how the combinations of the environment with turbidity (T), background (B), and distractors (D) affect the tracking performance. We fine-tune the base models for 10 epochs on the eight different sets of synthetic sequences. The synthetic sequences only contain the *small fish* class, therefore, we evaluate the fine-tuned models on a sub-set of the BrackishMOT test split consisting of the sequences with the *small fish* class. We name this sub-set the 'small fish split' and it contains eight sequences (the list of sequences is presented in Table 4 as part of another evaluation).

Table 2. Performance of CenterTrack models trained strictly on variations of the synthetic dataset. The models have been evaluated on the small fish split.

CT-COCO	HOTA↑	MOTA↑	IDF1↑	**CT-ImNet**	HOTA↑	MOTA↑	IDF1↑
Synth	0.08	−0.17	0.07	Synth	0.08	−0.90	0.06
Synth$_B$	0.12	−0.21	0.12	Synth$_B$	0.14	0.05	0.17
Synth$_T$	0.08	−42.02	0.06	Synth$_T$	0.06	−0.93	0.04
Synth$_D$	0.09	−0.12	0.09	Synth$_D$	0.08	0.03	0.08
Synth$_{TD}$	0.12	−0.14	0.12	Synth$_{TD}$	0.06	0.02	0.05
Synth$_{BT}$	0.19	0.16	0.21	Synth$_{BT}$	**0.19**	−0.24	**0.21**
Synth$_{BD}$	0.15	0.08	0.16	Synth$_{BD}$	0.17	**0.13**	0.19
Synth$_{BTD}$	**0.21**	**0.18**	**0.24**	Synth$_{BTD}$	0.13	0.00	0.14
CT-COCO-Brack	0.37	0.47	0.43	CT-ImNet-Brack	0.39	0.50	0.46

The results of the synthetically trained models evaluated on the small fish split are presented in Table 2 along with the results of the CT-COCO-Brack and CT-ImNet-Brack models for comparison. Although the synthetic models only perform up to half as well as the models trained on real data, we see that it is in fact possible to train CenterTrack to be able to detect and track the *small fish* class without ever seeing real images of the class. The feature that seems to increase the tracking performance the most is by adding background videos whereas the turbidity and distractors give mixed results. We see that the CT-COCO model performs the best when fine-tuned on the $Synth_{BTD}$ sequences while it is more unclear what benefits the CT-ImNet model the most, however, a good compromise seems to be the $Synth_{BD}$ sequences.

Two Strategies for Training on both Synthetic and Real Data. Previously, we found the synthetic sequences best suited for teaching the base models to track the *small fish* class. Now, we examine whether the CT-COCO model fine-tuned on $Synth_{BTD}$ and the CT-ImNet model fine-tuned on $Synth_{BD}$ provide better foundations compared to the base models. We fine-tune on top of these models for 30 epochs on the BrackishMOT train sequences and name these two-step fine-tuned models CT-COCO-Synth$_{BTD}$ and CT-ImNet-Synth$_{BD}$. Additionally, we examine the potential benefits of combining real and synthetic data in a single training step by fine-tuning the base models for 30 epochs on a combination of the BrackishMOT train and $Synth_{BTD}$ sequences. We name these the CT-COCO-Mix and CT-ImNet-Mix models.

Baseline results for the models are presented in Table 3 for both the regular test split and the small fish split. We use the small fish split to examine whether the models trained on the synthetic data overfits to the *small fish* class. Generally, we see a tendency that the ImageNet pre-trained models perform better than the models pre-trained on the MS COCO dataset, which indicates that the ImageNet dataset lays a stronger foundation for detecting the objects of the BrackishMOT dataset. Furthermore, the CT-COCO models do not seem to benefit from the synthetic data, which is in contrast to the results presented in Table 2 that showed that fine-tuning the CT-COCO model on the $Synth_{BTD}$ sequences gave the best performing purely synthetically trained tracker.

For the CT-ImNet-Synth$_{BT}$ model we see a slight decrease in MOTA and IDF1 when evaluating on the test split, but an increase in the *small fish* sequences, this indicates that the model learns from the synthetic data to better track the *small fish* objects but at the expense of some of the other classes. The CT-ImNet-Mix model exhibits similar performance as the CT-ImNet-Synth$_{BT}$ model on the *small fish* sequences. However, the performance is also increased when looking at all the test sequences, which indicates that the ability to track the other classes is maintained using this training strategy.

Table 3. Baseline tracking results for the BrackishMOT test split and the small fish split.

Model		HOTA↑	MOTA↑	IDF1↑		HOTA↑	MOTA↑	IDF1↑
CT-COCO-Brack		0.36	0.37	0.39		0.37	0.47	0.43
CT-COCO-Synth$_{BTD}$		0.36	0.38	0.39		0.39	0.47	0.44
CT-COCO-Mix	Test split	0.36	0.37	0.39	Small fish split	0.37	0.46	0.43
CT-ImNet-Brack		0.38	0.43	0.44		0.39	0.50	0.46
CT-ImNet-Synth$_{BT}$		0.38	0.42	0.41		**0.41**	**0.52**	0.48
CT-ImNet-Mix		**0.40**	**0.44**	**0.45**		**0.41**	**0.52**	**0.49**

5.1   Qualitative Evaluation

In the previous evaluation, we found the overall best-performing model to be CT-ImNet-Mix. In this section, we analyse how the model performs on each of the eight sequences from the small fish split. The qualitative results of the CT-ImNet-Mix model when evaluated on the small fish split are presented in Table 4. When we inspect the brackishMOT-93 and brackishMOT-95 sequences we see that they have 45 and 1 GT IDs, respectively. However, both sequences score a HOTA performance of 0.44 indicating that a higher number of objects does not seem to have a significantly negative impact on the tracking performance. If we look at BrackishMOT-67 it has four GT IDs but the tracker only manages to get a HOTA score of 0.18. Inspecting the BrackishMOT-67 sequence visually shows that it contains a single medium-sized object of the *small fish* class, which the model tracks well throughout the sequence, however, there are also three tiny objects of the *small fish* class near the seabed that the model largely fails to detect and this penalizes the tracking performance greatly.

Table 4. Performance of CT-ImNet-Mix model on the sequences of the small fish split.

Sequence	HOTA↑	MOTA↑	IDF1↑	Dets	GT dets	IDs	GT IDs	IDSW
brackishMOT-50	0.40	0.46	0.47	1785	2129	105	17	50
brackishMOT-55	0.35	0.46	0.43	2318	3192	187	37	112
brackishMOT-56	0.53	0.41	0.75	80	87	7	1	4
brackishMOT-67	0.18	0.11	0.10	173	636	31	4	7
brackishMOT-90	0.61	0.53	0.74	401	426	18	3	7
brackishMOT-93	0.44	0.51	0.51	1450	1567	160	45	82
brackishMOT-95	0.44	0.38	0.39	619	148	28	1	0
brackishMOT-98	0.49	0.58	0.60	1728	1930	137	36	80

6   Conclusion

We propose a new underwater multi-object tracking dataset named BrackishMOT, which is an extension of the Brackish dataset captured in turbid waters in Denmark. This is the first and only dataset of its kind and it is a necessary step towards increasing the capability of underwater trackers as there currently only exist very few underwater tracking datasets and they have all been captured in clear tropical waters. Furthermore, we propose a framework for generating synthetic underwater MOT sequences and present baseline results based on fine-tuning CenterTrack using three different training strategies. We show that tracking performance can be increased by including sequences generated by the proposed synthetic framework in the training procedure.

Acknowledgements. This work has been funded by the Independent Research Fund Denmark under the case number 9131-00128B.

References

1. Bergmann, P., Meinhardt, T., Leal-Taixe, L.: Tracking without bells and whistles. In: 2019 IEEE/CVF International Conference on Computer Vision (ICCV). IEEE (Oct 2019). https://doi.org/10.1109/iccv.2019.00103
2. Bernardin, K., Stiefelhagen, R.: Evaluating multiple object tracking performance: The CLEAR MOT metrics. EURASIP J. Image Video Process. **2008**(1), 1–10 (2008). https://doi.org/10.1155/2008/246309
3. Bewley, M., et al.: Australian sea-floor survey data, with images and expert annotations. Sci. Data **2**(1) (Oct 2015). https://doi.org/10.1038/sdata.2015.57
4. Bochkovskiy, A., Wang, C.Y., Liao, H.Y.M.: Yolov4: Optimal speed and accuracy of object detection (2020). https://doi.org/10.48550/ARXIV.2004.10934
5. Community, B.O.: Blender - a 3D modelling and rendering package. Blender Foundation, Stichting Blender Foundation, Amsterdam (2018), http://www.blender.org
6. Dendorfer, P., et al.: Mot20: A benchmark for multi object tracking in crowded scenes (2020). https://doi.org/10.48550/ARXIV.2003.09003
7. Deng, J., Dong, W., Socher, R., Li, L.J., Li, K., Fei-Fei, L.: ImageNet: A large-scale hierarchical image database. In: 2009 IEEE Conference on Computer Vision and Pattern Recognition. IEEE (Jun 2009). https://doi.org/10.1109/cvpr.2009.5206848
8. Fisher, R.B., Chen-Burger, Y.-H., Giordano, D., Hardman, L., Lin, F.-P. (eds.): ISRL, vol. 104. Springer, Cham (2016). https://doi.org/10.1007/978-3-319-30208-9
9. Geiger, A., Lenz, P., Urtasun, R.: Are we ready for autonomous driving? the KITTI vision benchmark suite. In: 2012 IEEE Conference on Computer Vision and Pattern Recognition. IEEE (Jun 2012). https://doi.org/10.1109/cvpr.2012.6248074
10. Giordano, D., Palazzo, S., Spampinato, C.: Fish tracking. In: Fish4Knowledge: Collecting and Analyzing Massive Coral Reef Fish Video Data, pp. 123–139. Springer International Publishing (2016). https://doi.org/10.1007/978-3-319-30208-9_10
11. Hartman, C., Beneš B.: Autonomous boids. Comput. Animation Virtual Worlds **17**(3–4), 199–206 (2006). https://doi.org/10.1002/cav.123
12. Ishiwaka, Y., et al.: Foids. ACM Trans. Graph. **40**(6), 1–15 (2021). https://doi.org/10.1145/3478513.3480520
13. Jäger, J., Wolff, V., Fricke-Neuderth, K., Mothes, O., Denzler, J.: Visual fish tracking: Combining a two-stage graph approach with CNN-features. In: OCEANS 2017 - Aberdeen. IEEE (Jun 2017). https://doi.org/10.1109/oceanse.2017.8084691
14. Kano, Y., et al.: An online database on freshwater fish diversity and distribution in mainland southeast asia. Ichthyol. Res. **60**(3), 293–295 (2013). https://doi.org/10.1007/s10228-013-0349-8
15. Kezebou, L., Oludare, V., Panetta, K., Agaian, S.S.: Underwater object tracking benchmark and dataset. In: 2019 IEEE International Symposium on Technologies for Homeland Security (HST). IEEE (Nov 2019). https://doi.org/10.1109/hst47167.2019.9032954
16. Krizhevsky, A., Sutskever, I., Hinton, G.E.: Imagenet classification with deep convolutional neural networks. In: Pereira, F., Burges, C., Bottou, L., Weinberger, K. (eds.) Advances in Neural Information Processing Systems. vol. 25. Curran Associates, Inc. (2012), https://proceedings.neurips.cc/paper/2012/file/c399862d3b9d6b76c8436e924a68c45b-Paper.pdf
17. Leal-Taixé, L., Milan, A., Reid, I., Roth, S., Schindler, K.: Motchallenge 2015: Towards a benchmark for multi-target tracking (2015). https://doi.org/10.48550/ARXIV.1504.01942
18. Lin, T.Y., Goyal, P., Girshick, R., He, K., Dollar, P.: Focal loss for dense object detection. IEEE Trans. Pattern Anal. Mach. Intell. **42**(2), 318–327 (2020). https://doi.org/10.1109/tpami.2018.2858826

19. Lin, T.-Y., et al.: Microsoft COCO: common objects in context. In: Fleet, D., Pajdla, T., Schiele, B., Tuytelaars, T. (eds.) ECCV 2014. LNCS, vol. 8693, pp. 740–755. Springer, Cham (2014). https://doi.org/10.1007/978-3-319-10602-1_48
20. Liu, T., Li, P., Liu, H., Deng, X., Liu, H., Zhai, F.: Multi-class fish stock statistics technology based on object classification and tracking algorithm. Eco. Inform. **63**, 101240 (2021). https://doi.org/10.1016/j.ecoinf.2021.101240
21. Luiten, J., et al.: HOTA: A higher order metric for evaluating multi-object tracking. International Journal of Computer Vision **129**(2), 548–578 (Oct 2020). https://doi.org/10.1007/s11263-020-01375-2
22. Madsen, N., Pedersen, M., Jensen, K.T., Møller, P.R., Andersen, R.E., Moeslund, T.B.: Fishing with c-tucs (cheap tiny underwater cameras) in a sea of possibilities. J. Ocean Technol.**16**(2), 19–30 (2021), https://www.thejot.net/article-preview/?show_article_preview=1250
23. Mahmood, A., et al.: Automatic detection of western rock lobster using synthetic data. ICES J. Mar. Sci. **77**(4), 1308–1317 (2019). https://doi.org/10.1093/icesjms/fsz223
24. Mandel, T., et al.: Detection confidence driven multi-object tracking to recover reliable tracks from unreliable detections. Pattern Recogn. **135**, 109107 (2023). https://doi.org/10.1016/j.patcog.2022.109107
25. Martija, M.A.M., Naval, P.C.: SynDHN: Multi-object fish tracker trained on synthetic underwater videos. In: 2020 25th International Conference on Pattern Recognition (ICPR). IEEE (Jan 2021). https://doi.org/10.1109/icpr48806.2021.9412291
26. Milan, A., Leal-Taixé, L., Reid, I., Roth, S., Schindler, K.: Mot16: A benchmark for multi-object tracking (2016). https://doi.org/10.48550/ARXIV.1603.00831
27. Musić, J., Kružić, S., Stančić, I., Alexandrou, F.: Detecting underwater sea litter using deep neural networks: An initial study. In: 2020 5th International Conference on Smart and Sustainable Technologies (SpliTech). IEEE (Sep 2020). https://doi.org/10.23919/splitech49282.2020.9243709
28. de Oliveira Barreiros, M., de Oliveira Dantas, D., de Oliveira Silva, L.C., Ribeiro, S., Barros, A.K.: Zebrafish tracking using YOLOv2 and kalman filter. Sci. Reports **11**(1) (Feb 2021). https://doi.org/10.1038/s41598-021-81997-9
29. Panetta, K., Kezebou, L., Oludare, V., Agaian, S.: Comprehensive underwater object tracking benchmark dataset and underwater image enhancement with GAN. IEEE J. Oceanic Eng. **47**(1), 59–75 (2022). https://doi.org/10.1109/joe.2021.3086907
30. Pedersen, M., Bruslund Haurum, J., Gade, R., Moeslund, T.B.: Detection of marine animals in a new underwater dataset with varying visibility. In: Proceedings of the IEEE/CVF Conference on Computer Vision and Pattern Recognition Workshops, pp. 18–26 (2019)
31. Pedersen, M., Haurum, J.B., Bengtson, S.H., Moeslund, T.B.: 3d-ZeF: A 3d zebrafish tracking benchmark dataset. In: 2020 IEEE/CVF Conference on Computer Vision and Pattern Recognition (CVPR). IEEE (Jun 2020). https://doi.org/10.1109/cvpr42600.2020.00250
32. Pedersen, M., Haurum, J.B., Dendorfer, P., Moeslund, T.B.: MOTCOM: The multi-object tracking dataset complexity metric. In: Lecture Notes in Computer Science, pp. 20–37. Springer Nature Switzerland (2022). https://doi.org/10.1007/978-3-031-20074-8_2
33. Pedersen, M., Madsen, N., Moeslund, T.B.: No machine learning without data: Critical factors to consider when collecting video data in marine environments. J. Ocean Technol. 16(3), (2021)
34. Podila, S., Zhu, Y.: Animating escape maneuvers for a school of fish. In: Proceedings of the 21st ACM SIGGRAPH Symposium on Interactive 3D Graphics and Games. ACM (Feb 2017). https://doi.org/10.1145/3023368.3036845
35. Ren, S., He, K., Girshick, R., Sun, J.: Faster r-CNN: Towards real-time object detection with region proposal networks. IEEE Trans. Pattern Anal. Mach. Intell. **39**(6), 1137–1149 (2017). https://doi.org/10.1109/tpami.2016.2577031

36. Reynolds, C.W.: Flocks, herds and schools: A distributed behavioral model. In: Proceedings of the 14th annual conference on Computer graphics and interactive techniques - SIGGRAPH '87. ACM Press (1987). https://doi.org/10.1145/37401.37406
37. Ristani, E., Solera, F., Zou, R., Cucchiara, R., Tomasi, C.: Performance measures and a data set for multi-target, multi-camera tracking. In: Hua, G., Jégou, H. (eds.) ECCV 2016. LNCS, vol. 9914, pp. 17–35. Springer, Cham (2016). https://doi.org/10.1007/978-3-319-48881-3_2
38. Romero-Ferrero, F., Bergomi, M.G., Hinz, R.C., Heras, F.J.H., de Polavieja, G.G.: idtracker.ai: tracking all individuals in small or large collectives of unmarked animals. Nature Methods 16(2), 179–182 (Jan 2019). https://doi.org/10.1038/s41592-018-0295-5
39. Simonyan, K., Zisserman, A.: Very deep convolutional networks for large-scale image recognition (2014). https://doi.org/10.48550/ARXIV.1409.1556
40. Stephens, K., Pham, B., Wardhani, A.: Modelling fish behaviour. In: Proceedings of the 1st international conference on Computer graphics and interactive techniques in Australasia and South East Asia. ACM (Feb 2003). https://doi.org/10.1145/604471.604488
41. Technologies, U.: Unity (2005), https://www.unity.com, Accessed 21 Mar 2023
42. Tobin, J., Fong, R., Ray, A., Schneider, J., Zaremba, W., Abbeel, P.: Domain randomization for transferring deep neural networks from simulation to the real world. In: 2017 IEEE/RSJ International Conference on Intelligent Robots and Systems (IROS). IEEE (Sep 2017). https://doi.org/10.1109/iros.2017.8202133
43. Tremblay, J., et al.: Training deep networks with synthetic data: Bridging the reality gap by domain randomization. In: 2018 IEEE/CVF Conference on Computer Vision and Pattern Recognition Workshops (CVPRW). IEEE (Jun 2018). https://doi.org/10.1109/cvprw.2018.00143
44. UnitedNations: Life below water. https://www.un.org/sustainabledevelopment/goal-14-life-below-water/ (2021) Accessed 21 Mar 2023
45. Wang, H., Zhang, S., Zhao, S., Wang, Q., Li, D., Zhao, R.: Real-time detection and tracking of fish abnormal behavior based on improved YOLOV5 and SiamRPN++. Comput. Electron. Agric. 192, 106512 (2022). https://doi.org/10.1016/j.compag.2021.106512
46. Xu, Y., Ošep, A., Ban, Y., Horaud, R., Leal-Taixé, L., Alameda-Pineda, X.: How to train your deep multi-object tracker. In: 2020 IEEE/CVF Conference on Computer Vision and Pattern Recognition (CVPR). IEEE (Jun 2020). https://doi.org/10.1109/cvpr42600.2020.00682
47. Yu, F., Wang, D., Shelhamer, E., Darrell, T.: Deep layer aggregation. In: 2018 IEEE/CVF Conference on Computer Vision and Pattern Recognition. IEEE (Jun 2018). https://doi.org/10.1109/cvpr.2018.00255
48. Yu, H., Li, G., Zhang, W., Huang, Q., Du, D., Tian, Q., Sebe, N.: The unmanned aerial vehicle benchmark: object detection, tracking and baseline. Int. J. Comput. Vision 128(5), 1141–1159 (2019). https://doi.org/10.1007/s11263-019-01266-1
49. Zhang, Y., et al.: ByteTrack: Multi-object tracking by associating every detection box. In: Lecture Notes in Computer Science, pp. 1–21. Springer Nature Switzerland (2022). https://doi.org/10.1007/978-3-031-20047-2_1
50. Zhang, Y., Wang, C., Wang, X., Zeng, W., Liu, W.: FairMOT: on the fairness of detection and re-identification in multiple object tracking. Int. J. Comput. Vision 129(11), 3069–3087 (2021). https://doi.org/10.1007/s11263-021-01513-4
51. Zhou, X., Koltun, V., Krähenbühl, P.: Tracking objects as points. In: Vedaldi, A., Bischof, H., Brox, T., Frahm, J.-M. (eds.) ECCV 2020. LNCS, vol. 12349, pp. 474–490. Springer, Cham (2020). https://doi.org/10.1007/978-3-030-58548-8_28

Camera Calibration Without Camera Access - A Robust Validation Technique for Extended PnP Methods

Emil Brissman[1,2]([✉]) [iD], Per-Erik Forssén[1] [iD], and Johan Edstedt[1] [iD]

[1] Computer Vision Laboratory, Department EE, Linköping University, Linköping, Sweden
{emil.brissman,per-erik.forssen,johan.edstedt}@liu.se
[2] Saab, Linköping, Sweden

Abstract. A challenge in image based metrology and forensics is intrinsic camera calibration when the used camera is unavailable. The unavailability raises two questions. The first question is how to find the *projection model* that describes the camera, and the second is to detect incorrect models. In this work, we use off-the-shelf extended PnP-methods to find the model from 2D-3D correspondences, and propose a method for model validation. The most common strategy for evaluating a projection model is comparing different models' residual variances—however, this naive strategy cannot distinguish whether the projection model is potentially underfitted or overfitted. To this end, we model the residual errors for each correspondence, individually scale all residuals using a predicted variance and test if the new residuals are drawn from a standard normal distribution. We demonstrate the effectiveness of our proposed validation in experiments on synthetic data, simulating 2D detection and Lidar measurements. Additionally, we provide experiments using data from an actual scene and compare non-camera access and camera access calibrations. Last, we use our method to validate annotations in MegaDepth.

1 Introduction

Intrinsic camera calibration is a fundamental computer vision problem. It involves finding the parameters that allow the conversion of pixel coordinates to bearing angles [12]. It is possible to use the camera for *metrology* using a calibration. In the single-view case, metrology means measuring the lengths and angles of objects depicted in an image. As an extension, it is the underpinning of single view 3D reconstruction [4]. Metrology has many applications, including non-contact measurements, sensor fusion, and forensic analysis.

Traditionally, intrinsic calibration is a semi-automatic process, which involves imaging of calibration objects [28,30]. Such calibration allows controlled accuracy; however, access to the camera is required. In forensic analysis, the camera is only sometimes available, depending on the received material. Therefore, we aim to facilitate measurements in an image when the camera is unavailable.

R. Gade et al. (Eds.): SCIA 2023, LNCS 13885, pp. 34–49, 2023.
https://doi.org/10.1007/978-3-031-31435-3_3

Using a calibration profile from a camera of the same model often works well, but the accuracy is unknown in this approach and should thus be avoided in forensics.

In the Perspective-n-Point (PnP) problem, the goal is to estimate the camera pose given a set of 2D-3D point correspondences. Early methods assume a calibrated camera, and only estimate translation and rotation parameters [7]. More recent variants of PnP also estimate the intrinsic camera parameters [14,21]. These *extended PnP methods* (xPnP) do not require the camera to be available, in contrast to calibration pattern methods [28,30]. However, they introduce new challenges such as 2D-3D matching and validation.

In this work, we attend to the validation of camera calibration for forensic metrology applications [2]. Usually, a model is assumed to be validated if it, on average, has low residuals. However, this approach will not provide any measure of uncertainty in the image plane. Moreover, deriving the uncertainty is challenging because the amount of distortion scales non-linearly with the distance to the camera centre. Thus, we treat noise modelling as a robust regression problem and predict a residual scaling for each 2D-3D correspondence. When the model is correct, we assume the scaled residuals to follow a standard normal distribution (Fig. 1). Next, to verify this assumption, we use a hypothesis test. Simulated data, an indoor scene and MegaDepth [18], with annotated cameras depicting different scenes, demonstrate our proposed validation.

Contributions. Our contributions are as follows: **(i)** We propose a method for testing residuals based on variance predictions and standardisation. **(ii)** We suggest using xPnP methods for unavailable cameras as input to our method, given 2D-3D correspondences. **(iii)** An empirical estimate of the variance scales residuals poorly. Instead, we propose a predictive noise model to scale individual residuals over the 2D detector and projected 3D noise. **(iv)** We analyse

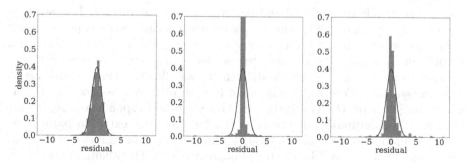

Fig. 1. Left: Standardised residuals for a correct model with one distortion parameter using our robust scale estimate. I.e. the residuals are not affected by the model error. Middle: Standardised residuals from images under more distortion, for an incorrect model using a non-robust scale estimate. Right: Standardised residuals for an incorrect model, using a robust scale estimate.

the effectiveness of our method in quantitative and qualitative experiments and demonstrate its ability to significantly predict incorrect models, also when the mean of the residuals is low.

1.1 Background Motivation

At the Swedish National Forensic Center (NFC), the task is to collect information linked to crimes without the possibility of misinterpretations when used in the Swedish court system. At the time of writing, NFC uses the Zhang method [30] for metrology. However, this commonly accepted practice only applies to images where the camera is available. Lidar scanning is a standard technique in forensic investigations, and in many legal cases, the depicted location is revisited for scanning. Using this working methodology, Olsson [22] first investigated the validation of xPnP methods that forms the basis of this work. In [22], model correctness is assessed by checking if the empirical mean of the re-projected sample distances is within the two centre quartiles. However, this decision will prefer incorrect models since outliers will expand the decision range.

1.2 Ethical Consideration

This work does not concern any police investigations or legal cases. Instead, the method we propose analyses residuals using synthetic data, available benchmark data, and a snapshot of a fictional crime scene provided by a police agency. These are all free from apparent ethical dilemmas.

2 Related Work

Semi-Automatic Calibration. Camera calibration is a broad subject found in many areas of industry and research. However, the most common camera calibration practice is to use a printed pattern on a planar surface. This strategy was proposed by Zhang [30], who suggested using a checkerboard pattern with equidistant squares of black and white. The inner corners of the pattern form unambiguous features that are easy to find. Detecting several of these features, also called saddle points, between different views allows camera parameters to be estimated. Each detected saddle point in each picture is assigned to its corresponding point on the checkerboard. This set of correspondences represents a series of homographies, determining the intrinsic and extrinsic parameters for one or more cameras. In the case of Tsai [28], camera calibration depends only on one view of a co-planar checkerboard pattern. The Zhang method [30], instead depends on at least three pictures of a planar checkerboard pattern. More recently, deep methods like Li *et al.* [17] take a single image as input and jointly learn to predict distortion coefficients and optical flow from images with lens-type annotations. However, this problem only concerns visual quality and provides no model accuracy assessment.

Perspective-n-Point. The PnP problem [7] refers to finding a rigid transformation from 2D-3D point matches. That is, to estimate the rotation matrix $\mathbf{R}$ and the translation vector $\mathbf{t}$ describing the camera pose in the coordinate system of the 3D points, assuming that the intrinsic parameters never changed during the sampling of a scene. The minimal but ambiguous case, P3P [9,19,23] is not considered in this work.

Lepetit *et al.* [16] (EPnP), reduced the computational complexity to $O(n)$ operations. Although the convergence is fast, the solution depends on initialization and global convergence is not guaranteed. Later works recognized the need to include intrinsic parameters to generalize application tasks [14,21]. Nakano [21] extends the PnP problem by including intrinsic parameters and dividing the parameter estimation into different stages. Radial distortion and equal focal length horizontally and vertically are assumed, as well as fixating the principal point to the image center. Larsson *et al.* [14] instead require a minimal correspondence set and add a local optimization step [15]. In our work, we propose a method to validate xPnP methods, for application in forensic analysis, by *Goodness-of-Fit* (GoF) testing between distributions. That is, we do not improve the methods [21] and [14].

Empirical Performance Evaluation. Works by Wang *et al.* [29] and Thai *et al.* [27] are related to our work. Wang *et al.* [29] propose a method to test hypotheses about the effect of conceptual changes in deep classification models. That is, if the difference is the probable reason behind the increase in accuracy. Thai *et al.* [27] propose to identify cameras by raw pixel intensities. Similar to our approach, two quantities parametrize the intensity variance–analog gain, controlled by the camera ISO setting, and electronic noise caused during sensor readout. For an unknown camera the parameters are first estimated and secondly tested against known parameter values (null hypothesis).

Goodness of Fit. The goodness of fit testing is one of the fundamental tasks in statistics. In this work, we focus on normality testing due to normal distributions being a good model for uncertainty in projective geometry [8,13]. Still, our approach could easily be generalized to GoF tests for arbitrary distributions.

There are a large variety of proposed statistics for normality testing, of which the Kolmogorov-Smirnov (KS) [20], D'Agostino-Pearson (DAP) [6], and Shapiro-Wilk (SW) [26] tests are well known. These tests all seek to maximize the power, i.e., minimizing the risk of the null hypothesis being accepted, given that an alternative hypothesis is correct. We discuss those further in Sect. 3.5 and test all three in Sect. 4.

Single View Metrology. Metrology is the study of measurement. In the context of computer vision, single view metrology [5] involves estimating, e.g., angles and lengths, from a single image. In all metrology, an accurate measure of uncertainty is crucial, and in particular in the forensic setting. Previous work in single view metrology has focused on the undistorted (but often uncalibrated) case [5], with model uncertainty assumed to be normally distributed [3,5,8,13]. At inference time, these uncertainties can be propagated by first order propagation or by Monte Carlo simulation. However, in those works, both the uncertainty estimation and propagation requires *a priori* knowledge of the noise levels and estimation method, and implicitly assume the estimated model is approximately correct. In contrast to those methods, our approach

1. Is estimation agnostic, i.e., we can treat the estimators as black boxes.
2. Generalises to arbitrary projection models.
3. Does not implicitly assume that the estimated model is approximately correct.

In particular, perturbation theory, as used in previous work, does not provide a reliable measure of the trustworthiness of the estimated model, it simply provides an approximate measure of the estimation sensitivity to the input. In contrast, our method directly measures trustworthiness by testing the hypothesis of the matches being generated from the estimated model.

3 Method

We propose a method that compares observed and expected noise levels. The method takes residual values as input, given a calibration computed from an xPnP method and 2D-3D correspondences. We decompose the residual error for each correspondence as three additive terms: (i) 2D detector noise, (ii) 3D detector noise projected into the image, and (iii) model noise. The expected model noise is zero for the correct model, not affecting the residual distribution in any direction. We describe this in Sect. 3.3. To handle unexpected model noise, Sect. 3.4 details robust regression over (i) and (ii) to obtain a scale value for each 2D point. Finally, we assume the scaled residuals are drawn from a standard normal distribution and test this using a GoF test. We motivate our preferred choice of test in Sect. 3.5. We consider all points to influence the validation decision and believe this to improve applications in forensic analysis. We begin with an example to get a good intuition of our approach.

3.1 Motivating Example

Consider a correct data model $y = x$ and an (incorrect) hypothesis $h_{bad} : y = x + 0.5x^5$. Under the assumption that y is observed with some Gaussian noise, the residuals r of the true model will be distributed as $\mathcal{N}(0, \sigma_y^2)$. In contrast, the residuals of the incorrect hypothesis are typically *significantly* different from the expected distribution (as shown in Fig. 1). Thus, if σ_y is known, a simple hypothesis test is whether $\frac{r}{\sigma_y} \sim \mathcal{N}(0, 1)$. However, in real world scenarios σ_y is typically not known and needs to be estimated. Since incorrect hypotheses typically contain outliers, it is important with a robust estimate of the noise level. We show these steps in Fig. 1. It is clear that h_{bad} produces a tailed residual distribution that does not follow the expected Gaussian curve. Hence we can use the KS test [20] to validate the produced models.

Underfittning and Overfitting. It is common for a complex model to be optimized to fit the data y perfectly. We can describe overfitting and underfitting as a constant multiplication of σ_y^2, yielding residuals distributed as $\mathcal{N}(0, a\sigma_y^2)$. When $a < 1$ the model is overfitted, and when $a > 1$ it is incorrect (underfittning). The following sections describe how we can apply this intuition to validate a camera calibration.

3.2   Camera Calibration

Calibration fundamentally depends on correspondences of point coordinates. An arbitrary camera, c, observes a set of K 3D points $\{\mathbf{X}_k\}_{k=1}^{K}$, and a set of corresponding image points $\{\mathbf{x}_k^c\}_{k=1}^{K}$. Point sets and correspondences are known $\forall k$, and for each camera. In this work, we consider xPnP based camera calibration using the methods proposed by [21] and [14]. Both extrinsic (rotation and translation) and intrinsic (focal length and distortion) parameters in (3) are computed to enable measurement of length and angles in the camera image.

Distortion. Depending on the optical system of a camera, small or large displacements of image coordinates can be introduced, called image distortion. Unlike the focal length, which scales the image uniformly, distortion is characterised as scaling the image differently depending on the distance to a distortion centre. The farther the pixels are from the centre of distortion, the more they are distorted. We let the same point represent the distortion and optical centra, which is assumed to be fixed and in the centre of the image.

$$\mathbf{y}' = g(\mathbf{y}, \boldsymbol{\theta}) \tag{1}$$

We model the distortion as in (1), and let $\boldsymbol{\theta} = [\theta_1, \theta_2, \theta_3]$ specify the non-linear distortion terms. When g uses one distortion term, it will be denoted as D(1,0) and as D(3,0) when all three terms are used, according to [14].

Correspondences. The calibration uses Lidar measurements, which map physical features with high precision by emitting narrow laser beams that are reflected back. Even if the image, whose camera we want to calibrate, and the lidar map are recorded at separate times, there should be enough overlapping features left for calibration. That is, consistent physical properties. Such properties, which are more likely to be consistent, are, for example, those found on buildings, vegetation, paintings, furniture, etc. In practice, correspondences can be of varying quality, making robust estimation a critical importance, when computing an xPnP solution [14]. Therefore, we use only the residuals from correspondences marked as inliers by the model estimator for model validation.

3.3   Residual Error Model

Regardless of whether the corresponding coordinates are found by an interest point detector, or whether they are manually annotated by a human, they will suffer from *detection noise*. This means that a location estimate $\tilde{\mathbf{x}}$ has a residual $\epsilon_{detector}$, compared to the ideal point location $\hat{\mathbf{x}}$. This residual is typically modelled as a 2D normal distribution:

$$\epsilon_{detector} = \tilde{\mathbf{x}} - \hat{\mathbf{x}} \sim \mathcal{N}(\mathbf{0}, \sigma_d^2 \mathbf{I}) \,. \tag{2}$$

For a successful calibration, the residual between a detected point, and the projection of the corresponding 3D point using the estimated parameters, should also satisfy (2). In other words:

$$\hat{\mathbf{x}} = \mathrm{proj}_{\Theta,\mathbf{P}}(\hat{\mathbf{X}}) = \mathbf{K}g(\pi(\mathbf{R}\hat{\mathbf{X}} + \mathbf{t}), \boldsymbol{\theta}) \,. \tag{3}$$

Here $\hat{\mathbf{X}}$ is the ideal 3D point, and π is the pinhole projection. The *intrinsic calibration*, $\Theta = (\mathbf{K}, \boldsymbol{\theta})$, and the *extrinsic calibration*, $\mathbf{P} = (\mathbf{R}, \mathbf{t})$ (the camera pose) are of course estimates in practice. We summarize the error caused by the estimation in an additive *modelling noise* term ϵ_{model}. We intend to explain the residuals by detection noise in the image and in the Lidar, and test whether the explanation holds using a test on the residual data, e.g. by the DAP test [6], or by the KS test [20], testing the GoF.

For 2D-3D matches, the 3D points are also affected by noise ϵ_{3D}. Thus, (3) should be replaced by:

$$\hat{\mathbf{x}} + \epsilon_{lidar} + \epsilon_{model} = \mathrm{proj}_{\Theta,\mathbf{P}}(\hat{\mathbf{X}} + \epsilon_{3D}) \,, \tag{4}$$

where ϵ_{3D} is the detection noise in 3D, and ϵ_{lidar} its projection. By combining (4) with (2) we obtain the following residual model:

$$\epsilon = \epsilon_{detector} + \epsilon_{lidar} + \epsilon_{model} \,. \tag{5}$$

We model the detection error as in (2), and describe the lidar error model in detail below.

Lidar Error Model. For a Lidar sensor, the 3D noise has both angular and depth components. However, when the camera and 3D-sensor are close to being co-axial, and point in roughly the same direction (i.e. $\mathbf{t}$ is small, and $\mathbf{R} \approx \mathbf{I}$ in (3)), the depth error becomes irrelevant, and the projection in the image ϵ_{lidar} is dominated by $\mathbf{K}$, which is affine. This means that the shape of ϵ_{lidar} is a simple, but location dependent scaling.

We thus model the projection of the Lidar error ϵ_{lidar} as:

$$\epsilon_{lidar} \sim \mathcal{N}(0, \sigma_l^2 \mathrm{diag}(a_x^2, a_y^2)) \,, \tag{6}$$

where σ_l^2 is a noise variance, and a_x, a_y are the noise scalings in horizontal and vertical directions. These depend on the location in the image. To estimate a_x, a_y, we can project the current 3D point and its neighbours in pan and tilt directions to obtain:

$$a_{x,k} = \|\mathrm{proj}(\mathbf{X}_k) - \mathrm{proj}(\mathbf{X}_k^P)\| \tag{7}$$

$$a_{y,k} = \|\mathrm{proj}(\mathbf{X}_k) - \mathrm{proj}(\mathbf{X}_k^T)\| \,, \tag{8}$$

where $\mathbf{X}_k$ is the current 3D point, and $\mathbf{X}_k^P$, and $\mathbf{X}_k^T$ are its neighbours in pan and tilt directions.

3.4 Noise Estimation

The parameters of the detector errors in the model (5) can be fitted to the observed residuals using robust linear regression. However, estimating the variance of ϵ_{model} is neglected since its observed values are those that will remain in order to test whether the model is incorrect. I.e. we assume:

$$E\left\{\epsilon^2\right\} = \sigma_d^2 + \sigma_l^2. \tag{9}$$

By using a common σ_d for x, and y image residuals, and the aspect ratio model in (6) we obtain:

$$E\left\{\begin{pmatrix}\epsilon_x^2\\\epsilon_y^2\end{pmatrix}\right\} = \begin{pmatrix}1 & a_x^2\\1 & a_y^2\end{pmatrix}\begin{pmatrix}\sigma_d^2\\\sigma_l^2\end{pmatrix}. \tag{10}$$

We fit these to the observed K residuals, for each camera, c, separately, to obtain the regressor parameters (σ_d, σ_l).

$$\left(\epsilon_{x,1}^2\ \epsilon_{y,1}^2\ \cdots\ \epsilon_{x,K}^2\ \epsilon_{y,K}^2\right)^{\mathsf{T}} = \begin{pmatrix}1 & 1 & \cdots & 1 & 1\\a_{x,1}^2 & a_{y,1}^2 & \cdots & a_{x,K}^2 & a_{y,K}^2\end{pmatrix}^{\mathsf{T}}\begin{pmatrix}\sigma_d^2\\\sigma_l^2\end{pmatrix} \tag{11}$$

In practice, we do not use linear regression by solving the normal equations to (11), but use robust regression using IRLS [11] with initial weights $1/p(\epsilon_k|\sigma = 5.0)$. We can now obtain standardised residuals:

$$\tilde{\epsilon}_k = \begin{pmatrix}\epsilon_{x,k}/\sigma_{k,x}\\\epsilon_{y,k}/\sigma_{k,y}\end{pmatrix} = \begin{pmatrix}\epsilon_{x,k}/\sqrt{\sigma_d^2 + a_{x,k}^2\sigma_l^2}\\\epsilon_{y,k}/\sqrt{\sigma_d^2 + a_{y,k}^2\sigma_l^2}\end{pmatrix}. \tag{12}$$

3.5 Hypothesis Testing

When the modelling error is low, the standardised residuals in (12) should pass a statistical test, such as, e.g., the KS test. We can thus use the test to check whether the calibration worked for a particular set of 2D-3D correspondences. More formally, we test the **null hypothesis** $\mathcal{H}_0$: *The standardised residuals* (12) *are distributed explicitly according to a standard normal*, against $\mathcal{H}_1$: *at least one value does not match that distribution*. Related to this classical approach is that the data we are testing is random, so the test decision is random too, which means there is still a small probability of an incorrect decision. Nevertheless, tests are useful to detect low model errors and thus further validate the calibration.

Evidence. The approach involves comparing the samples (residuals) with a statistical model under $\mathcal{H}_0$, where a test statistic measures the discrepancy between the data and the model. To this end, we use the KS test [20] and compute a *p-value*, measuring the error size of rejecting $\mathcal{H}_0$. Commonly, when the *p-value* is below 5%, $\mathcal{H}_0$ can be rejected in favour of $\mathcal{H}_1$. That is, the error probability is sufficiently low. However, this probability does not directly infer confidence for the data distributed as a standard normal.

Other tests also calculate a *p-value* to test the normality of data. For example, the DAP test [6] sums the discrepancies from a skewness test and a kurtosis test into a single *p-value*. Skewness is the asymmetry about the mean, and kurtosis is the measure of the "tailedness". Although parametric tests are preferable to non-parametric ones, and the SW test is one of the more powerful [10], we believe their null hypotheses to be non-directional, where a broader chance of normality is possible, leading to unstable decisions.

4   Experiments

We first evaluate the proposed method for testing a calibration using 2D-3D correspondences on synthetic data simulating detection and Lidar errors. Next, we provide results on a real scenario using Lidar measurements and compare this with a semi-automatic calibration. Last, we analyse a large-scale dataset using our method.

4.1   Synthetic Data

We implemented the simulator in OpenCV [1] and provide code at https:// github.com/emibr948/ccwca. We aim to render a fictitious checkerboard pattern with equidistant squares of black and white into a camera c with a small angular rotation maintaining the image centre as its viewpoint at a distance t. The pattern contains equally many saddle points (inner checkerboard corners) vertically as horizontally (15×15). The simulator iterates three main tasks to render each image, which is presented next. **(i)** The projection model has fixed intrinsic parameters according to D(3,0) [14]. That is, $f = 800$ and $\theta = [-0.0684, 0.0100, 0.0006]$. We let the distortion centre coincide with the image centre. Rotation parameters are randomly sampled in the range $\pm 15°$ relative to the z-axis. The translation can also be random, but in our generated

Table 1. Projection models used with the Zhang method [30]. Models M_1 and M_3 use the same number of parameters as the PnP methods D(1,0) [14], D(3,0) [14] and D(3,0) [21]. The standard deviation on a set of test images, determines how accurately the two cameras have been calibrated, *axis223m* and *axisp3364*. M_5 is the most accurate model. Models M_1-M_4 (approximately) share the same standard deviation, although models M_4 and M_5 only differ with one parameter. Our method instead decides M_1 as incorrect but M_3 as a plausible model, still usable for metrology.

Model	f_x	f_y	c_x	c_y	k_1	k_2	k_3	std(*axis223m*)	std(*axisp3364*)
M_1	✓				✓			2.99	2.91
M_2	✓		✓	✓	✓			2.96	2.73
M_3	✓				✓	✓	✓	3.03	2.95
M_4	✓		✓	✓	✓	✓	✓	3.19	2.52
M_5	✓	✓	✓	✓	✓	✓	✓	0.81	0.19

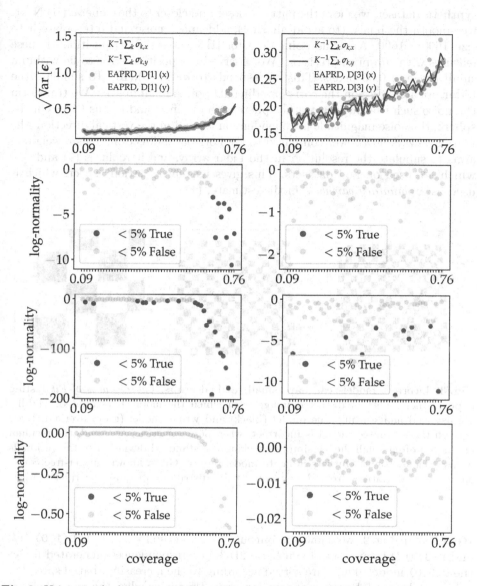

Fig. 2. Using residuals computed from 56 sets of simulated 2D-2D correspondences, we show the results for the incorrect D(1,0) model (left) and the correct D(3,0) model (right). We sort the correspondence sets in ascending order, using the area of the 2D points' convex hull (*coverage*). The second, third and fourth rows show the outcome of the KS [20], DAP [6] and SW [26] tests at level 5%. When the model is correct, the standard deviation is low (first row), and our predicted variance follows the corresponding empirical value. The KS test rejects images under an incorrect model, while accepting images under the correct model. In contrast, [6] is too strict, while [26] is too permissive.

synthetic dataset, we move the pattern closer and closer to the camera. **(ii)** Next, we smooth the image (to avoid aliasing), add image noise and interpolate it to size 1600×1600. A saddle point detector [1] locates the 2D position of these features with sub-pixel precision. We observed the position error to lie within a small range of 0.03 pixels. This corresponds to σ_d in (9).**(iii)** To simulate the Lidar, we add noise on the corresponding 3D points in all images. We transform the noise such that it lives on the sphere with origin t^c and radius $||t^c||$. On the sphere, the noise magnitude is dependent on t^c, and in the vertical direction, the noise is always 90% lower compared to the horizontal direction. This reweighing aims to simulate the resolution in the Lidar array, and it replaces (7) and (8), which are used on real datasets. This gives us the resolution aspect, which is used as *explanatory variables* in the estimate, (11).

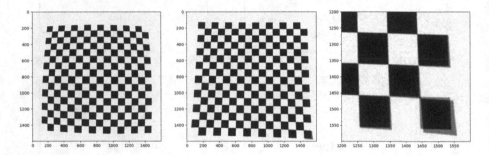

Fig. 3. Incorrect models can have small residual errors, but *Godness-of-Fit* testing exposes them. Left: A distorted image from which the model is estimated. Middle: Undistorted image using true model (black) and wrong model (grey) overlaid. Even though the estimated model is incorrect, when detector errors are small the residual errors are often small, hence simply checking the standard deviation of the residuals is insufficient. Right: Outliers indicate model failure. Given known outlier-free correspondences, deviations from the expected noise distribution expose incorrect models.

Results In Fig. 2, we show the output of our method on models D(1,0) [14] and D(3,0) [14] for a set of synthetic 2D-3D correspondences, generated under the D(3,0) model. These are sorted according to the increasing spatial spread of correspondences. The closer the points are to the image edge, the more they are affected by the distortion (Fig. 3). The first row in Fig. 2 shows the empirical and estimated standard deviations over all points per image, and rows 2–4 whether the scaled residuals are sufficient evidence to reject the null hypothesis for three different tests. Given in the second row of Fig. 2 is the output of the KS test for both D(1,0) [14] and D(3,0) [14]. To the left, in the same row, our method rejects the images with correspondences more uniformly spread over the entire image at level 5% using D(1,0) [14]. Making the same test using a projection model with more parameters fits the data more accurately, as indicated by both the low standard deviation and the test. We also test the scaled residuals in

Table 2. The results for the *axis223m* camera. For each PnP method, with distortion type DT, we report the empirical standard deviation and whether $\mathcal{H}_0$ can be rejected at level 5% using the KS, DAP or SW test.

PnP Method	DT	$\sqrt{\mathrm{Var}\,[\epsilon]}$	KS [20]	DAP [6]	SW [26]
EAPRD [14]	D(1,0)	2.23	✓	✓	
EAPRD [14]	D(3,0)	0.98		✓	
PNPRF [21]	D(3,0)	1.05		✓	

Table 3. The results for the *axisp3364* camera. For each PnP method, with distortion type DT, we report the empirical standard deviation and whether $\mathcal{H}_0$ can be rejected at level 5% using the KS, DAP or SW test.

PnP Method	DT	$\sqrt{\mathrm{Var}\,[\epsilon]}$	KS [20]	DAP [6]	SW [26]
EAPRD [14]	D(1,0)	1.71	✓	✓	
EAPRD [14]	D(3,0)	1.47			
PNPRF [21]	D(3,0)	1.08			

the third and fourth rows using [6], and [26], respectively. While these tests are parametric, they test for any normal distribution. Compared to the proposed KS test, this leads to false positives and negatives, see rows 3–4 of Fig. 2. In the second column, the stronger D(3,0) projection model is tested. In most images, $\mathcal{H}_0$ can not be rejected as expected due to the simulated data conforming to D(3,0) [14]. However, the output does not reveal the tests' differences in this case. Thus, [20] generally leads to more accurate decisions.

4.2 Lidar Measurements

Next, we compare our method using images from two real cameras, *axis223m* and *axisp3364*, respectively, and a 3D point cloud from a *Leica RTC360* scanner, with semi-automatic calibration. The second camera offers lower-quality images than the first, which is visible in Fig. 4. There is also no verified annotation for the cameras; in practice, there is none, and the camera can be inaccessible. The cameras instead depict a scene such that their optical axes have a relatively small angle to the 3D point cloud coordinate system's z-axis, similar to the simulations in Sect. 4.1.

Semi-Automatic. For semi-automatic calibration, we collect images for both cameras using a checkerboard pattern. The pattern has 6×6 saddle points. We split the set of images and estimate the model parameters on the first set using the Zhang method [30] implemented in [1]. Then, we calibrate using different projection models where M_5, in Table 1, achieves the lowest standard deviation on the second set of images (test). A factor of almost 4 differs between the accuracy of models M_4 and M_5. In Table 1, the number of model parameters differs by one.

Fig. 4. The first column shows two images of two cameras, *axis223m* and *axisp3364*, respectively. In the second and third columns, the undistortion looks to be visually removed for both D(1,0) [14], and D(3,0) [14]. Our method correctly detects D(1,0) [14] as incorrect for both cameras (Tables 2 and 3). The undistorted images in the fourth column are visually similar to their original, but this is not detected. For more details, see Sect. 4.2.

Without Camera Access. We show in Tables 2 and 3 that D(1,0) [14] is incorrect compared to D(3,0) [14] and D(3,0) [21] using the KS test on 24 manually annotated correspondences. Each of the annotated 3D points is visible in both cameras and projected to consistent features, e.g. corners. Similar to simulation, we observe that the parametric tests contradict each other and are thus infeasible for our application. While models M_1 to M_4, in Table 1, obtain higher standard deviations using [30], models M_1 and M_3 are equvivalent to the models used from [14] and [21], and thus there is possibility that M_5 is overfitted.

Finally, we found that the computed distortion parameters of D(3,0) [21] were all zero, shown in the rightmost column of Fig. 4. To our knowledge, [21] divides the xPnP problem into subproblems. In the subproblem that solves distortion, we can't find a condition on θ preventing the *normal equations* from giving the trivial solution. Thus, our method can not make the correct decision to either reject $\mathcal{H}_0$ or not based on residuals from D(3,0) [21].

4.3 Structure-from-Motion

In this experiment, we use our proposed method on annotations computed from a Structure-from-Motion (SfM) pipeline to get a broader insight into its effectiveness. To this end, we use 1000 images from each scene of MegaDepth [18]. This dataset contains many scenes with 2D-3D correspondences and camera intrinsic and extrinsic parameters given. The SfM pipeline, COLMAP [24,25], estimates the annotation parameters of the widely used benchmark for state-of-the-art comparison. In the dataset, the assumed projection model, a *simple radial*, models a single focal length, one distortion parameter, the distortion centre, rotation and translation. The histogram to the left in Fig. 5 shows that residuals are overall low. However, in 70 out of 100 images, our method rejects the null hypothesis at level 5%. The two images in the middle and to the right, in Fig. 5, show when $\mathcal{H}_0$ can not be rejected at level 5% and when $\mathcal{H}_0$ is rejected in favour of $\mathcal{H}_1$. We can thus assume mostly overfitted projection models in [18].

Fig. 5. Left: Density plot of residuals from 1000 images in all scenes, on which the annotation in MegaDepth depends. It is unlikely residuals will be high for images in [18] measuring a good performance. However, our proposed method tests each image and rejects the null hypothesis, $\mathcal{H}_0$, on 70 out of 100 images. Middle: Example of when $\mathcal{H}_0$ can not be rejected, and the *simple radial* projection model is suitable. Right: Example of when our method rejects $\mathcal{H}_0$. As can be seen, e.g. on the flagpole to the right, the images are distorted.

5   Conclusion

We suggested that metrology applications in forensic analysis use xPnP methods and use our proposed method to validate the calibration without camera access. The method formulation processes a single image, estimating a robust scaling of each correspondence and tests if the scaled set of residuals is drawn from a standard normal distribution. We demonstrate via qualitative and quantitative experiments that the KS test is most suitable and provide further insight from an extensive collection of annotated cameras.

Although we are sufficiently confident that the test can determine models as incorrect with a small margin of error, the challenge remains to infer confidence in the image measurements. A test is not a classification, and the *p-value* does not imply measurement confidence. However, when rejection of the null hypothesis is not possible at the acceptable error level, our error model explicitly provides the expected measurement errors over the image. Depending on the number of correspondences, we can get local estimates of expected measurement error from our assumptions of normally distributed residuals. Therefore, our method is a useful tool for xPnP camera calibration.

Acknowledgement. *This work was partially supported by the Wallenberg AI, Autonomous Systems, and Software Program (WASP) funded by the Knut and Alice Wallenberg Foundation; and the computations were enabled by the Berzelius resource provided by the Knut and Alice Wallenberg Foundation at the National Supercomputer Centre; and a point cloud of a realistic scene was provided by the Swedish National Forensic Centre (NFC).*

References

1. Bradski, G.: The OpenCV Library. Dr. Dobb's Journal of Software Tools (2000)
2. Bramble, S., Compton, D., Klasén, L.: Forensic image analysis. In: Proceedings of the 13th INTERPOL Forensic Science Symposium (2001)
3. Brandner, M.: Bayesian uncertainty evaluation in vision-based metrology. In: Gallegos-Funes, F. (ed.) Vision Sensors and Edge Detection, chap. 5. IntechOpen, Rijeka (2010). https://doi.org/10.5772/10135, https://doi.org/10.5772/10135
4. Criminisi, A.: Single-view metrology: Algorithms and applications. In: Pattern Recognition, 24th DAGM Symposium. Zurich, Switzerland (January 2002)
5. Criminisi, A., Reid, I., Zisserman, A.: Single view metrology. Int. J. Comput. Vision **40**(2), 123–148 (2000)
6. D'Agostino, R., Pearson, E.S.: Tests for departure from normality. Biometrika **60**, 613–622 (1973)
7. Fischler, M.A., Bolles, R.C.: Random sample consensus: A paradigm for model fitting with applications to image analysis and automated cartography. Commun. ACM **24**(6), 381–395 (jun 1981)
8. Förstner, W.: Uncertainty and projective geometry. In: Handbook of Geometric Computing, pp. 493–534. Springer (2005)
9. Gao, X.S., Hou, X.R., Tang, J., Cheng, H.F.: Complete solution classification for the perspective-three-point problem. IEEE Transactions on Pattern Analysis and Machine Intelligence (Volume: 25, Issue: 8, Aug. 2003) (2003)
10. Ghasemi, A., Zahediasl, S.: Normality tests for statistical analysis: a guide for non-statisticians. Int. J. Endocrinol. Metab. **10**(2), 486–489 (2012)
11. Green, P.J.: Iteratively reweighted least squares for maximum likelihood estimation, and some robust and resistant alternatives. J. Royal Stat. Society. Series B (Methodological) **46**(2), 149–192 (1984)
12. Hartley, R., Zisserman, A.: Multiple View Geometry in Computer Vision, 2nd ed. Cambridge University Press (2003)
13. Heuel, S.: Uncertain Projective Geometry. LNCS, vol. 3008. Springer, Heidelberg (2004). https://doi.org/10.1007/b97201
14. Larsson, V., et al.: Revisiting radial distortion absolute pose. In: International Conference on Computer Vision (ICCV) (2019)
15. Lebeda, K., Matas, J., Chum, O.: Fixing the locally optimized ransac. British Machine Vision Conference (BMVC) (2012)
16. Lepetit, V., Moreno-Noguer, F., Fua, P.: Epnp: An accurate o(n) solution to the pnp problem. Int. J. Comput. Vision **81**, 155–166 (2009)
17. Li, X., Zhang, B., Sander, P.V., Liao, J.: Blind geometric distortion correction on images through deep learning. In: Conference on Computer Vision and Pattern Recognition (CVPR) (2019)
18. Li, Z., Snavely, N.: Megadepth: Learning single-view depth prediction from internet photos (2018)

19. Lu, X.X.: A review of solutions for perspective-n-point problem in camera pose estimation. J. Phys.: Conf. Ser. **1087**, 052009 (2018)
20. Massey, F.J.: The kolmogorov-smirnov test for goodness of fit. J. Am. Stat. Assoc. **46**(253), 68–78 (1951)
21. Nakano, G.: A versatile approach for solving PnP, PnPf, and PnPfr Problems. In: Leibe, B., Matas, J., Sebe, N., Welling, M. (eds.) ECCV 2016. LNCS, vol. 9907, pp. 338–352. Springer, Cham (2016). https://doi.org/10.1007/978-3-319-46487-9_21
22. Olsson, E.: Lens Distortion Correction Without Camera Access. Master's thesis, Linköping University, Sweden (2022)
23. Persson, M., Nordberg, K.: Lambda twist: an accurate fast robust perspective three point (P3P) Solver. In: Ferrari, V., Hebert, M., Sminchisescu, C., Weiss, Y. (eds.) ECCV 2018. LNCS, vol. 11208, pp. 334–349. Springer, Cham (2018). https://doi.org/10.1007/978-3-030-01225-0_20
24. Schönberger, J.L., Frahm, J.M.: Structure-from-motion revisited. In: Conference on Computer Vision and Pattern Recognition (CVPR) (2016)
25. Schönberger, J.L., Zheng, E., Frahm, J.-M., Pollefeys, M.: Pixelwise view selection for unstructured multi-view stereo. In: Leibe, B., Matas, J., Sebe, N., Welling, M. (eds.) ECCV 2016. LNCS, vol. 9907, pp. 501–518. Springer, Cham (2016). https://doi.org/10.1007/978-3-319-46487-9_31
26. Shapiro, S.S., Wilk, M.B.: An analysis of variance test for normality (complete samples)†. Biometrika **52**(3–4), 591–611 (1965). https://doi.org/10.1093/biomet/52.3-4.591
27. Thai, T.H., Cogranne, R., Retraint, F.: Camera model identification based on the heteroscedastic noise model. IEEE Trans. Image Process. **23**(1), 250–263 (2014)
28. Tsai, R.Y.: A versatile camera calibration technique for high-accuracy 3d machine vision metrology using off-the-shelf tv cameras and lenses. IEEE (1987)
29. Wang, Q., Alexander, W., Pegg, J., Qu, H., Chen, M.: Hypoml: visual analysis for hypothesis-based evaluation of machine learning models. IEEE Trans. Visual Comput. Graphics **27**(2), 1417–1426 (2021)
30. Zhang, Z.: A flexible new technique for camera calibration. IEEE Trans. Pattern Anal. Mach. Intell. **22**(11), 1330–1334 (2000)

CHAD: Charlotte Anomaly Dataset

Armin Danesh Pazho[✉], Ghazal Alinezhad Noghre, Babak Rahimi Ardabili,
Christopher Neff, and Hamed Tabkhi

University of North Carolina at Charlotte, Charlotte, NC 28223, USA
{adaneshp,galinezh,brahimia,cneff1,htabkhiv}@uncc.edu

Abstract. In recent years, we have seen a significant interest in data-driven deep learning approaches for video anomaly detection, where an algorithm must determine if specific frames of a video contain abnormal behaviors. However, video anomaly detection is particularly context-specific, and the availability of representative datasets heavily limits real-world accuracy. Additionally, the metrics currently reported by most state-of-the-art methods often do not reflect how well the model will perform in real-world scenarios. In this article, we present the Charlotte Anomaly Dataset (CHAD). CHAD is a high-resolution, multi-camera anomaly dataset in a commercial parking lot setting. In addition to frame-level anomaly labels, CHAD is the first anomaly dataset to include bounding box, identity, and pose annotations for each actor. This is especially beneficial for skeleton-based anomaly detection, which is useful for its lower computational demand in real-world settings. CHAD is also the first anomaly dataset to contain multiple views of the same scene. With four camera views and over 1.15 million frames, CHAD is the largest fully annotated anomaly detection dataset including person annotations, collected from continuous video streams from stationary cameras for smart video surveillance applications. To demonstrate the efficacy of CHAD for training and evaluation, we benchmark two state-of-the-art skeleton-based anomaly detection algorithms on CHAD and provide comprehensive analysis, including both quantitative results and qualitative examination. The dataset is available at https://github.com/TeCSAR-UNCC/CHAD.

Keywords: Anomaly Detection · Dataset · Computer Vision · Deep Learning

1 Introduction

Video anomaly detection, which requires understanding if a video contains anomalous behaviors, is a popular but challenging task in computer vision. In addition to substantial research interest, many real-world applications greatly benefit from being able to determine if such anomalous behaviors are present. Parking lot surveillance is one such application, where being able to determine the presence of an anomalous action (e.g. fighting, theft, fainting) is paramount.

R. Gade et al. (Eds.): SCIA 2023, LNCS 13885, pp. 50–66, 2023.
https://doi.org/10.1007/978-3-031-31435-3_4

Current state-of-the-art (SotA) deep learning solutions take one of two approaches. The first is an appearance-based method, where the algorithm works directly on video frames. The second is the skeleton-based methodology, in which algorithms rely on extracted human pose data to understand human behaviors. Both methods require large amounts of quality data. Anomaly detection is particularly context-specific, so training data must also be representative of both the environment and the context of the target application. This need is amplified for unsupervised approaches, which try to learn the normal behaviors of a specific context and need many example frames to do so.

There are currently only a limited number of datasets for video anomaly detection. These datasets, while seeing continual growth in the amount of data provided, also tend to fall short regarding the number of normal frames per context (i.e., per scene). Additionally, no current video anomaly dataset provides the detection, tracking, and pose information required by skeleton-based methods, leaving them to rely on external algorithms to generate this data. Since there is no standard for this, it is difficult to determine how much of an approach's error is due to the noise in this generated data or from the algorithm itself. This is further obfuscated by the inconsistency of the metrics used in reporting performance. Of the three main metrics for anomaly detection, discussed in Sect. 6, most SotA approaches only report one. However, all of them are necessary for a full understanding of an algorithm's performance, especially in the real-world.

In this paper, we present the Charlotte Anomaly Dataset (CHAD), a high-resolution, multi-camera anomaly detection dataset in a parking lot setting. CHAD is designed to address the most challenging issues facing current video anomaly detection datasets. The first video anomaly dataset with multiple camera views of a single scene, CHAD has over 1.15 million frames capturing the same context. With over 1 million normal frames, CHAD places itself as the premiere video anomaly dataset for unsupervised methods, providing human detection, tracking, and pose annotations. Thanks to these annotations, CHAD allows for a more accurate standard, positioning itself as the best-in-class dataset for skeleton-based anomaly detection.

We also propose a new standard in the benchmarking and evaluation of real-world video anomaly detection. Included is a detailed discussion on metrics, the benefits and disadvantages of each, and how the use of all three is needed to truly understand an algorithm's performance. To demonstrate the efficacy of CHAD, we train two SotA skeleton-based approaches, report both single camera and multi-camera performance, and compare to those methods trained on other datasets. Additionally, we perform cross-validation on CHAD, and the ShanghaiTech Campus Dataset [17], demonstrating CHAD's suitability for enabling generalization and revealing it to be more challenging than its peers.

In summary, this paper has the following contributions:

- We introduce CHAD, a high resolution, multi-camera video anomaly detection dataset in a parking lot setting. With over 1.15 million frames of a single context and detection, tracking, and pose annotations, CHAD positions itself

as the best-in-class dataset for both unsupervised and skeleton-based anomaly detection methods.
- We propose a new standard in real-world video anomaly detection benchmarking and evaluation. We provide a detailed discussion on the metrics used, including the insights they provide.
- To validate the efficacy of CHAD, we train and evaluate two SotA skeleton-based models with our proposed methodology. We further perform cross-validation on CHAD and ShanghaiTech Campus [17], demonstrating that CHAD is robust enough for generalization while being more challenging.

2   Related Work

Anomaly Detection Algorithms. Appearance-based methods utilize appearance and motion features generated directly from pixel data for detecting anomalies [6,11,12,18,27,30,32,35]. These methods generally achieve high accuracies in their context at the cost of high computation. Skeleton-based methods utilizes high-level, low-dimensional human pose skeletons [14,20,23,24,28]. These skeletons are informative in the context of human behavior while requiring far less computation than working with raw video data. They are more privacy preserving, and they remove demographic biases. As such, researchers have found significant success in skeleton-based anomaly detection.

Anomaly Detection Datasets. The CUHK Avenue Dataset [19] consists of nearly 31K frames captured from a single camera. Abnormal objects, walking in the wrong direction, and sudden movements are examples of anomalous behaviors in this dataset.

The UCSD Anomaly Detection Dataset [22] consists of 19K frames overlooking pedestrian walkways. UCSD has been categorized into two subsets, each one covering a different view. UCSD Ped1 sees pedestrian movement perpendicular to the camera, while UCSD Ped2 sees movement parallel to the camera. UCSD contains positional information for localizing anomalies.

The Subway dataset [2] consists of two surveillance videos, the subway entrance, and exit. With a combined total of 139 min of video, this dataset counts behaviors such as running, loitering, and walking in the opposite direction of the crowd as anomalous behaviors.

Street Scene [26] is a single scene anomaly detection dataset captured from a bird's eye view of a two lane street. Compared to most other datasets, Street Scene is relatively large at over 200K frames. Street Scene also contains non-human anomalies, such as illegally parked cars, dogs on the sidewalk, and cars making u-turns.

The ShanghaiTech Campus dataset [17] contains 13 different scenes taken from a campus setting. With over 317K frames, ShanghaiTech is one of the largest and most popular anomaly detection datasets available. However, ShanghaiTech has relatively few frames per context.

IITB-Corridor [28] was the largest single-stationary-camera anomaly detection dataset that existed before CHAD. It contains nearly 440K frames in a campus setting. Recorded in high-resolution 1080p, it is the only continuous video anomaly detection dataset with a resolution comparable to CHAD.

The ADOC dataset [25] is captured from a single high-resolution camera over 24 h in a campus setting. ADOC consists of 260K frames and adopts an approach of considering any low-frequency behavior to be anomalous. Assuming only walking is normal, they consider all other behaviors as anomalous, even relatively commonplace activities like walking with a briefcase, having a conversation, or a bird flying through the air. While this categorization works for ADOC's context, it is inconsistent with how other datasets define anomalous behaviors.

Specifically for supervised anomaly detection, UBnormal [1] is comprised entirely of synthetically generated videos. With a total of 236,902 frames, UBnormal is moderately large compared to other anomaly datasets, though with 29 scenes the average number of frames per scene is fairly low.

The NOLA dataset [9] is another new dataset. Collected over an entire week, NOLA contains over 1.4 million frames including both day and night scenes. In contrast to most other anomaly datasets, NOLA uses a single moving camera instead of stationary cameras. The rapid movement of the camera introduces a massive change of context, making the video anomaly detection more challenging. Due to the way annotations are presented in the dataset and the lack of clarifying documentation, it is impossible to ascertain what constitutes an anomaly in the context of NOLA. As such, it is difficult to determine the efficacy of this dataset for anomaly detection, and fair comparison to other datasets is not feasible.

UCF Crime [30] and X-D Violence [34] collect video clips from many different sources in varying contexts, as opposed to continuous recordings. This allows them to be enormous by anomaly dataset standards but is so fundamentally different in problem formulation that it could be considered a different task altogether. XD-Violence provides both video and audio, making it unique among video anomaly datasets.

All of these datasets bring their own benefits and have helped advance the field of video anomaly detection. However, while they all have their own strengths, each of them also provides its own challenges when it comes to training networks for the real-world. Some datasets are too small, either in overall frames or frames per scene. Some of them have strict definitions of normal behaviors that would be undesirable in a real-world context. Some have to contend with domain shift, either from taking a large amalgamation of clips from entirely different contexts or from training with synthetic actors and moving to real persons and objects when used in a real-world context. And while many of these datasets provide multiple contexts, none of them provide different views of the same context, as would be fairly common in a surveillance setting. Further, none of these datasets provide the human detection, tracking, and pose annotations needed for skeleton-based anomaly detection. It is impossible for a single dataset to fit every possible scenario.

3   Data Collection and Setup

Since anomaly detection is such a context-specific task, it is important that the data used to train algorithms is representative of their real-world environments. Often the disconnect between training data, and inference data leads to unsatisfying performance in the real-world [3]. CHAD was designed to accurately mimic a real-world parking lot surveillance setting. The four cameras, as seen in Fig. 1, were positioned to cover the same general scene, though their perspectives give them each a unique context compared to the others. Each video is recorded in full HD (1920 × 1080, 30fps), except camera 4 which is in standard HD (1280 × 720, 30fps), as seen in Table 3.

There are thirteen actors present in CHAD. The actors represent diverse demographics (gender, age, ethnicity, etc.) and each participates in both normal and anomalous clips. There are 22 classes of anomalous behaviors in CHAD, which can be seen in Table 1. This list has been curated in line with other state-of-the-art datasets [2,17,19,22]. All other actions present in CHAD (e.g. walking, waving, talking, etc.) are considered normal.

Table 1. Anomalous behaviors present in CHAD.

Type of Anomalous Behavior			
Group Activities		Individual Activities	
Fighting	Punching	Throwing	Running
Kicking	Pushing	Riding	Falling
Pulling	Slapping	Littering	Jumping
Strangling	Body Hitting	Hopping	Sleeping
Theft	Pick-Pocketing		
Tripping	Playing with Ball		
Chasing	Playing with Racket		

4   Annotation Methodology

CHAD contains four types of annotations: frame-level anomaly labels, person bounding boxes, person ID labels, and human keypoints.

4.1   Anomaly Annotations

We annotate anomalous behaviors at the frame level. This is, we mark the frame where the anomalous behavior begins, the frame where it ends, and every frame in between. This is done by hand, accounting for all the behaviors defined in Sect. 3. These frame-level labels are needed for both appearance-based and skeleton-based approaches. CHAD does not include anomaly localization labels.

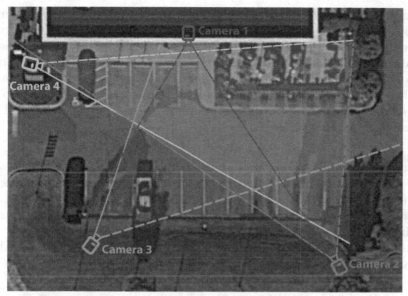

(a) Birds-eye View

(b) Camera View

Fig. 1. Approximate position and the views of the cameras.

4.2 Person Annotations

One of the innovations that sets CHAD above its peers is the inclusion of person annotations. In real-world scenarios, there is no access to hand-annotated data. The annotations must be generated through available tools and are not always perfect. We include generated person-annotations to ensure they are more representative of a real-world situation. This is by design, as a certain amount of noise is desirable in the dataset to assist models in learning how to deal with unclean data inherent in real-world situations [5]. It also allows skeleton-based anomaly detection methods to have access to the processed data they need without having to spend time extracting it themselves. We hope this will make skeleton-based anomaly detection more accessible to researchers, leading to more innovation. It also sets a standard previously unavailable for how to generate this human detection, tracking, and pose information. With this standard, the variability based on the quality of input data is removed, leading to more precise and fair comparisons between approaches.

Bounding Boxes. The bounding box of a person refers to the upper and lower x and y coordinate limits they occupy in an image. Having quality bounding boxes for each individual and for every frame is doubly important for CHAD, as this localization is needed for the extraction of both person ID labels and human keypoints as well. For this reason, CHAD utilizes the popular object detection algorithm YOLOv4 [4] for generating quality bounding boxes. Since CHAD is focused on anomalous human behavior, only the bounding boxes for people are used.

Person ID Labels. Anomaly detection algorithms often utilize temporal information to understand the behaviors of people. Particularly for skeleton-based methods, it is necessary to be able to associate the different poses of a person to that specific person across frames. Person ID labels provide this information, allowing for temporal tracking of individual persons in each video clip. Given the bounding box information generated previously, DeepSORT [33] was utilized to provide tracking for persons through frames, generating unique person ID labels for each person in a video clip. For label stability, a three frame warm-up is used by DeepSORT before providing person ID labels. As such, the first two frames of each video clip are absent of person annotations.

Human Keypoints. CHAD contains pose information in the form of human pose skeletons. These skeletons are made up of human keypoints, or points of interest on the human body. While there are several methods for defining what keypoints to use, CHAD follows the 17 keypoint methodology proposed by MS COCO [16]. Using the localization provided by the previously generated bounding boxes, keypoints are extracted using HRNet [31], a prolific algorithm for human pose estimation used by many. To ensure we only provide quality keypoint annotations, we remove any person with low confidence (<50%) for at least half of their keypoints (9+). While this leads to some frames where people are not detected, it helps reduce the overall noise of the data that is present.

4.3   Annotation Smoothing

The algorithms used to annotate CHAD are imperfect, and there are instances where people are completely missed at either the object detection or keypoint extraction stage. Combined with our purposeful removal of overly noisy data, this results in an undesirable number of missed persons. To compensate for this, we introduce annotation smoothing to CHAD, using high confidence annotations to help fill in the missing information.

Given the relatively high frame rate of CHAD at 30 frames per second, it is a reasonable assumption that the positions and skeletons of a person will not drastically change between consecutive frames. As such, we can use linear interpolation to approximate the bounding box coordinates of each individual, assuming we have accurate detection at the start and end of the missing frames, and the number of missing frames is not too large. We choose 15 frames, or half a second, as a qualitative analysis showed this to be long enough to provide a significant benefit to annotation consistency, but not so long that the data it produced became unreliable. We apply the same smoothing technique to the keypoint annotations, with the same frame limitations. The details of smoothing are provided in the following equation:

$$\mathbb{X}_i = (\frac{\mathbb{X}_{\mathcal{N}} - \mathbb{X}_{\mathcal{M}}}{\mathcal{N} - \mathcal{M}}) \times i + \mathbb{X}_{\mathcal{M}} \tag{1}$$

where $\mathbb{X}_i$ refers to a missing point (either bounding box or keypoint coordinate) at frame i, $\mathbb{X}_{\mathcal{M}}$ and $\mathbb{X}_{\mathcal{N}}$ refer to the two nearest matching points at frames $\mathcal{M}$ and $\mathcal{N}$ respectively, and where $\mathcal{M} < i < \mathcal{N}$ and $\mathcal{N} - \mathcal{M} + 1 \leq 15$.

The added consistency in annotations created by this smoothing is particularly useful in the context of unsupervised learning. However, the confidence scores of keypoints generated by this smoothing are set to Null, so they can be easily discarded if undesired.

5   CHAD Statistics

With over 1.15 million frames, CHAD is the largest anomaly detection dataset available that is recorded from continuous videos captured from stationary cameras, and includes person annotations. As shown in Table 4, CHAD has more than 2× the number of frames as the next largest dataset, providing a substantial amount of learnable data. Additionally, CHAD has over 1 million frames of purely normal behaviors, which are required for unsupervised methods that rely on learning the normal to understand the anomalous. This is nearly 3× more than can be found in other datasets. The 59K anomalous frames in CHAD are comprised of the 22 anomalous behaviors presented in Table 1. To facilitate supervised, unsupervised, and semi-supervised approaches, CHAD includes two splits for training and testing. The *unsupervised split* has a training set comprised only of normal behaviors, while the test set contains both normal and anomalous behaviors. The details of the *unsupervised split* can be found in Table 4. For the *supervised split*, the normal and anomalous frames were distributed uniformly between the training and test sets, with 60% of each belonging to the training set and 40% to the test set.

Table 2. Annotation availability in ShanghaiTech [17], CUHK [19], UCSD [22], Subway [2], IITB [28], Street Scene [26], UBnormal [1], and CHAD (Ours). * partially annotated, − not annotated.

Dataset	Anomaly Annotations		Person Annotations		
	Frame-level	Pixel-level	Bounding Box	ID Number	Keypoints
ShanghaiTech	✓	✓	−	−	−
CUHK	✓	✓	−	−	−
UCSD	✓	*	−	−	−
Subway	✓	✓	−	−	−
IITB	✓	✓	−	−	−
Street Scene	✓	✓	−	−	−
UBnormal	✓	✓	✓	−	−
CHAD (Ours)	✓	−	✓	✓	✓

More than just the amount of data, CHAD benefits from having high quality image data. As discussed in Sect. 3, CHAD was recorded from four high-resolution cameras with an overlapping view of a scene. Recorded at 30 FPS, CHAD not only boasts a higher resolution and frame rate than other datasets, shown in Table 3, but also presents data in a format representative of modern real-world surveillance systems. While resolution and frame rate are indicators of overall video quality and the amount of data present in each frame, they can not convey how much of that data is actually useful for learning. Difference of Gaussian [7] is an image processing method that has been used to simulate how the human eye extracts visual details of an image for neural processing [21]. More simply, it creates a visual illustration of the density and richness of the features in an image. This allows us to visually analyze the quality of the data present in each dataset by comparing the Difference of Gaussian between them.

We visualize the Difference of Gaussian for a single frame of each dataset in Fig. 2. We set a Gaussian blur radius of one pixel to maximize the precision of the resulting representation. Looking at the images, CHAD very clearly presents the most detail. This was anticipated due to its high resolution, but the amount by which it surpasses the other datasets far exceeded expectations. Fine details in the persons, clothing, vehicles, and the environment are clear, granting an accurate perception of the original image. IITB-Corridor [28] is the only other dataset with 1080p images. However, the Difference of Gaussian tells a different story. While there are details present in the environment, they are comparably indistinct. Even in the brightened image, it is difficult to tell there is a person in the image. This demonstrates a surprising lack of rich features in the IITB-Corridor, despite the resolution.

Street View, at the next highest resolution, shows much more detail and clarity than IITB-Corridor, though nowhere near the level of CHAD. What is most interesting is that while the building, car, and street boundaries are clear,

Table 3. Resolution and frame rate in Shanghai [17], CUHK [19], UCSD [22], Subway [2], IITB [28], Street Scene [26], UBnormal [1], and CHAD (Ours). N/A means Not Available.

Dataset		Resolution (Pixels)	Frame Rate (FPS)
Shanghai		856*480	N/A
CUHK		640*360	25
UCSD	Ped1	238*158	N/A
	Ped2	360*240	N/A
Subway		N/A	N/A
IITB		1920*1080	25
Street Scene		1280*720	15
UBnormal		varies	30
CHAD (Ours)	Scene 1–3	1920*1080	30
	Scene 4	1280*720	30

it is difficult to notice the two people in the bottom left of the image. This is perhaps due to their relative size compared to the other objects mentioned and not necessarily indicative of a lack of features. Unsurprisingly, the lower resolution datasets, UCSD and CUHK Avenue, show sharp focal points (bright white pixels) but very little overall detail. Interestingly for ShanghaiTech [17], despite its slightly higher resolution, it presents a similar level of detail as Street Scene. However, due to the different camera perspectives, this translates into Shanghai providing better features for people, which is beneficial for its context.

Overall, we can see that CHAD not only has the best-in-class resolution and frame rate among anomaly detection datasets but also that the videos in CHAD are extremely feature rich, unrivaled among its peers. Additionally, there is a significant amount of background information irrelevant to person behaviors. The brightest spot in the Difference of Gaussian for CHAD is the foliage in the bottom left. This is noise - a distractor from information pertinent to anomaly detection. This means CHAD is not only more informative than other datasets but also suggests that it is more challenging as well. This level of challenge is needed if algorithms are to perform well in real-world scenarios, which are notorious for being more demanding than dataset benchmarks.

6 Metrics and Measurements

There are three main metrics used for evaluating performance on anomaly detection datasets: Area Under the Receiver Operating Characteristic Curve, Area Under the Precision-Recall Curve, and Equal Error Rate. While none of these metrics are truly representative of overall performance, they each have their strengths and weaknesses, and, taken together, they can provide a comprehensive understanding of how an algorithm truly performs.

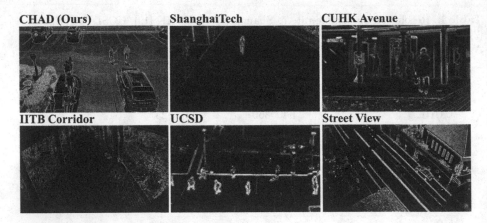

Fig. 2. Visualization of Difference of Gaussian in Shanghai [17], CUHK [19], UCSD [22], IITB [28], Street Scene [26], and CHAD (Ours). UCSD cropped to fit. All brightened for readability.

Table 4. Dataset comparison for ShanghaiTech [17], CUHK [19], UCSD [22], Subway [2], IITB [28], Street Scene [26], UBnormal [1], and CHAD (Ours). CHAD uses unsupervised split. N/A means Not Available.

Dataset	Number of Frames					Scene(s)	Camera(s)
	Total	Train	Test	Normal	Anomalous		
ShanghaiTech	317,398	274,515	42,883	300,308	17,090	13	13
CUHK	30,652	15,328	15,324	26,832	3,820	1	1
UCSD	18,560	9,050	9,210	12,919	5,641	2	2
Subway	208,925	27,500	181,425	205,805	3120	2	2
IITB	483,566	301,999	181,567	375,288	108,278	1	1
Street Scene	203,257	56,847	146,410	N/A	N/A	1	1
UBnormal	236,902	116,087	28,175	147,887	89,015	29	-
CHAD (Ours)	1,152,649	1,026,174	126,475	1,093,477	59,172	1	4

6.1   Receiver Operating Characteristic Curve

The Area Under the Receiver Operating Characteristic Curve (AUC-ROC) is simply the area under the curve when plotting the True Positive Rate (TRP) over the False Positive Rate (FPR) over various thresholds. This metric is specific to binary classification, such as determining if a video does or does not contain anomalous behavior. Generally, a higher AUC-ROC indicates that the model is better at separating inputs into their corresponding classes. The ROC curve itself also helps give insight into the trade-off between TPR and FPR at different thresholds [10]. However, AUC-ROC is not indicative of the final decisions of a model. The metric reports a final calculated number, and concluding useful information about the actual amount of False Negative Rate (FNR), when an anomaly is classified as normal is almost unfeasible. FNR is particularly important for real-world applications, and reporting it separately is crucial. Additionally, AUC-ROC is very sensitive to imbalances in data [13], making it sub-optimal if one class is over represented, as is often the case with normal behaviors in anomaly datasets [8].

6.2 Precision-Recall Curve

Precision is the fraction of correct positive guesses over all positive guesses, while Recall is the fraction of correct positive guesses over all positive samples. The Precision-Recall Curve (PR) is useful for understanding how to balance Precision and Recall, while the area under this curve summarizes all the information represented in it. While AUC-PR heavily focuses on the positive class, it still accounts for the False Negative Rate (FNR) – that is when the model classifies an anomaly as normal. As such, AUC-PR is a better metric for understanding the prediction ability of a model when compared to AUC-ROC [29]. Additionally, AUC-PR is better suited for highly imbalanced data [29], making it better at evaluation the minority class [13]. As the minority class in anomaly detection usually refers to the anomalous behaviors, this is an important quality for this context. However, AUC-PR is a final calculated number and it does not provide direct insight into the correct classification of negative samples, nor does it provide a measure for the number of incorrect decisions a model makes. Thus, much like AUC-ROC, AUC-PR provides an incomplete understanding of a model's performance.

6.3 Equal Error Rate

Another useful metric is the Equal Error Rate (EER) [15]. Plotting the FNR and FPR over various thresholds produces two curves that intersect at one point. The value at the intersection is the EER and shows what threshold value allows the model to achieve a balance between FNR and FPR. In the context of video anomaly detection, the EER illustrates how many false alarms a model will raise and how many anomalous frames it will miss when at equilibrium. On its own, this metric offers little insight into the overall performance of a model [30]. However, when used as a complement to AUC-ROC and AUC-PR, a more complete understanding can be achieved.

7 Evaluation

All experiments were conducted on a server containing two Intel Xeon Silver 4114, one V100 GPU, and 256 GB of RAM. We performed each experiment (training and testing) five times, averaging the results to remove any potential skew due to variability. For each model, training is performed exactly as described in their respective papers unless otherwise specified.

7.1 Standard Validation

To demonstrate CHAD's viability as an anomaly detection dataset, we train and evaluate two state-of-the-art skeleton-based models using the *unsupervised split*. We select Graph Embedded Pose Clustering (GEPC) [23] and Message-Passing Encoder-Decoder Recurrent Neural Network (MPED-RNN) [24] for their high accuracy and model availability. GEPC utilizes a spatio-temporal graph autoencoder, while MPED-RNN uses a two-headed structure with reconstruction and prediction.

Both models were trained on each of CHAD's four camera views individually, the results reported in Table 5. The most obvious observation is that both models were able to learn on CHAD. GEPC achieved an average AUC-ROC of 0.663 and AUC-PR of 0.619, while MPED-RNN achieved an average AUC-ROC of 0.718 and AUC-PR of 0.635. For both models, the AUC-ROC is noticeably higher than the AUC-PR. This is largely due to the overwhelming majority of normal frames in the data, which if properly classified will a significant boost to the AUC-ROC. AUC-PR, on the other hand, does not count True Negatives, and as such gives a more measured result for the imbalanced data. Additionally, GEPC achieved an EER of 0.378 and MPED-RNN an EER of 0.339. This means that, given the threshold at equilibrium, both models can expect to see between 34% and 38% of both normal frames and anomalous frames to be misclassified. This is important to understand when targeting real-world applications, where misclassification rates are more important than class separability.

Table 5. Evaluation of GEPC [23] and MPED-RNN [24] on CHAD (Ours).

Model	Camera	AUC-ROC	AUC-PR	EER
GEPC	1	0.673	0.636	0.363
	2	0.660	0.566	0.382
	3	0.661	0.586	0.384
	4	0.656	0.688	0.382
MPED-RNN	1	0.747	0.715	0.303
	2	0.691	0.567	0.349
	3	0.771	0.584	0.331
	4	0.662	0.674	0.372

7.2 Cross Validation

To illustrate CHAD's ability to train models that can generalize, we perform cross validation experiments with another anomaly dataset in the same domain. We choose the popular ShanghaiTech Campus Dataset [17] for its relatively large size, its similar context to CHAD, and its proven track record in anomaly

Table 6. Cross-validation of GEPC [23] on ShanghaiTech [17] and CHAD (Ours).

Model	Train	Test	AUC-ROC	AUC-PR	EER
GEPC	CHAD (Ours)	CHAD (Ours)	0.649	0.587	0.385
		ShanghaiTech	0.728	0.637	0.326
	ShanghaiTech	CHAD (Ours)	0.639	0.572	0.399
		ShanghaiTech	0.741	0.657	0.315

detection research. For these experiments, we use GEPC, as its multi-camera training methodology allows for a simple conversion to cross validation. For both CHAD and ShanghaiTech, a single model is trained for all cameras in one dataset, then tested on both datasets. The results can be seen in Table 6.

The first thing to notice is that models trained on CHAD perform well on ShanghaiTech, and models trained on ShanghaiTech perform well on CHAD. This is logical, as the contexts for the two datasets (i.e. setting, camera views, anomalous behaviors) are quite similar. In all metrics, the validation of models across datasets performs within 1–2% of models validated on their parent datasets, showing that models trained on either can generalize quite well given their similar contexts.

Another trend seen in Table 6 is that for all metrics, models tend to achieve lower scores (or higher in the case of EER) on CHAD than they do on ShanghaiTech. Since both models performed equally well in cross validation, the logical assumption is that CHAD's test set is more challenging than ShanghaiTech's. This is in part due to the additional noise and distractors present in CHAD, as explained in Sect. 5. The other major factor is the inclusion of very subtle and complex anomalies in CHAD. Pick-pocketing is subtle by design, as most pick-pockets are trying not to be seen. Littering is also quite complex to learn, especially for a model that relies solely on human keypoints. Combined with the sheer size of CHAD's test set (3× that of ShanghaiTech's), this makes for a very challenging dataset for current anomaly detection algorithms.

8 Conclusion

This paper presented the Charlotte Anomaly Dataset (CHAD). Consisting of more than $1.15\,m$ high-resolution frames of a single scene, CHAD is the largest available anomaly detection dataset consisting of continuous video from stationary cameras. In addition to frame-level anomaly labels, CHAD goes further than other datasets and provides bounding-box, person ID, and human keypoints annotations, enabling a unified benchmarking standard for both skeleton and appearance-based anomaly detection. Additionally, this paper assesses three metrics for anomaly detection and proposes their use in combination as a new standard for real-world video anomaly detection.

Acknowledgements. This research is supported by the National Science Foundation (NSF) under Award No. 1831795 and NSF Graduate Research Fellowship Award No. 1848727. In addition, this research is IRB approved under Document Number IRBIS-17-0307.

References

1. Acsintoae, A., et al.: Ubnormal: New benchmark for supervised open-set video anomaly detection. In: Proceedings of the IEEE/CVF Conference on Computer Vision and Pattern Recognition (CVPR) (June 2022)
2. Adam, A., Rivlin, E., Shimshoni, I., Reinitz, D.: Robust real-time unusual event detection using multiple fixed-location monitors. IEEE Trans. Pattern Anal. Mach. Intell. **30**(3), 555–560 (2008)
3. Alinezhad Noghre, G., Danesh Pazho, A., Sanchez, J., Hewitt, N., Neff, C., Tabkhi, H.: Adg-pose: Automated dataset generation for real-world human pose estimation. In: International Conference on Pattern Recognition and Artificial Intelligence. pp. 258–270. Springer (2022). https://doi.org/10.1007/978-3-031-09282-4_22
4. Bochkovskiy, A., Wang, C.Y., Liao, H.Y.M.: Yolov4: Optimal speed and accuracy of object detection. arXiv preprint arXiv:2004.10934 (2020)
5. Chandra, R., Bhattacharya, U., Roncal, C., Bera, A., Manocha, D.: Robusttp: End-to-end trajectory prediction for heterogeneous road-agents in dense traffic with noisy sensor inputs. In: ACM Computer Science in Cars Symposium, pp. 1–9 (2019)
6. Chu, W., Xue, H., Yao, C., Cai, D.: Sparse coding guided spatiotemporal feature learning for abnormal event detection in large videos. IEEE Trans. Multimedia **21**(1), 246–255 (2019). https://doi.org/10.1109/TMM.2018.2846411
7. Crowley, J.L., Parker, A.C.: A representation for shape based on peaks and ridges in the difference of low-pass transform. IEEE Trans. Pattern Anal. Mach. Intell. **2**, 156–170 (1984)
8. Davis, J., Goadrich, M.: The relationship between precision-recall and roc curves. In: Proceedings of the 23rd International Conference on Machine learning, pp. 233–240 (2006)
9. Doshi, K., Yilmaz, Y.: Rethinking video anomaly detection-a continual learning approach. In: Proceedings of the IEEE/CVF winter Conference on Applications of Computer Vision, pp. 3961–3970 (2022)
10. Fernández, A., García, S., Galar, M., Prati, R.C., Krawczyk, B., Herrera, F.: Learning from Imbalanced Data Sets. Springer, Cham (2018). https://doi.org/10.1007/978-3-319-98074-4
11. Ganokratanaa, T., Aramvith, S., Sebe, N.: Anomaly event detection using generative adversarial network for surveillance videos. In: 2019 Asia Pacific Signal and Information Processing Association Annual Summit and Conference (APSIPA ASC), pp. 1395–1399 (2019). https://doi.org/10.1109/APSIPAASC47483.2019.9023261
12. Goodfellow, I., et al.: Generative adversarial nets. In: Ghahramani, Z., Welling, M., Cortes, C., Lawrence, N., Weinberger, K. (eds.) Advances in Neural Information Processing Systems. vol. 27. Curran Associates, Inc. (2014), https://proceedings.neurips.cc/paper/2014/file/5ca3e9b122f61f8f06494c97b1afccf3-Paper.pdf
13. He, H., Ma, Y.: Imbalanced learning: foundations, algorithms, and applications. Wiley-IEEE Press (2013)

14. Li, N., Chang, F., Liu, C.: Human-related anomalous event detection via spatial-temporal graph convolutional autoencoder with embedded long short-term memory network. Neurocomputing **490**, 482–494 (2022)

15. Li, W., Mahadevan, V., Vasconcelos, N.: Anomaly detection and localization in crowded scenes. IEEE Trans. Pattern Anal. Mach. Intell. **36**(1), 18–32 (2013)

16. Lin, T.-Y., et al.: Microsoft COCO: common objects in context. In: Fleet, D., Pajdla, T., Schiele, B., Tuytelaars, T. (eds.) ECCV 2014. LNCS, vol. 8693, pp. 740–755. Springer, Cham (2014). https://doi.org/10.1007/978-3-319-10602-1_48

17. Liu, W., W. Luo, D.L., Gao, S.: Future frame prediction for anomaly detection - a new baseline. In: 2018 IEEE Conference on Computer Vision and Pattern Recognition (CVPR) (2018)

18. Liu, W., Luo, W., Lian, D., Gao, S.: Future frame prediction for anomaly detection - a new baseline. In: Proceedings of the IEEE Conference on Computer Vision and Pattern Recognition (CVPR) (June 2018)

19. Lu, C., Shi, J., Jia, J.: Abnormal event detection at 150 fps in matlab (2013)

20. Luo, W., Liu, W., Gao, S.: Normal graph: Spatial temporal graph convolutional networks based prediction network for skeleton based video anomaly detection. Neurocomputing 444, 332–337 (2021). https://doi.org/10.1016/j.neucom.2019.12.148, https://www.sciencedirect.com/science/article/pii/S0925231220317720

21. Lv, Y., Jiang, G., Yu, M., Xu, H., Shao, F., Liu, S.: Difference of gaussian statistical features based blind image quality assessment: A deep learning approach. In: 2015 IEEE International Conference on Image Processing (ICIP), pp. 2344–2348. IEEE (2015)

22. Mahadevan, V., Li, W., Bhalodia, V., Vasconcelos, N.: Anomaly detection in crowded scenes. In: 2010 IEEE Computer Society Conference on Computer Vision and Pattern Recognition, pp. 1975–1981 (2010). https://doi.org/10.1109/CVPR.2010.5539872

23. Markovitz, A., Sharir, G., Friedman, I., Zelnik-Manor, L., Avidan, S.: Graph embedded pose clustering for anomaly detection. In: Proceedings of the IEEE/CVF Conference on Computer Vision and Pattern Recognition, pp. 10539–10547 (2020)

24. Morais, R., Le, V., Tran, T., Saha, B., Mansour, M., Venkatesh, S.: Learning regularity in skeleton trajectories for anomaly detection in videos. In: Proceedings of the IEEE/CVF Conference on Computer Vision and Pattern Recognition (CVPR) (June 2019)

25. Pranav, M., Zhenggang, L., K, S.S.: A day on campus - an anomaly detection dataset for events in a single camera. In: Proceedings of the Asian Conference on Computer Vision (ACCV) (November 2020)

26. Ramachandra, B., Jones, M.J.: Street scene: A new dataset and evaluation protocol for video anomaly detection. In: 2020 IEEE Winter Conference on Applications of Computer Vision (WACV), pp. 2558–2567 (2020). https://doi.org/10.1109/WACV45572.2020.9093457

27. Ravanbakhsh, M., Sangineto, E., Nabi, M., Sebe, N.: Training adversarial discriminators for cross-channel abnormal event detection in crowds. In: 2019 IEEE Winter Conference on Applications of Computer Vision (WACV), pp. 1896–1904 (2019). https://doi.org/10.1109/WACV.2019.00206

28. Rodrigues, R., Bhargava, N., Velmurugan, R., Chaudhuri, S.: Multi-timescale trajectory prediction for abnormal human activity detection. In: Proceedings of the IEEE/CVF Winter Conference on Applications of Computer Vision (WACV) (March 2020)

29. Saito, T., Rehmsmeier, M.: The precision-recall plot is more informative than the roc plot when evaluating binary classifiers on imbalanced datasets. PLoS ONE **10**(3), e0118432 (2015)
30. Sultani, W., Chen, C., Shah, M.: Real-world anomaly detection in surveillance videos. In: Proceedings of the IEEE Conference on Computer Vision and Pattern Recognition (CVPR) (June 2018)
31. Sun, K., Xiao, B., Liu, D., Wang, J.: Deep high-resolution representation learning for human pose estimation. In: Proceedings of the IEEE/CVF Conference on Computer Vision and Pattern Recognition, pp. 5693–5703 (2019)
32. Tian, Y., Pang, G., Chen, Y., Singh, R., Verjans, J.W., Carneiro, G.: Weakly-supervised video anomaly detection with robust temporal feature magnitude learning. In: Proceedings of the IEEE/CVF International Conference on Computer Vision (ICCV), pp. 4975–4986 (October 2021)
33. Wojke, N., Bewley, A., Paulus, D.: Simple online and realtime tracking with a deep association metric. In: 2017 IEEE International Conference on Image Processing (ICIP), pp. 3645–3649. IEEE (2017)
34. Wu, P., et al.: Not only look, but also listen: learning multimodal violence detection under weak supervision. In: Vedaldi, A., Bischof, H., Brox, T., Frahm, J.-M. (eds.) ECCV 2020. LNCS, vol. 12375, pp. 322–339. Springer, Cham (2020). https://doi.org/10.1007/978-3-030-58577-8_20
35. Zhou, S., Shen, W., Zeng, D., Fang, M., Wei, Y., Zhang, Z.: Spatial-temporal convolutional neural networks for anomaly detection and localization in crowded scenes. Signal Processing: Image Communication 47, 358–368 (2016). https://doi.org/10.1016/j.image.2016.06.007, https://www.sciencedirect.com/science/article/pii/S0923596516300935

iDFD: A Dataset Annotated for Depth and Defocus

Saqib Nazir[1]([✉])[iD], Zhouyan Qiu[2,3][iD], Daniela Coltuc[1][iD],
Joaquín Martínez-Sánchez[2][iD], and Pedro Arias[2][iD]

[1] CEOSpaceTech, University POLITEHNICA of Bucharest (UPB),
Bucharest, Romania
{saqib.nazir,daniela.coltuc}@upb.ro
[2] CINTECX, Universidade de Vigo, Applied Geotechnology Group, Vigo, Spain
[3] ICT and Innovation Department, Ingeniería Insitu, Vigo, Spain

Abstract. Depth estimation and image deblurring from a single defocused image are fundamental tasks in Computer Vision (CV). Many methods have previously been proposed to solve these two tasks separately, using Deep Learning (DL) powerful learning capability. However, when it comes to training the Deep Neural Networks (DNN) for image deblurring or Depth from Defocus (DFD), the mentioned methods are mostly based on synthetic training datasets because of the difficulty of densely labeling depth and defocus on real images. The performance of the networks trained on synthetic data may deteriorate rapidly on real images. In this work, we present Indoor Depth from Defocus (iDFD), a Depth And Defocus Annotated dataset, which contains naturally defocused, All-in-Focus (AiF) images and dense depth maps of indoor environments. iDFD is the first public dataset to contain natural defocus and corresponding depth obtained using two appropriate sensors, DSLR and MS-Kinect camera. This dataset can support the development of DL based methods for depth estimation from defocus and image deblurring by providing the possibility to train the networks on real data instead of synthetic data. The dataset is available for download at iDFD.

Keywords: RGBD-Dataset · Defocus Deblurring · Deep Learning

1 Introduction

The advancement of DL techniques has revolutionized the domain of CV, where the quality of tasks such as classification, image segmentation, restoration, augmented reality etc. has been substantially improved due to the more accurate models learned with DNNs. Over the last few years, there has been a lot of

---

This project has received funding from the European Union's Horizon 2020 research and innovation programme under the Marie Skłodowska-Curie grant agreement No. 860370.

Fig. 1. Sample from our iDFD dataset. Each sample contains the All-in-Focus, Defocus, Raw Depth, and Depth after in-painting.

Table 1. iDFD Overview.

Sensors	Scenes	Range	Data	Total no. Images	Depth Maps	Out-of-Focus DSLR Settings
DLSR-Nikon, MS-KINECT	Indoor	0–10 m	RGB/Depth	764	Raw/In-painted	Lens: 14 mm $f/2.8$

interest in Monocular Depth Estimation (MDE). MDE is critical for scene under-standing, $3D$ reconstruction in robotics, augmented reality or autonomous driv-ing and flight [1]. Many methods were proposed in the past for solving MDE problem from single images or sequences of frames [1]. When dealing with single images, the DL relies mainly on the scene geometry and learns the depth models generally, in a supervised way. The success of supervised DL methods how-ever depends on the existence of large diverse training datasets such as KITTI, Cityscapes or Make3D that contain RGB images and their corresponding Ground Truth (GT) consisting in either dense depth maps or cloud of points. The depth GT is captured with $3D$ range sensors that may be noisy, sparse, expensive, and above all these, laborious to obtain. A list of the various modalities used to cap-ture depth for indoor and outdoor scenes is in Table 2. The Kinect camera from Microsoft and other devices from the family of Time-of-Flight (ToF) cameras,

Table 2. Depth sensors and common criteria in comparing them.

Modality	Scenes	Compactness	Response Time	Depth Accuracy	Power Consumption	Range
Stereo Vision	Outdoor, Indoor	Low	Medium	Low	Low	Unlimited
Structured Light	Indoor	High	Slow	High	Medium	Scalable
Time-of-Flight	Indoor	Low	Fast	Medium	Scalable	Scalable
LIDAR	Outdoor, Indoor	Low	Fast	Medium	Scalable	Scalable

were used to acquire depth GT for several indoor datasets, such as the NYU-v2 [2], Sun3D [3], and ScanNet [4].

One of the cues in MDE is the defocus blur, which is naturally present in every image due to the camera lens. Defocus blur manifests more visibly in images taken with a shallow Depth of Field (DoF), which is caused by large apertures. The blur strength depends on the distance to the camera. Although extensively studied in the past and rigorously modelled, the defocus blur remained difficult to use in MDE because of the uncertainty by respect to the focal plane and its almost absence in DoF. To alleviate these drawbacks, supplementary constraints like coded apertures [5], dual images [6], multi focus stack [7] etc. have to be added.

In depth DL, the defocus blur is naturally associated with the scene geometry and can be exploited without any other constraints. Despite its valuable content, there were few attempts to use defocused images in MDE by DL [8,9]. One of the obstacles has been the unavailability of real defocused datasets annotated for depth. In their works, the authors of [8,10] have introduced synthetic blur into the RGB images and fed them into the network to estimate DFD. Although these works obtained promising results for DFD on synthetic datasets, their performance decreases drastically when tested on real scenes.

Defocus blur is the main actor in applications of image restoration, where it is seen as a distortion that varies not only with the distance to the object but also spatially. Image deblurring techniques aim to reduce the blur and to restore a sharp image from its defocused version. Many DL based methods have been developed to this purpose, and the results were remarkable [11].

Image restoration and DFD are related problems due to the mere presence of defocus blur. Image deblurring can be assisted by depth while the depth accuracy can improved by exploiting the complementary information carried by blur. Despite this strong relationship, image deblurring and DFD were treated separately by the overwhelming majority of the DL based methods. Only recently, the authors of [12] proposed a multi-task network called 2HDED:NET, where DFD and deblurring support each other to better accomplish these tasks. But again, 2HDED:NET is trained on a synthetically defocused dataset without confirming its effectiveness on naturally defocused images because of the lack of such dataset [13].

A consistent dataset with defocused images annotated for depth and defocus would boost the supervised DL methods for MDE and image deblurring and improve their accuracy. Even when the focus is on training unsupervised learning for MDE or image deblurring, the need for diverse GT depth and AiF images is still necessary for evaluating the models. To fill in this gap, we introduce iDFD, a Depth and Defocus Annotated dataset, with naturally defocused images and corresponding depth and AiF GT. iDFD is a large-scale dataset with various indoor scenes. The dataset was acquired with an MS-Kinect camera and a DSLR camera tightly coupled to each other. The DSLR camera was adopted to capture the naturally defocused images, since the RGB camera of MS-Kinect does not have the facility to work with different apertures. Table 1 gives an overview of iDFD, and Fig. 1 shows some examples from the iDFD dataset illustrating the variety of scenes and the quality of the depth maps as well as the natural defocus images along with the corresponding AiF images.

The primary contribution of this paper is iDFD, a new dataset consisting in a collection of naturally defocused images and their GT AiF images, as well as corresponding depth maps. The dataset is addressed to DFD and image deblurring applications. To the best of our knowledge, iDFD is the first consistent dataset with natural defocus and depth annotations. The detailed description of the dataset is given in the next sections, as follows: in Sect. 2, we provide an overview of the main datasets used for MDE and image deblurring by DL, in Sect. 3 we present our dataset along with the undergone pre-processing, then the dataset is validated by tests with 2HDED:NET in Sect. 4, and finally, conclusions and future work are presented in Sect. 5.

2   Related Work

A series of RGB-D (RGB images paired with Depth maps) and defocused datasets (RGB images paired with defocus maps) have been proposed over the years. With very rare exceptions, these datasets have been dedicated to the exclusive use of MDE, $3D$ scene understanding or image restoration. This section briefly reviews several *indoor* datasets used by these applications.

2.1   RGB-D Datasets

NYU-Depth V2 [2] is the most commonly used dataset for MDE. It was captured with the MS-Kinect RGB-D camera and features a wide range of indoor scenes. As GT, it proposes low resolution raw images and their inpainted versions with the invalid pixels filled in. $Sun3D$ [3] is a dataset dedicated to $3D$ scene understanding. It contains full $3D$ models with semantics: RGB-D images captured in infrared with a ASUS Xtion PRO LIVE sensor, camera poses, object segmentations, and point clouds registered into a global coordinate frame. ScanNet [4] is a more recent RGB-D video dataset that, likewise $Sun3D$, is annotated with $3D$ camera poses, surface reconstructions, and instance-level semantic segmentation. It was obtained with an easy-to-use and scalable RGB-D capture system that

Table 3. Datasets for Monocular Depth Estimation and Image deblurring

Depth Estimation

Dataset Name	Real/ Synthetic	Data Modalities	Sensor type/name	Total no. Images	Scene	Defocus
NYU [2]	Real	RGB, Depth	MS-Kinect	1.4K	Indoor	×
Sun3D [3]	Real	RGB, Depth, 3D-Polygon	Intel RealSense 3D Camera, Asus Xtion LIVE PRO, Kinect V1 and V2	10.3K	Indoor	×
Matterport3D [14]	Real	RGB, Depth, Panoramas	Matterport Camera, Stereo Camera	194K	Indoor	×
Scannet [4]	Real	RGB, Depth	Occipital Structure Sensor, Similar to Kinect V1	5M	Indoor	×
DIODE [15]	Real	RGB, Depth, Normal-Maps	FARO Focus S350	18K	Indoor/ outdoor	×
DSLR-DFD [8]	Real	RGB, Depth	Xtion, DSLR	120	Indoor/ outdoor	✓

Image Deblurring

Dataset Name	Real/ Synthetic	Data Modalities	Sensor type/name	Total no. Images	Scene	Defocus
Shi's [16]	Real	RGB	–	200	Outdoor	✓
DBD [17]	Real	RGB	DSLR	1.1k	Outdoor	✓
RTF [18]	Synthetic	RGB, Blur Map	DSLR	22	Outdoor	✓
LFDOF [19]	Synthetic	RGB	DSLR	12k	Indoor/ outdoor	✓
SYNDOF [20]	Synthetic	RGB, Blur Map	DSLR	9.2k	Indoor	✓
DPD [21]	Real	RGB	DSLR	500	Indoor/ outdoor	✓
RealDOF [22]	Real	RGB	DSLR	50	Indoor/ outdoor	✓
iDFD	**Real**	**RGB, Depth**	**DSLR, MS-Kinect**	**764**	**Indoor**	✓

included automated surface reconstruction and crowd-sourced semantic annotation. *Matterport3D* [14] is also from the category of datasets dedicated to training RGB-D scene understanding algorithms. It contains a large number of panoramic views of home environments. Specific annotations include surface reconstructions, camera poses, and *2D* and *3D* semantic segmentations. The setup to capture panoramas consisted of a tripod-mounted camera rig with three HDR RGB cameras and 3 Matterport depth cameras. DIODE [15] is a dense indoor/outdoor dataset captured with the same sensor -FARO Focus *S*350 scanner- in both environments. The utilization of such datasets for training enables the possibility of achieving generalization across diverse indoor and outdoor domains. The dataset is annotated for depth and normals to the surface and is dedicated to the MDE application. The small dataset DSLR DFD for MDE of Carvalho et al. [8] is the sole to propose defocused and AiF RGB images together with dense depth maps. The RGB images were captured with a DSLR Nikon camera by tuning the aperture such to have defocused and AiF versions of the same scene. For depth, they used an Xtion PRO sensor. Unfortunately, the dataset is very small. It consists of only 110 scenes. A list of datasets proposed over the years for indoor MDE can be found in the upper half of Table 3.

2.2   Datasets for Image Deblurring

The training of a DNN in view of defocus deblurring requires both sharp images with full focus and spatially varying defocused versions. The existing models solve the problem of procuring such datasets either by generating synthetically defocused images or by capturing them with DSLR cameras. Table 3 summarizes some of the existing datasets containing defocused images. Shi's and DBD datasets [16,17] were developed for blur detection/estimation, which can be further used in deblurring with classic methods. They provide as GT binary blur masks.

RTF [18] is a small synthetic dataset of 22 images, used also to estimate the defocus map. Defocused images and GT defocus maps are generated from light field data captured with a plenoptic camera. From the same category is LFDOF [19] based on the Stanford Multiview Light Field Dataset. It is a large defocus dataset generated with the open-sourced software Lytro Power Tool. A total of 11,986 defocused and AiF image pairs are included in this dataset.

SYNDOF [20] is a synthetic dataset, where each image is synthetically blurred using a GT depth map. The images come from three existing datasets, two with game scenes and road rendering and one with real stereo images.

DPD [21] and RealDOF [22] are datasets with naturally defocused images. The GT consists of AiF images captured with a narrower aperture, while the defocused ones are obtained with large apertures.

2.3   Datasets for Joint DFD and Image Deblurring

It is worth mentioning several DL applications where depth estimation and deblurring are used jointly to support each other. In the context of training a network for Joint DFD and Image Deblurring, a major drawback of previously proposed datasets is that they are either deficient in naturally defocused images or GT depth maps (Table 3). This deficiency poses a challenge to effectively train the network. In [10], two concatenated DNN estimate the depth map from a defocused image and later, this intermediate result is used to deblur the image by deconvolution. The method is tested on NYU-v2 [2] Make3D [23] datasets that contain only AiF and depth GT, the defocussed images being generated synthetically. A similar application is in [19], where instead of the depth is estimated a defocus map but with the same purpose of restoring an AiF image. The dataset

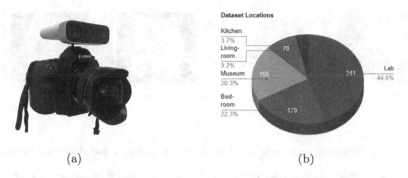

(a)   (b)

Fig. 2. (a). Experimental setup: DSLR-Nikon camera is coupled with MS-Kinect camera using a Camera Shoe Mount Adapter. Two cameras are fixed on a tripod to avoid motion. The DSLR is used to capture AiF and Defocus (RGB) images and Kinect is used to capture the GT depth for RGB images. (b). Data acquisition locations and the total number of images per location.

used to train the network is LFDOF that contains synthetic images generated from light field data. In [8], a DNN called D3-Net, is used to estimate the depth from defocused images. The network is trained first on NYU v2 dataset by using depth GT and synthetic defocused images. Afterwords, the network is fine tuned on a small proprietary dataset containing naturally defocused images captured with a DSLR camera. A similar situation is encountered in [12], where the multi-task network 2HDED-NET is trained with synthetically defocused images and depth GT from NYU v2 and Make3D datasets, in order to estimate both the depth map and AiF image. These examples show how necessary are the datasets with naturally defocused, AiF images and depth GT. Our iDFD comes to fill in this gap in the series of existing datasets.

3   iDFD Dataset for DFD and Image Deblurring

The dataset contains a variety of indoor scenes captured in 5 different environments: bedroom, living room, labs, kitchen, and museum (Fig. 1). During data acquisition, we paid attention to the diversity of scenes, so we selected challenging locations with complicated contents e.g. scenes from museum (Fig. 1). In total, we captured 764 scenes, most of them coming from the labs. Figure 2b shows the distribution of scenes over the 5 environments. Some scenes, such as those from bedrooms and living rooms, are much simpler as content than the scenes coming from labs and museum. The goal has been to acquire three images for each scene: defocused, AiF and depth.

3.1   Data Acquisition

We used a DSLR-Nikon camera with a $14mm$ lens at two different aperture sizes to capture AiF and defocused images. For AiF images we used a narrow

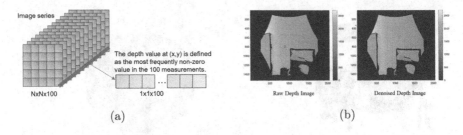

(a)   (b)

Fig. 3. (a) Denoising approach. (b) Example of raw depth and denoisied depth.

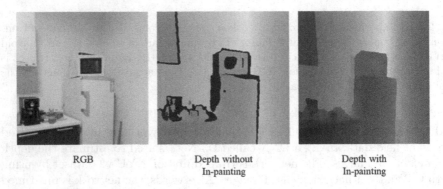

RGB   Depth without   Depth with
                       In-painting   In-painting

Fig. 4. Depth in-painting: recovery of the missing depth values. The invalid pixels in depth raw image are marked in black.

aperture with $f/10$, and for the out-of-focus ones a wider aperture with $f/2.8$. The wide aperture creates a shallow depth of field resulting in a blurred image. While capturing scenes with the DSLR camera, we needed to make sure that there is no motion blur, but it was impossible to completely avoid the occasional motion in the scenes.

To acquire the GT depth, we used the MS-Kienct camera in the Narrow Field-of-View (NFOV) depth mode with an operating range of 0.5–3.8 m, and the maximum Field of Interest (FoI) of $75° \times 65°$. The depth can also be provided outside the operating range, but it depends entirely on the reflectivity of the objects. In selecting the scenes, we avoided locations where depth information cannot be correctly captured, such as shadows, etc. Our experimental setup is shown in Fig. 2a. The Kinect camera was mounted on a DSLR-Nikon using a shoe mount adapter, and our experimental setup is similar to that of Carvalho et al. [8].

3.2 Data Preprocessing

For pre-processing the data, we followed the procedure of Qiu et al. [24], consisting in denoising, inpainting (optional), image registration, normalization and cropping. The overall workflow is depicted in Fig. 5.

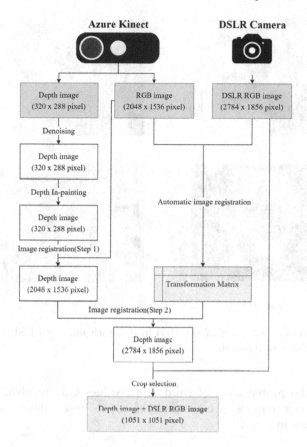

Fig. 5. Overall workflow of data pre-processing. The blue rectangles represent the input images, yellow is the output and white are the intermediate results after each step. The dimension of the data at each step is displayed and each process is represented by an arrow. (Color figure online)

Denoising. The images captured with the ToF camera are very noisy. The noise is primarily caused by the false phase shift occurring as a result of the differences in distances within a solid angle. This phenomenon is also known as flying pixels [25]. Considering that flying pixels are randomly distributed in each frame, we removed the noise by repeating the measurements, as shown in Fig. 3a. Similar to the procedure of Qiu et al. [24], we collected 100 depth images during each capture, and the value of each depth pixel was determined by the most frequent non-zero pixel in the 100 measurements. An example of raw depth and denoised depth is shown in Fig. 3b.

Depth In-painting. The dataset offers the in-painted depth in addition to the raw depth maps. To recover the invalid pixels in the raw depth maps, we use the iterative in-painting method proposed by Pertuz et al. [26]. This method works

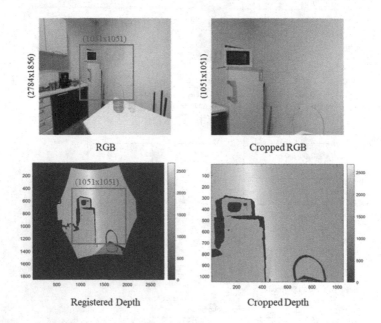

Fig. 6. Crop selection: selection of the ROI from depth map and DSLR RGB image, respectively. (Color figure online)

by restricting the propagation of valid depth values only to highly consistent, independent image regions. An example of depth image before and after inpainting is shown in Fig. 4.

Image Registration. Being captured from slightly different perspectives, the depth from Kinect and the DSLR image need to be registered. The registration is done by means of Kinect RGB image. First, the depth is registered to Kinect RGB and then, the two RGB images (from Kinect and DSLR) are again registered. This entail the registration of depth to the DSLR image. The process is depicted in Fig. 5.

Azure Kinect Sensor SDK provides a function for transforming the depth map from the perspective of depth camera to the perspective of built-in RGB camera. By warping a triangle mesh, it is possible to convert the depth camera geometry into the color camera geometry. The triangle mesh is used to prevent holes from being generated in the transformed depth image [27]. By using this function, we generated the so called RGB-D images, in which D represents an additional channel that records depth information.

To register the Kinect RGB to DSLR image, we followed Oriented FAST and Rotated BRIEF (ORB) method from Ethan et al. [28], having the following specific steps:

1. Detect the ORB keypoints on the Azure Kinect RGB image and the DSLR AiF image.

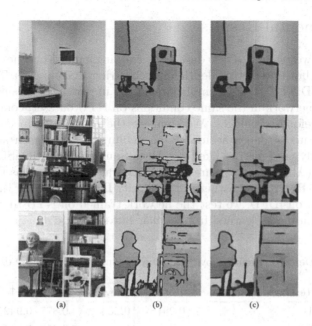

Fig. 7. Results with 2HDED:NET trained on raw depth: a) RGB image, b) Raw GT depth, c) Estimated depth.

2. Compute the ORB descriptors from the keypoints.
3. Match the two sets of ORB descriptors by calculating the pairwise distance between feature descriptors [28].
4. Estimate projective transformation matrix from matching point pairs [29].
5. Apply the transformation matrix for depth and DSLR image registration.

Image Normalization. We normalize the depth and DSLR images such to fit the pixels values in the range using $cv2.NORM - MINMAX$ function from OpenCV. The image normalization process is often used by DNNs as a data pre-processing step, in order to homogenize the entire dataset in terms of size and pixel values.

Crop Selection. Since we are using the NFOV depth mode, the depth is represented in a smaller hexagonal area (Fig. 6). We need to select a Region of Interest (ROI) from the depth map, and the correspondent of the DSLR RGB, such to avoid the blank area. We manually curated the registered depth maps and DSLR RGB to remove the unwanted regions (blue region around the ROI in the depth maps). We cropped patches with the resolution of 1051×1051 pixels from both deph and RGB images.

4   Experimental Results

In this section, we evaluate iDFD dataset by training a multi-task network that is challenging for both predicting depth and deblurring RGB images. The network is called 2HDED:NET [12]. To emphasize on the importance of using real data, we retrain the network in the same conditions on NYU dataset, completed with synthetically defocused images. We compare the results obtained on the two datasets, in terms of depth accuracy and deblurring. For depth accuracy, we utilize commonly used metrics like Root Mean Squared Error (RMSE), Absolute Relative Error (Abs. rel), and Thresholded Accuracy (δ), as well as Peak Signal-to-Noise Ratio (PSNR) and Structural Similarity Index (SSIM) for the deblurred images. Finally, on iDFD, we test the network trained on NYU to evaluate the performance of a model learned from similar but synthetic data.

Table 4. Depth Estimation by 2HDED:NET on NYU and iDFD datasets.

Training/Testing dataset	$RMSE \downarrow$	$Abs.\ rel \downarrow$	$\delta 1 \uparrow$	$\delta 2 \uparrow$	$\delta 3 \uparrow$
NYU/NYU	0.281	0.266	0.877	0.942	0.958
NYU/iDFD	0.401	0.284	0.759	0.813	0.839
iDFD/iDFD (raw depth)	0.312	0.194	0.727	0.793	0.807
iDFD/iDFD (in-painted depth)	0.248	0.145	0.799	0.854	0.959

2HDED:NET is a recently proposed architecture consisting of one encoder and two decoders [12]. The encoder is fed in with a defocused RGB image, while one decoder outputs the scene depth map and the other deblurred image. The network has to be trained under the joint supervision of the GT depth and AiF RGB images. Until now, it was trained only on synthetically defocused images from NYU indoor dataset.

In our tests with NYU and iDFD, we train 2HDED:NET by using AiF images as GT for the deblurring head. To have a fair comparison with the iDFD dataset, we change the blur specification to generate synthetic blur in the NYU dataset than [12]. In case of iDFD dataset, for GT depth, we have two options: the raw depth and the in-painted depth. We give results for both cases. The iDFD dataset consists of 764 RGB images and corresponding depth maps. To train the 2HDED:NET on the iDFD dataset, we divide the dataset into 500 images for training and 264 images for testing, the test set contains images from all scenes. For NYU dataset we use the same dataset split as [8,10,12]. The simulated defocus is similar as dB to the natural blur on iDFD images. For training the 2HDED:NET on the iDFD dataset, we use the same training parameters as [12], with a batch size of 4 images for 500 epochs and a Stochastic Gradient Descent (SGD) optimizer with an initial learning rate of 0.0002. The initial learning rate is reduced 10 times after the first 300 epochs. For data augmentation, we followed the same procedure as in [8,12,30]. Table 4 presents the errors in estimating the

(a)   (b)   (c)

Fig. 8. Results with 2HDED:NET trained on in-painted depth: a) RGB image, b) Depth GT, c) Estimated Depth.

depth for both NYU v2 and iDFD datasets. When training on NYU dataset, the RMSE is lower than that of the raw depths from the iDFD dataset. The figures show better accuracy when in-painted depth is used. This results must be taken under the reserve of a GT which is not 100% measured. The worst results are obtained for the network trained on NYU and tested on iDFD dataset. Figures 7 and 8 depict three examples of depth maps obtained after training with raw depth and in-painted depth, respectively. The visual inspection shows that when raw images are used, the network tends to wipe small regions like the empty spaces on the shelves in the second example. These details are preserved by the depth maps obtained after training on in-painted depth. This suggests that in-painting should be performed rather before training then after training with raw depth.

Table 5 illustrates the deblurring results of 2HDED:NET on NYU v2 and iDFD datasets. The 2HDED:NET has shown promising results for deblurring of synthetically defocused images from NYU, where the PSNR was improved from 26.09 dB to 32.68 dB (Table 5). On iDFD naturally defocused images, 2HDED:NET is able to improve the PSNR from 25.83 dB to 36.25 dB, which means a gain of 10.43 dB, significantly higher than the 6.59 dB obtained in the case of NYU. The initial PSNR of 25.83 dB has been calculated by taking as reference the AiF image obtained with aperture $f/10$.

Finally, we tested on iDFD the network trained on NYU. The deblurring has been more modest with a final PSNR of only 30.09 dB. This result shows that a

Table 5. Deblurring results obtained with 2HDED:NET trained or iDFD and NYU datasets.

Dataset	PSNR ↑	SSIM ↑
NYU after deblurring (training on NYU)	32.68	0.91
NYU defocused	26.09	0.49
Gain	6.59	0.42
iDFD after deblurring (training on iDFD)	36.25	0.99
iDFD defocused	25.83	0.51
Gain	10.43	0.43
iDFD after deblurring (training on NYU)	30.09	0.86

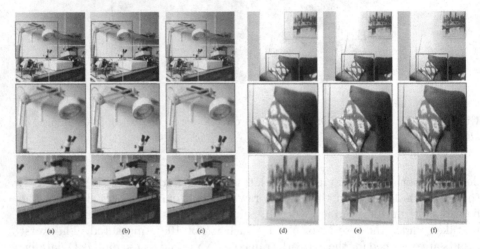

Fig. 9. 2HDED:NET results for image deblurring: from left to right in the first-row a) Defocus (input), b) AiF (GT), and c) Deblurred. Similarly, d) Defocus (input), e) AiF (GT), and f) Deblurred. The second row shows the zoomed-in regions to illustrate the deblurring results. Two examples are taken from the electronics lab and living room.

network trained on a dataset with natural defocus and high quality RGB images can be more effective for image deblurring than the same network trained on a large synthetically defocused dataset, such as that obtained from NYU.

Figure 9 depicts two images with a complex content, restored by 2HDED:NET trained on iDFD dataset. Crops representing the microscope in the first image or the pillow and the picture in the second image, are magnified in order to show how details sinked into blur, emerge after restauration with 2HDED:NET.

5 Conclusion

The iDFD dataset, proposed in this paper, is a collection of naturally defocused indoor scenes that has the novelty of being supplemented by both depth and AiF GT. The tests on iDFD with the multi-task network 2HDED:NET, which simultaneously estimates the depth and deblurs the image, have proved that training a network on real rather than simulated data like NYU synthetically defocused, is by far more effective. On iDFD, for deblurring, the PSNR gain has been higher by about 5 dB and the RMSE of estimated depth was 12% lower on in-painted depth maps. The same network trained on NYU and tested on iDFD gave worse results for both depth and deblurring, showing once again how the performance of a model trained on synthetic data degrades on real data. iDFD is useful for DL based applications that take into account the connection between depth estimation and image deblurring, two tasks inherently bridged by the defocus blur, present in all real images. A limitation of the iDFD dataset is the fixed camera settings used to capture defocused images that impose a certain range for defocus. In our future work, we plan to evaluate the iDFD dataset quality by training and testing it on various MDE and DFD networks.

References

1. Ming, Y., Meng, X., Fan, C., Yu, H.: Deep learning for monocular depth estimation: a review. Neurocomputing **438**, 14–33 (2021)
2. Silberman, N., Hoiem, D., Kohli, P., Fergus, R.: Indoor segmentation and support inference from RGBD images. In: Fitzgibbon, A., Lazebnik, S., Perona, P., Sato, Y., Schmid, C. (eds.) ECCV 2012. LNCS, vol. 7576, pp. 746–760. Springer, Heidelberg (2012). https://doi.org/10.1007/978-3-642-33715-4_54
3. Xiao, J., Owens, A., Torralba, A.: Sun3d: a database of big spaces reconstructed using SFM and object labels. In: Proceedings of the IEEE International Conference on Computer Vision, pp. 1625 1632 (2013)
4. Dai, A., Chang, A.X., Savva, M., Halber, M., Funkhouser, T., Nießner, M.: Scannet: richly-annotated 3d reconstructions of indoor scenes. In: Proceedings of the IEEE Conference on Computer Vision and Pattern Recognition, pp. 5828–5839 (2017)
5. Ikoma, H., Nguyen, C.M., Metzler, C.A., Peng, Y., Wetzstein, G.: Depth from defocus with learned optics for imaging and occlusion-aware depth estimation. In: 2021 IEEE International Conference on Computational Photography (ICCP), pp. 1–12, IEEE (2021)
6. Song, G., Lee, K.M.: Depth estimation network for dual defocused images with different depth-of-field. In: 2018 25th IEEE International Conference on Image Processing (ICIP), pp. 1563–1567. IEEE (2018)
7. Maximov, M., Galim, K., Leal-Taixé, L.: Focus on defocus: bridging the synthetic to real domain gap for depth estimation. In: Proceedings of the IEEE/CVF Conference on Computer Vision and Pattern Recognition, pp. 1071–1080 (2020)
8. Carvalho, M., Le Saux, B., Trouvé-Peloux, P., Almansa, A., Champagnat, F.: Deep depth from defocus: how can defocus blur improve 3d estimation using dense neural networks? In: Leal-Taixé, L., Roth, S. (eds.) ECCV 2018. LNCS, vol. 11129, pp. 307–323. Springer, Cham (2019). https://doi.org/10.1007/978-3-030-11009-3_18

9. Gur, S., Wolf, L.: Single image depth estimation trained via depth from defocus cues. In: Proceedings of the IEEE/CVF Conference on Computer Vision and Pattern Recognition, pp. 7683–7692 (2019)
10. Anwar, S., Hayder, Z., Porikli, F.: Deblur and deep depth from single defocus image. Machine Vision Appl. **32**(1), 1–13 (2021). https://doi.org/10.1007/s00138-020-01162-6
11. Zhang, K., et al.: Deep image deblurring: a survey. Int. J. Comput. Vision **130**(9), 2103–2130 (2022)
12. Nazir, S., Vaquero, L., Mucientes, M., Brea, V.M., Coltuc, D.: 2hded: net for joint depth estimation and image deblurring from a single out-of-focus image. In: 2022 IEEE International Conference on Image Processing (ICIP), pp. 2006–2010. IEEE (2022)
13. Nazir, S., Vaquero, L., Mucientes, M., Brea, V.M., Coltuc, D.: Depth estimation and image restoration by deep learning from defocused images. arXiv preprint arXiv:2302.10730 (2023)
14. Chang, A., et al.: Matterport3d: learning from RGB-D data in indoor environments. arXiv preprint arXiv:1709.06158 (2017)
15. Vasiljevic, I., et al.: Diode: a dense indoor and outdoor depth dataset. arXiv preprint arXiv:1908.00463 (2019)
16. Shi, J., Xu, L., Jia, J.: Just noticeable defocus blur detection and estimation. In: Proceedings of the IEEE Conference on Computer Vision and Pattern Recognition, pp. 657–665 (2015)
17. Zhao, W., Zhao, F., Wang, D., Lu, H.: Defocus blur detection via multi-stream bottom-top-bottom fully convolutional network. In: Proceedings of the IEEE Conference on Computer Vision and Pattern Recognition, pp. 3080–3088 (2018)
18. D'Andrès, L., Salvador, J., Kochale, A., Süsstrunk, S.: Non-parametric blur map regression for depth of field extension. IEEE Trans. Image Process. **25**(4), 1660–1673 (2016)
19. Ruan, L., Chen, B., Li, J., Lam, M.-L.: Aifnet: all-in-focus image restoration network using a light field-based dataset. IEEE Trans. Comput. Imaging **7**, 675–688 (2021)
20. Lee, J., Lee, S., Cho, S., Lee, S.: Deep defocus map estimation using domain adaptation. In: Proceedings of the IEEE/CVF Conference on Computer Vision and Pattern Recognition, pp. 12222–12230 (2019)
21. Abuolaim, A., Brown, M.S.: Defocus deblurring using dual-pixel data. In: Vedaldi, A., Bischof, H., Brox, T., Frahm, J.-M. (eds.) ECCV 2020. LNCS, vol. 12355, pp. 111–126. Springer, Cham (2020). https://doi.org/10.1007/978-3-030-58607-2_7
22. Lee, J., Son, H., Rim, J., Cho, S., Lee, S.: Iterative filter adaptive network for single image defocus deblurring. In: Proceedings of the IEEE/CVF Conference on Computer Vision and Pattern Recognition, pp. 2034–2042 (2021)
23. Saxena, A., Sun, M., Ng, A.Y.: Make3d: Learning 3d scene structure from a single still image. IEEE Trans. Pattern Anal. Mach. Intell. **31**(5), 824–840 (2008)
24. Qiu, Z., Martínez-Sánchez, J., Brea, V.M., López, P., Arias, P.: Low-cost mobile mapping system solution for traffic sign segmentation using azure kinect. Int. J. Appl. Earth Obs. Geoinf. **112**, 102895 (2022)
25. Lindner, M., Schiller, I., Kolb, A., Koch, R.: Time-of-flight sensor calibration for accurate range sensing. Comput. Vis. Image Underst. **114**(12), 1318–1328 (2010)
26. Pertuz, S., Kamarainen, J.: region-based depth recovery for highly sparse depth maps. In: 2017 IEEE International Conference on Image Processing (ICIP), pp. 2074–2078. IEEE (2017)

27. Microsoft. Use azure Kinect sensor SDK image transformations
28. Rublee, E., Rabaud, V., Konolige, K., Bradski, G.: Orb: an efficient alternative to sift or surf. In: 2011 International Conference on Computer Vision, pp. 2564–2571. IEEE (2011)
29. Hartley, R., Zisserman, A.: Multiple View Geometry in Computer Vision. Cambridge University Press, Cambridge (2003)
30. Nazir, S., Coltuc, D.: Edge-preserving smoothing regularization for monocular depth estimation. In: 2021 26th International Conference on Automation and Computing (ICAC), pp. 1–6. IEEE (2021)

TBPos: Dataset for Large-Scale Precision Visual Localization

Masud Fahim[1], Ilona Söchting[1], Luca Ferranti[1], Juho Kannala[2],
and Jani Boutellier[1(✉)]

[1] University of Vaasa, Vaasa, Finland
{masud.fahim,ilona.sochting,luca.ferranti,jani.boutellier}@uwasa.fi
[2] Aalto University, Espoo, Finland
juho.kannala@aalto.fi

Abstract. Image based localization is a classical computer vision challenge, with several well-known datasets. Generally, datasets consist of a visual 3D database that captures the modeled scenery, as well as query images whose 3D pose is to be discovered. Usually the query images have been acquired with a camera that differs from the imaging hardware used to collect the 3D database; consequently, it is hard to acquire accurate ground truth poses between query images and the 3D database. As the accuracy of visual localization algorithms constantly improves, precise ground truth becomes increasingly important. This paper proposes TBPos, a novel large-scale visual dataset for image based positioning, which provides query images with fully accurate ground truth poses: both the database images and the query images have been derived from the same laser scanner data. In the experimental part of the paper, the proposed dataset is evaluated by means of an image-based localization pipeline.

Keywords: Visual localization · 6DoF pose · Dataset · Computer vision

1 Introduction

Image based localization is one of the enabling technologies for autonomous vehicles [12], augmented reality [5], and robotics [15]. Driven by advances in deep learning [1,6,16,17], the precision of image based localization has progressed in significant leaps during the recent years: for example, using the classical Aachen Day-Night dataset, the 2016 state-of-the-art visual localization approach Active Search [18] yielded 43.9% accuracy for night images, whereas HFNet [16] proposed in 2019 yielded[1] already 72.4% accuracy [16].

To increase the level of challenge for state-of-the-art visual localization algorithms, two main directions of development can be identified: 1) introducing more challenging datasets (for example, by visual data sparsity, occlusions, lighting changes), such as InLoc [23], and 2) reducing the pose correctness threshold to, e.g., 0.5 m and 2 ° [7]. However, the prerequisite for pose correctness assessment

---

[1] 5.0 m distance and 10 ° orientation threshold.

R. Gade et al. (Eds.): SCIA 2023, LNCS 13885, pp. 84–94, 2023.
https://doi.org/10.1007/978-3-031-31435-3_6

Fig. 1. Overview of the laser-scanned building, acquired by registering the 182 point clouds that consist the 3D dataset.

is accurate ground truth (GT); a recent work [4] pointed out that with most current visual positioning datasets the ground truth poses have been acquired by a reference algorithm such as SfM (e.g., [21]) or SLAM (e.g., [22]). Consequently, the authors of [4] conclude that benchmarks, which rely on such *pseudo ground truth*, measure how well visual localization algorithms are able to reproduce the output of the reference algorithm (SfM or SLAM), instead of absolute pose accuracy.

This paper proposes a new large-scale dataset for visual positioning and provides exact ground truth by the use of *synthetic query images*. In the proposed dataset the 3D structure of the environment (see Fig. 1) has been captured using a high-resolution 3D laser scanner, and formatted into *database images* that describe the 3D environment for visual localization methods. Then, instead of relying on secondary camera equipment for capturing *query images*, the query images are generated algorithmically from the same 3D laser scanner data, yielding exact ground truth poses. Based on our experiments, there clearly is a need for exact ground truth: the tested visual localization algorithm [23] was able to reach a localization accuracy of less than 0.02 m and 0.3 °, which is more than an order of magnitude less than the presently dominant threshold of 0.25 m and 10 °. Besides ground truth accuracy, synthetic query image generation allows producing an arbitrary number of new query images, which can be beneficial for future developments in visual localization.

Obviously, the proposed approach of using synthetic images risks creating either a) too easy or b) unrealistic query images. However, our claim is that the generated query images are sufficiently realistic, because they have been formed from measured visual data, and their degree of challenge can be increased by introducing a set of irreversible distortions, such as occlusion or nonlinear lighting variation. In the experimental section of the paper we apply the InLoc [23] visual localization pipeline to the proposed dataset to assess its level of challenge compared to the well-known InLoc dataset [23].

Table 1. Statistics of selected large-scale localization datasets: number of locations used to acquire database images, number of provided database and query images, means for acquiring ground truth (GT) poses; P3P: Perspective-three-point [10], BA: Bundle adjustment, SfM: Structure-from-Motion [21], VS: View synthesis [23], ICP: Iterative closest point [3], SIFT: Scale-invariant feature transform [13]. (*) The number of query images synthesized for experiments presented in this work.

Dataset	#Locations	#Database	#Query	GT
InLoc [23]	277	9972	329	P3P+BA
Aachen Day-Night [19,20]	n/a	4328	922	SfM+P3P
Aachen Day-Night v1.1 [19,24]	n/a	6697	1015	SfM+VS
RobotCar Seasons [14,20]	8707	26121	11934	ICP
CMU Seasons [2,20]	17	7159	75335	SIFT+BA
TBPos (proposed)	182	6552	338*	VS

The contributions of this work are as follows:

1. TBPos, a novel open[2] 3D dataset for large-scale visual precision localization,
2. Proposed approach of view synthesis for query image and exact ground truth generation, and
3. Benchmarking of the proposed dataset using a visual localization pipeline.

Furthermore, our results point out that the use of view synthesis enables generating challenging query images from viewpoints, where traditional query image acquisition could not provide reliable ground truth. For the development of practical image based localization applications (e.g., autonomous vehicles) such challenging real-life cases need to be considered as well.

2   Related Work

2.1   Datasets

Several well-known datasets have been published for large-scale visual localization, some of which are shown in Table 1. Related to this work, the most significant differences between the datasets can be seen in the means how ground truth poses for query images have been acquired. For the RobotCar Seasons dataset [14,20], query image ground truth poses were acquired from LIDAR-based point clouds after ICP [3] alignment and some manual adjustment [20]. For the CMU Seasons dataset [20], bundle-adjusted SIFT [13] features were used to construct local 3D scenery models, and to acquire GT poses. In the InLoc dataset [23], GT poses were acquired using the P3P-RANSAC and bundle adjustment, with manual matching for difficult cases. The Aachen Day-Night dataset [20] exists in two versions, for both of which GT poses were estimated from an SfM model

---

[2] https://gitlab.com/jboutell/tbpos; https://doi.org/10.5281/zenodo.7466448.

using a P3P solver. The recent v1.1 version [24] of the Aachen Day-Night dataset comes closest to the proposed work in the sense that it uses view synthesis [23]; however, whereas [24] uses view synthesis for refinement of GT poses related to separately acquired query images, in our proposed dataset view synthesis is used to *render the query images*. Whereas the approach taken in [24] enables more accurate reference pose acquisition for manually acquired query images, the proposed approach enables automatic generation of query images with exact ground truth. A further advantage of the proposed approach is its applicability to the training of machine learning based visual localization techniques that require numerous query images, such as [9].

2.2 Algorithms

Algorithms for image based localization can be classified into *direct matching*, *retrieval-based* and *learning-based* approaches [8].

The InLoc [23] work presented both an extensive dataset for visual localization, and an image retrieval-based localization pipeline. In retrieval based pipelines, the main stages are *feature extraction, image retrieval, dense matching* and *pose estimation* [9]. Whereas the earlier pipeline stages between feature extraction and dense matching are based on learned image features, the pose estimation step leverages classical geometric computer vision for estimating the 6DoF camera pose for a given query image. InLoc appended this generic pipeline with the *pose verification* step that further increases localization accuracy. Recently, the work PCLoc [11], built on InLoc, also added a further *pose correction* step between the stages of pose estimation and pose verification.

The learning-based work HFNet [16] proposed a monolithic CNN for simultaneous keypoint detection, as well as local and global keypoint extraction, which provides significant runtime computation savings. Despite computational efficiency, results reported in [11] show that on average HFNet yields a similar level of accuracy as the InLoc [23] visual localization algorithm.

3 Visual Data Acquisition Procedure

The visual data for TBPos was acquired by laser-scanning an industrial building (see Fig. 1), which consists of inside and outside areas. Typical to such buildings, both repeating and absent texture are commonplace on surfaces. The 3D structure contains a lot of recurring shapes: identically-shaped rooms, tie beams, etc. Compared to the related datasets of Table 1, the proposed TBPos dataset is characteristically most similar to the InLoc dataset, with the most significant difference that TBPos covers also outside areas.

The laser scanning was performed using a Faro Focus 3D scanner, providing colored point clouds that mostly contain 27 million points each. Individual point clouds were registered against each other using the Faro SCENE software that required some degree of manual correspondence annotation.

Fig. 2. Six database images extracted from a single point cloud scan. Each image has a 60° field-of-view (hor.), and the images have been taken with a 30° stride.

To ease the adoption of TBPos, the data structure of InLoc [23] was used for point clouds, database images, query images and supporting data. For database images, each laser scan was sliced into 36 perspective RGBD images with 30° sampling stride and 60° field-of-view. The chosen resolution for database images was 1024×768. An example of six consecutive database images is shown in Fig. 2.

4   Synthesizing Query Images

The adopted procedure of query image synthesis (see Fig. 4) is similar to the procedure of database image generation. Starting from the original laser scanner position $\mathbf{t}$, the virtual camera position is randomly perturbed along X, Y and Z axes, providing a new position $\mathbf{t}^1$. Similarly, starting from initial laser scanner pointing direction $\mathbf{R}$, a random view direction $\mathbf{R}^1$ is generated, forming the virtual camera 6DOF pose $\mathbf{R}^1\mathbf{t}^1$ *that we also adopt as the ground truth for pose estimation benchmarking.* After acquiring $\mathbf{R}^1\mathbf{t}^1$, the 3D environment can be modified for the purpose of generating query images such that they reflect challenges relevant to practical pose estimation, e.g., lighting variation, displaced objects, etc.

Provided $\mathbf{R}^1\mathbf{t}^1$, a predefined focal length f, and the desired query image resolution $\{r_x, r_y\}$, a novel view can be synthesized from the surrounding 3D point cloud. Due to various reasons, such as occluded view directions, a certain proportion of synthetic view pixels can remain absent of visual data. In our proposed view synthesis approach these missing pixels are filled using an iterative, clamping based interpolation procedure. However, in order to maintain high visual quality for the synthesized query images, randomly generated virtual camera orientations $\mathbf{R}^1\mathbf{t}^1$ that have too much missing visual information can be discarded by straightforward missing pixel counting. As an optional last step, the

Fig. 3. Synthesized query image samples. Left-hand images have been extracted from the laser scanner data without modification, whereas their right-hand counterparts are after lighting adjustment and addition of 2D occlusion (lower-right image).

generated 2D view can be further modified to increase the positioning challenge. Such modification can include occlusion, lighting variation or image noise.

In the practical implementation of the proposed view synthesis approach, the random view rotations $\mathbf{R}^1$ around the Y (horizontal) axis were restricted in order to avoid generating views that dominantly show floor or ceiling. Similarly, perturbation of camera location was limited to small translations of a couple of meters to avoid placing the virtual camera behind walls.

5   Experimental Results

For benchmarking purposes a series of 338 query image poses were generated using the procedure described in Sect. 4. The query images were chosen to be sampled with a view angle and resolution identical to the database images (60 °, 1024×768), although there was no technical restriction to select other values. In order to increase the pose estimation challenge, the data was modified both in 3D (point cloud format) and 2D in the query image synthesis process.

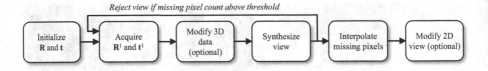

Fig. 4. The proposed procedure for query image synthesis.

The 3D point cloud data was lighting-adjusted to simulate a situation where an autonomous mobile platform navigates in a dark environment, lighting its surroundings by in-built lights. In Fig. 3, the two topmost images show the effect of this operation: brightness of 3D points decreases as a function of camera-point distance, simulating the effect of a flashlight close to the camera. In terms of pose estimation challenge, this effect causes the appearance of distant query image features to change considerably compared to their database counterparts. On the other hand, lightly-colored surfaces close to the camera tend to get overexposed, losing similarity with the respective database images.

Besides 3D lighting adjustment, also additional occlusions were introduced to most query images: an example of this can be seen in the lower-right image of Fig. 3. Each query image to be occluded was allowed to get a random quadrangle to cover between 1% and 50% of query image area, simulating the case where an object gets in between the surroundings and the autonomous platform's camera.

A significant benefit in using synthetic query images instead of manually acquired ones can be discovered by comparing the TBPos queries to the queries of the InLoc dataset [23]: whereas InLoc queries have exclusively been taken at human eye level with unnoticeable changes in camera pitch or roll, the TBPos query image locations range from close-to-floor level (Fig. 3) to higher altitudes, and include random rotations along all camera axes.

To assess the difficulty level of the query image set, the InLoc visual localization pipeline was used to acquire localization accuracy at the commonly applied translation thresholds of 0.25 m, 0.5 m and 1.0 m, and angular threshold of 10 °. The results are shown in Table 2(a) together with the respective results for the InLoc dataset. The numbers show that the proposed TBPos dataset with the 338 query images are significantly more challenging than the InLoc dataset, as localization success at 1.0 m and 10° is more than 30% lower for TBPos.

Table 2(b) details the TBPos pose estimation accuracy analysis at different stages of the InLoc pose estimation pipeline. It can be seen that the impact of the pose verification (PV) stage is minor, ranging between 0.9% and 2.4% in the success rate of $\leq$ 1.0 m accuracy. The 'Top10' column of Table 2(b), on the other hand, shows how often at least one database image *from the same point cloud scan as the query image*, appears in the list of top-10 database image candidates. For example, 66.6% means that for 225 out of the 338 query images, the InLoc pipeline has ranked at least one database image from the same point cloud scan as the query, into top-10 best candidates, and consequently this figure measures the success rate of image retrieval (IR).

Table 2. TBPos evaluation.

Dataset	0.25m	0.5m	1.0m
InLoc [23]	41.6%	56.5%	67.2%
TBPos	29.6%	34.3%	35.2%

(a) InLoc localization pipeline accuracy at various thresholds for the proposed and reference (InLoc) datasets using the 10° angular threshold.

TBPos datas.	0.25m	0.5m	1.0m	Top10
IR+DM+PV	29.6%	34.3%	35.2%	66.6%
IR+DM	27.2%	33.4%	34.3%	66.6%
IR	n/a	n/a	n/a	63.9%

(b) InLoc pose estimation success rate at different pipeline stages, for the TBPos dataset. IR: image retrieval, DM: dense matching, PV: pose verification.

With the success rate of 63.9%, the InLoc IR clearly has some challenges with the TBPos dataset. However, when the actual camera pose is estimated by dense matching (DM) for the best matching database image candidate using one of the metric thresholds, the success rate drops by around 30%. Consequently, it can be stated that for the TBPos set of 338 query images, the pose estimation difficulties are related both to IR and DM stages.

The precision of the InLoc pose estimation pipeline can be evaluated by computing the average deviation from ground truth for the 29.6% of cases where the query image is within 0.25 m and 10 ° from ground truth: for these success cases the average location deviation is 0.10 m and the average angular deviation is 2.26 ° from the ground truth for the TBPos set of 338 query images. In order to get better understanding of visual localization accuracy potential, an easier version of the 338 query image set was generated using the same ground truth poses, but without lighting variation or occlusions. With this considerably easier set of query images, the top-30% of pose estimates reached a localization accuracy below 0.02 m and angular accuracy below 0.3 °. Evidently, accuracy analysis like this requires exact ground truth that is available by the proposed approach of query image synthesis.

Still, also with synthesized query images, the achievable lower bound of accuracy is determined by the precision of the imaging device used for dataset collection, as well as the resolution of the database and query images.

5.1   Analysis of Pose Estimation Failure Cases

Figure 5 depicts representative samples from the set of 338 queries, where pose estimation by InLoc failed. The top-left image shows a query from outside environment, where distance-based darkening has rendered distant visual details almost invisible. For this query image, InLoc is not capable of computing any pose. However, if the brightness of the query image is manually increased using regular image processing (see Fig. 5, bottom-left), ample visual detail is revealed and manual localization would be possible. Hence, we consider the pose estimation failure of this case a shortcoming of the InLoc pose estimation pipeline.

Fig. 5. Examples of TBPos query images where the InLoc pose estimation pipeline completely fails.

The top-center and bottom-center cases of Fig. 5 show two further cases where the InLoc pipeline completely fails in pose estimation. The top-center image shows a large area without any texture, whereas the bottom case exhibits patterns that are repeated in several locations of the whole dataset, and successful localization would require paying attention to small details that discriminate the particular location from similar places elsewhere in the database.

The top-right case shows a close-up query image with very specific detail that would make manual pose estimation straightforward, but also in this case InLoc completely fails due to the overexposure that makes the query image's features differ significantly from the ones of the respective database image.

Finally, the failure case depicted in the bottom-right corner is particularly interesting, as it contains a high amount of distinct visual detail, yet the pose estimation pipeline fails here as well. Unlike the previous examples, this last example is without darkening or occlusion.

6   Conclusions

This paper proposes a novel open dataset, TBPos, for image based large-scale precision localization. In order to achieve exact ground truth for localization algorithm benchmarking, we have adopted the approach of query image synthesis. In the experimental section the proposed dataset has been benchmarked using the InLoc localization pipeline, and has been compared in terms of difficulty to the well-known InLoc dataset, showing that TBPos is significantly more challenging. In addition to measuring the conventional localization success rate, our approach also enables measuring the metric precision of image-base localization.

References

1. Arandjelovic, R., Gronat, P., Torii, A., Pajdla, T., Sivic, J.: NetVLAD: CNN architecture for weakly supervised place recognition. In: CVPR (2016)
2. Badino, H., Huber, D., Kanade, T.: Visual topometric localization. In: IEEE IV. IEEE (2011)
3. Besl, P.J., McKay, N.D.: Method for registration of 3-D shapes. In: Sensor fusion IV: Control Paradigms and Data Structures, vol. 1611. SPIE (1992)
4. Brachmann, E., Humenberger, M., Rother, C., Sattler, T.: On the limits of pseudo ground truth in visual camera re-localisation. In: ICCV, pp. 6218–6228 (2021)
5. Cavallari, T., et al.: Real-time RGB-D camera pose estimation in novel scenes using a relocalisation cascade. IEEE TPAMI $42(10)$, 2465–2477 (2019)
6. DeTone, D., Malisiewicz, T., Rabinovich, A.: Superpoint: Self-supervised interest point detection and description. In: CVPR workshops (2018)
7. Dusmanu, M., et al.: D2-Net: a trainable CNN for joint description and detection of local features. In: CVPR (2019)
8. Ferranti, L.: Confidence estimation in image-based localization. Master's thesis, Tampere University (2019)
9. Ferranti, L., Li, X., Boutellier, J., Kannala, J.: Can you trust your pose? confidence estimation in visual localization. In: ICPR. IEEE (2021)
10. Gao, X.S., Hou, X.R., Tang, J., Cheng, H.F.: Complete solution classification for the perspective-three-point problem. IEEE TPAMI $25(8)$, 930–943 (2003)
11. Hyeon, J., Kim, J., Doh, N.: Pose correction for highly accurate visual localization in large-scale indoor spaces. In: ICCV (2021)
12. Li, P., Qin, T., Shen, S.: Stereo vision-based semantic 3d object and ego-motion tracking for autonomous driving. In: Ferrari, V., Hebert, M., Sminchisescu, C., Weiss, Y. (eds.) ECCV 2018. LNCS, vol. 11206, pp. 664 679. Springer, Cham (2018). https://doi.org/10.1007/978-3-030-01216-8_40
13. Lowe, D.G.: Distinctive image features from scale-invariant keypoints. Int. J. Comput. Vis. $60(2)$, 91–110 (2004). https://doi.org/10.1023/B:VISI.0000029664.99615.94
14. Maddern, W., Pascoe, G., Linegar, C., Newman, P.: 1 year, 1000 km: the Oxford RobotCar dataset. Int. J. Robot. Res. $36(1)$, 3–15 (2017)
15. Naseer, T., Burgard, W., Stachniss, C.: Robust visual localization across seasons. IEEE Trans. Robot. $34(2)$, 289–302 (2018)
16. Sarlin, P.E., Cadena, C., Siegwart, R., Dymczyk, M.: From coarse to fine: robust hierarchical localization at large scale. In: CVPR (2019)
17. Sarlin, P.E., DeTone, D., Malisiewicz, T., Rabinovich, A.: Superglue: learning feature matching with graph neural networks. In: CVPR (2020)
18. Sattler, T., Leibe, B., Kobbelt, L.: Efficient & effective prioritized matching for large-scale image-based localization. IEEE TPAMI $39(9)$, 1744–1756 (2016)
19. Sattler, T., Weyand, T., Leibe, B., Kobbelt, L.: Image retrieval for image-based localization revisited. In: BMVC, vol. 1 (2012)
20. Sattler, T., et al.: Benchmarking 6DoF outdoor visual localization in changing conditions. In: CVPR (2018)
21. Schonberger, J.L., Frahm, J.M.: Structure-from-motion revisited. In: CVPR (2016)
22. Schops, T., Sattler, T., Pollefeys, M.: BAD SLAM: bundle adjusted direct RGB-D SLAM. In: CVPR (2019)

23. Taira, H., et al.: InLoc: indoor visual localization with dense matching and view synthesis. In: CVPR (2018)
24. Zhang, Z., Sattler, T., Scaramuzza, D.: Reference pose generation for long-term visual localization via learned features and view synthesis. Int. J. Comput. Vis. **129**(4), 821–844 (2021). https://doi.org/10.1007/s11263-020-01399-8

FinnWoodlands Dataset

Juan Lagos$^{(\boxtimes)}$, Urho Lempiö, and Esa Rahtu

Tampere University, Tampere, Finland
{juanpablo.lagosbenitez,urho.lempio,esa.rahtu}@tuni.fi

Abstract. While the availability of large and diverse datasets has contributed to significant breakthroughs in autonomous driving and indoor applications, forestry applications are still lagging behind and new forest datasets would most certainly contribute to achieving significant progress in the development of data-driven methods for forest-like scenarios. This paper introduces a forest dataset called *FinnWoodlands*, which consists of RGB stereo images, point clouds, and sparse depth maps, as well as ground truth manual annotations for semantic, instance, and panoptic segmentation. *FinnWoodlands* comprises a total of 4226 objects manually annotated, out of which 2562 objects (60.6%) correspond to tree trunks classified into three different instance categories, namely "Spruce Tree", "Birch Tree", and "Pine Tree". Besides tree trunks, we also annotated "Obstacles" objects as instances as well as the semantic stuff classes "Lake", "Ground", and "Track". Our dataset can be used in forestry applications where a holistic representation of the environment is relevant. We provide an initial benchmark using three models for instance segmentation, panoptic segmentation, and depth completion, and illustrate the challenges that such unstructured scenarios introduce. *FinnWoodlands* dataset is available at https://github.com/juanb09111/FinnForest.git.

Keywords: Machine Learning · Deep Learning · Forestry · Dataset

1 Introduction

Public datasets have contributed to attracting interest in research in the field of computer vision. Data availability has accelerated the development of new paradigms, techniques, and models, especially data-driven methods and deep learning models designed to solve different computer vision tasks such as image segmentation, object detection, depth estimation, flow estimation, and object tracking [32]. Some public datasets have even become evaluation benchmarks, allowing different methods to claim state-of-the-art status [4,5,7–9,11,13,15,16,22,25,26,28,36]. While most popular datasets are collected from urban environments, with clear benefits for the development of autonomous driving applications, other contexts are lagging behind in terms of the availability of data, more specifically, off-road landscapes like forests. Moving through the forest with a proper data collection setup is all the more challenging as compared to urban scenarios, and that explains, to some extent, the gap between the availability of forest datasets compared to indoor or urban datasets.

© The Author(s), under exclusive license to Springer Nature Switzerland AG 2023
R. Gade et al. (Eds.): SCIA 2023, LNCS 13885, pp. 95–110, 2023.
https://doi.org/10.1007/978-3-031-31435-3_7

Nonetheless, numerous applications would benefit from increasing the availability of forest datasets. For instance, in agricultural, farming, and exploration applications [2,3,23,30], objects are less structured and defined, boundaries between objects are less clear and the movement, interaction, and appearance are fuzzier as compared to urban and indoor scenarios. This makes tasks such as autonomous navigation in forest-like scenarios very challenging since it is a less controlled environment with no lane lines, navigation signs, or clear paths.

Moreover, navigation in such unstructured scenarios is less standardized. The variety of scenarios ranges from dirt and gravel roads in the middle of the forest, where larger vehicles and heavy machinery like harvesters and tractors used in agricultural applications [27] can be driven, to smaller trekking trails for hikers and explorers only. While forest datasets are of high relevance within a specific group of applications, more specifically in the forestry industry, there are common features in such datasets e.g. structureless nature, that other applications might possibly exploit.

Collecting data from forests is not a trivial task. Forests change significantly during different seasons, and the type of vegetation varies depending on the geographic location. Hence, providing an optimal forest dataset most certainly requires a collective effort. Motivated by this, we introduce *FinnWoodlands*, a dataset collected from trekking trails in the forests of Finland. It consists of 5170 stereo RGB frames, the corresponding LIDAR point clouds, and sparse depth maps for each frame. Besides, we provide semantic segmentation, instance segmentation, and panoptic segmentation annotations for 300 frames that contain 4226 objects annotated manually using CVAT [29], an open-source tool for image and video annotation. We provide guidelines for other scientists to extend FinnWoodlands with more frames which would also increase variability in our dataset with forest images from different parts of the world. In that sense, FinnWoodlands sets up a robust seed dataset for forestry applications, with which we expect to attract the attention of the community of data scientists.

We evaluated our dataset with three different models for instance segmentation, panoptic segmentation and depth completion, namely Mask R-CNN [18], EfficientPS [24], and FuseNet [10], respectively. We thus set an initial benchmark for our dataset. The major contributions of our work are the following:

- We provide a forest dataset named *FinnWoodlands* that consists of RGB stereo frames, point clouds, and sparse depth maps, as well as ground truth (GT) annotations for semantic segmentation, instance segmentation, and panoptic segmentation. To our best knowledge, no other dataset in the context of forestry applications provides panoptic segmentation GT annotations. Thus, we aim to facilitate the research in holistic scene understanding in forest environments.
- We illustrate how to collect data in scenarios where mobility and navigation are challenging with a simple data collection setup that consists of a LIDAR sensor and a stereo camera mounted on a backpack. Our setup can be easily replicated elsewhere to collect reliable data compared to other, more expensive solutions.

2   Related Works

Among numerous lists of public datasets, only a few are collected in forest scenarios, some of which are oriented to navigation tasks, while the rest focus on tree detection and segmentation for industrial forestry applications. Hereby, we present an overview of publicly available forest datasets and summary of the annotations provided by each one of the datasets shown in Table 1. Visualization of representative images and annotations for the listed datasets are depicted in Fig. 1 and Fig. 2.

CANATREE100 dataset [17] consists of 100 RGB and 100 depth images and approximately 920 annotated trees. It was collected in the forests of Canada and provides annotations with segmentation masks for the tree trunks.

ForTrunkDet [31] is a dataset for tree trunk detection collected from three different forest locations. It consists of manually annotated visible and thermal images comprising two tree species corresponding to eucalyptus and pinus. It contains 2029 images in the visible spectrum and 866 thermal images.

RELLIS-3D dataset [20] consists of multi-modal synchronized sensor data frames collected from off-road environments. It is composed of five sequences that consist of RGB images, IMU data, GPS data, LIDAR point clouds, and stereo images.

The Robot Unstructured Ground Driving dataset (RUGD) [35] is a dataset collected from a ground robot in semi-urban locations and unstructured scenarios like forests. It consists of video sequences containing objects with irregular and inconsistent geometric morphology. The robot moves around different types of terrain labeled as *"creek"*, *"park"*, *"trail"*, and *"village"*.

SYNTHTREE43K [17] is a synthetic dataset collected from a simulated forest environment for tree detection with as many as 17 different types of trees. It consists of 43k synthetic images produced using the Unity game engine [21].

TartanAir [34] is a large photo-realistic dataset for navigation tasks. It provides outdoor and indoor scenes and various types of environments, including two subsets that contain forest scenes during different seasons, namely, autumn and winter. Due to its synthetic nature, TartanAir dataset also provides multimodal GT labels such as semantic segmentation tags, depth, camera pose, optical flow, stereo disparity, synthetic LIDAR points, and synthetic IMU readings.

TimberSeg 1.0 dataset [14] consists of 220 RGB images collected during different seasons and environment illumination conditions. It provides bounding box and segmentation mask annotations for individual tree logs. There are a total of 2500 segmented logs in the *TimberSeg 1.0* dataset. This dataset targets especially forwarding and log-picking applications.

Table 1. Datasets and GT Annotations

Dataset	BBox	Instance Segm	Semantic Segm	Panoptic Segm	Point Cloud Segm	Optical Flow	Depth
CANATREE100 [17]	✓	✓	✗	✗	✗	✗	✓
ForTrunkDet [31]	✓	✗	✗	✗	✗	✗	✗
RELLIS-3D [20]	✗	✗	✓	✗	✓	✗	✗
RUGD [35]	✗	✗	✓	✗	✗	✗	✗
SYNTHTREE43K [17]	✓	✓	✗	✗	✗	✗	✓
TartanAir [34]	✗	✗	✓	✗	✓	✓	✓
TimberSeg 1.0 [14]	✓	✓	✗	✗	✗	✗	✗
FinnWoodlands (ours)	✓	✓	✓	✓	✗	✗	✗

Fig. 1. Semantic Segmentation Annotations. Sample RGB images and their corresponding semantic segmentation GT annotations from three forest datasets: RELLIS-3D [20], RUGD [35], and TartanAir [34].

3 Dataset Features

Our dataset, *FinnWoodlands*, comprises 5170 synchronized stereo RGB images, LIDAR point clouds, and sparse depth maps for every frame. Figure 4 shows a visualization of one sample frame with its corresponding point cloud (Fig. 4a), as well as the projection of the LIDAR points onto the image (Fig. 4b). The data was collected from three different locations near Tampere, Finland. The annotations were done manually, consisting of semantic segmentation images,

Dataset	RGB Frame	Instance Segmentation and Object Detection

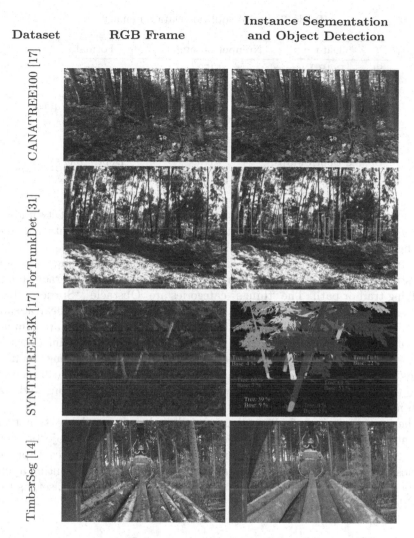

Fig. 2. Instance Segmentation Annotations. Sample RGB images and their corresponding instance segmentation GT annotations from four different forest datasets: CANA-TREE100 [17], ForTrunkDet [31], SYNTHTREE43K [17], TimberSeg [14].

bounding boxes, class labels, instance segmentation, and panoptic segmentation GT images. We also provide COCO format annotation files for instance segmentation and panoptic segmentation. Table 2 shows a summary list of the data contained in our dataset.

We used "stuff" and "things" objects as class categories in our dataset. In the context of computer vision, "stuff" classes refer to uncountable objects, for instance, "sky" or "ground", while "things" classes refer to countable objects such as "car" or "person" [1]. Objects under the category "things" can be annotated

Table 2. FinnWoodlands Data Summary

Data	Number of samples	Format
Stereo RGB Frames	5170	.jpg
LIDAR Point Clouds	5170	.pcd
Semantic GT	300	Label and RGB .png
Instance GT	300	RGB .png
Panoptic GT	300	RGB .png
Sparse Depth Maps	5170	Depth .png

with bounding boxes and instance segmentation masks, and objects under the category "stuff" are generally annotated with pixel-wise segmentation masks, and they cannot be confined within a single bounding box.

FinnWoodlands comprises three "stuff" categories and five "things" categories. The "stuff" categories are "Lake", "Ground", and "Track", where "Track" refers to a walking trail or path. The "things" categories are "Obstacle", "Spruce", "Birch", "Pine", and "Tree". The class "Obstacle" refers to obstacles on or near a walking trail. "Spruce", "Birch", and "Pine" refer to the tree species which are commonly encountered in Finnish forests. Any other tree that did not fall into these categories was labeled under the general category "Tree". By including "stuff" and "things" categories in our dataset, we aim to provide a more holistic 3D representation of forest scenarios, using panoptic segmentation annotations and sparse depth maps of the scenes.

The vast majority of objects annotated and segmented are tree trunks, accounting for approximately 60.6% of the total amount of objects. Table 3 shows the overall representation of classes. Within the tree species, "Spruce" is the most common type of tree encountered in our dataset, representing approximately 32.5% of the total amount of objects annotated in our dataset, as shown in Table 4.

Table 3. FinnWoodlands Object Classes

Object Class	Total Count	Total Area(px)	Representation
Lake	378	8.728M	8.9%
Obstacle	525	0.907M	12.4%
Ground	554	73.873M	13.1%
Track	207	7.778M	4.8%
Tree	2562	27.906M	60.6%

The annotations can be inspected visually in Fig. 3, where representative samples of *FinnWoodlands* dataset have been chosen to depict GT images. Every

Fig. 3. FinnWoodlands GT Annotations. Representative samples of our dataset *FinnWoodlands*. The first row displays three different RGB images. Gt annotations for semantic segmentation, instance segmentation and panoptic segmentation are shown from the second to the fourth row respectively.

Table 4. FinnWoodlands Tree Object Classes

Tree Type	Total Count	Total Area (px)	Representation
Spruce	1374	16.439M	32.5%
Birch	683	6.884M	16.1%
Pine	430	4.262M	10.1%
Tree (other type)	75	0.319M	1.7%

column corresponds to a different scene; the rows show the RGB frames and corresponding annotations. The first-row show three different scenes; the left-most frame is a typical Finnish forest with a walking trail. The frame in the middle column depicts a forest with a walking trail, obstacles, and a lake on the right side. Finally, the rightmost column shows a forest with no walking trail and with a high density of spruce trees. From the second to the fourth row, the corresponding semantic segmentation GT, instance segmentation GT, and panoptic segmentation GT are shown.

(a) Point Cloud

(b) Point Cloud Projection

Fig. 1. *FinnWoodlands* sample point cloud and sparse depth map visualization. The sparse depth map corresponds to the projection of the 3D point cloud onto to the 2D image.

4 Data Collection

For the *FinnWoodlands* dataset, a 64-beam Ouster OS1 LIDAR sensor and a ZED2 stereo camera were used to record point clouds and 720p resolution stereo images at 10 frames per second so that for each frame, there are stereo RGB images and one point cloud. The LIDAR and the camera are securely mounted on the backpack on a metal chassis, they are on top of the wearers head with OS1 above and ZED2 right below, as seen in Fig. 5. One Lenovo ThinkPad T440P laptop was used to collect and synchronize data from the cameras and sensor. The computer was running Ubuntu Linux (64-bit, version: 18.04), Robot Operating System (ROS, version: Melodic), Nvidia CUDA toolkit (version: 10.2), and Linuxptp software. PTP tools provided by linuxptp were used as a master clock for synchronizing the sensors. The data was then recorded using the ROS tool.

Recording sessions lasted about 50–100 s each. They were done by walking at a leisurely pace on a trekking path or in the forest. The locations were chosen from different types of forests near Tampere, Finland. These locations are Kyötikkälä, Hervantajärvi, and Suolijärvi. There is snow on the ground, and the lake is frozen since the recording sessions took place in the early spring. From the three locations, Hervantajärvi and Suolijärvi are surrounded by lakes with more birch trees, whereas Kyötikkälä has spruce trees primarily.

Forests are a difficult environment to work in compared to urban scenarios. The ground is not even, and there are many obstacles, e.g. rocks, trees, pits, and hills. Some places proved to be challenging to collect data by walking due to the snow on the ground and other obstacles. However, our data-collection setup, albeit simple, is flexible enough to collect reliable data under such conditions.

5 Experiments

In forestry applications, as in other applications, it is essential to understand the semantic meaning of a given image. Very often interacting with the objects present in the scene is also necessary. For that matter, semantic segmentation is not enough since it only provides pixel-wise classification, whereas instance segmentation allows for object detection and object segmentation. Moreover, it is possible to combine instance and semantic segmentation using panoptic segmentation, which provides a more holistic representation of the scenes. On the other hand, depth completion provides depth values for every pixel on the input frames, given incomplete sparse depth maps as input, which is relevant, especially for applications where it is needed to interact with objects in a 3D space.

Therefore, we conducted baseline experiments on three different models, namely, Mask R-CNN [18] for instance segmentation, EfficientPS [24] for panoptic segmentation and FuseNet [10] for depth completion based on sparse depth maps and RGB images. Our training set consists of 150 frames collected from two different locations, and the evaluation set contains 50 frames collected from a different location not used in our training set.

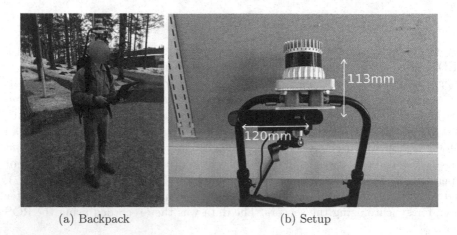

(a) Backpack   (b) Setup

Fig. 5. The backpack worn by a user and a close up of the setup. An OS1 LIDAR and a ZED2 stereo camera are mounted on top of a backpack. The ZED2 camera is placed right bellow the LIDAR, at a distance of 113 mm from the vertical center of the LIDAR. Both the LIDAR and the ZED2 camera are placed above the head of the user using an adjustable metal chassis. The baseline of the ZED2 camera is 120 mm.

In order to train the depth completion model *FuseNet* [10], we sub-sampled the sparse depth maps and used the resulting sparse depth maps as training data while keeping the original sparse depth maps as GT. The sub-sampled sparse depth maps contain 3500 depth values which account for approximately 25% of points in the original depth maps.

Mask R-CNN [18] reached the state-of-the-art of end-to-end instance segmentation in 2017 and has been broadly used as a reference model ever since. It takes RGB images as input and returns three different types of outputs. More specifically, it returns bounding boxes for object localization, class labels for every object detected, and segmentation masks. It consists of a backbone for feature extraction, a region proposal network (RPN), and three output heads in parallel that predict the corresponding class labels, bounding boxes, and segmentation masks for the given input image. We used Mask R-CNN with EfficientNet-B5 as the backbone [33], pre-trained on the ImageNet dataset [12]. We replaced the batch normalization layers [19] with synchronized Inplace Activated Batch Normalization [6] for GPU optimization.

EfficientPS [24] is a model for end-to-end panoptic segmentation. Similar to Mask R-CNN, it uses a backbone to extract multiple feature maps, fed to two output branches, and finally, it contains one panoptic fusion module. The first branch performs instance segmentation based on Mask R-CNN, using an RPN and three output heads for predicting bounding boxes, class labels, and segmentation masks. The second branch performs semantic segmentation, and finally, the panoptic fusion module fuses the semantic logits and the instance segmentation mask logits to produce a panoptic segmentation map. The name of the model makes reference to its backbone, which is based on the scalable *Efficient-Net* [33] architecture, wrapped in a two-way feature pyramid network (FPN) that allows for feature extraction at multiple scales.

FuseNet [10] uses RGB images, and sparse depth maps to perform end-to-end depth completion. It returns fully dense depth maps as output. FuseNet learns 2D and 3D features jointly using a building block that processes 2D tensors using 2D convolutional layers and 3D points using continuous convolution. The resulting features of the building block are then fused in 2D space.

5.1   Evaluation

We used the standard COCO evaluation metrics [22]. More specifically, we computed the mean Average Precision (mAP) for evaluating instance segmentation, Mean Intersection over Union (mIoU) for semantic segmentation, Panoptic Quality (PQ), Segmentation Quality (SQ), and Recognition Quality (RQ) for panoptic segmentation. We computed the Root Mean Square Error (RMSE) to evaluate the depth completion task.

Fig. 6. Mask R-CNN [18] Qualitative Results. The first column shows three different input frames. The first and second column, depict the instance segmentation output from Mask R-CNN [18] and the GT segmentation respectively.

5.2   Results

The quantitative performance results of every model are shown in Table 5. The overall performance of Mask R-CNN [18] as measured by the mAP@50 is 28%, and from Fig. 6, it is clear that, while it can detect tree trunks, the segmentation masks are not highly accurate, especially when the tree trunks are very close to each other. This reveals an opportunity for improvement in densely populated forests. Similarly, EfficientPS [24] detects tree trunks effectively, as shown in Fig. 7. Nonetheless, segmentation masks are not tightly bound to the corresponding trees in dense forest scenes. The overall PQ reached 27.8%; however, there is a significant difference when the PQ is evaluated separately for "things" and "stuff". EfficientPS performance is poorer when evaluated on "things" instances, which holds for the SQ and RQ metrics as well. These results again highlight the challenges that dense forest scenarios pose to machine learning methods and data-driven algorithms, especially segmenting single objects. We also computed the mAP@50 for EfficientPS, and it is noticeable that it outperforms Mask R-CNN in the same metric. On the other hand, FuseNet [10] generalizes reasonably well, even over the areas within the images where no sparse depth information is provided. However, fine structures and boundaries of objects like trees are lost, as shown in Fig. 8; hence there is also room for improvement in depth completion.

Table 5. Quantitative Results

Model	Metric	Value
Mask R-CNN	mAP@50	28%
EfficientPS	PQ	27.8
	SQ	32.5%
	RQ	34.1%
	PQ Stuff	41.5%
	SQ Stuff	41.5%
	RQ Stuff	50.0%
	PQ Things	18.7%
	SQ Things	26.5%
	RQ Things	23.5%
	mAP@50	50.0%
FuseNet	RMSE(mm)	489.96

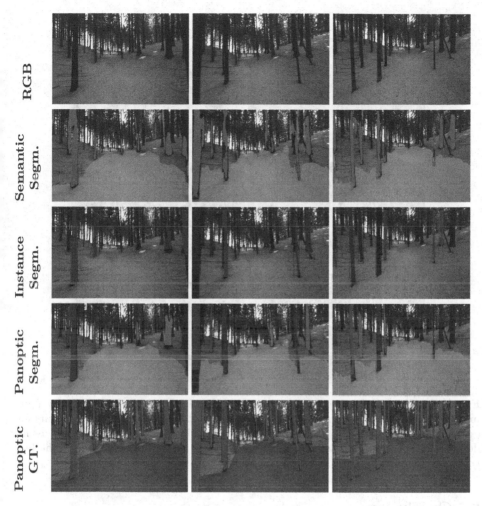

Fig. 7. EfficientPS [24] Qualitative Results. On the top row, three different RGB input images are shown. From the second to the fourth row, we present the segmentation results from EfficientPS for every task, more specifically, semantic segmentation, instance segmentation and panoptic segmentation respectively. The last row shows the GT segmentation.

RGB   Depth
                           Completion

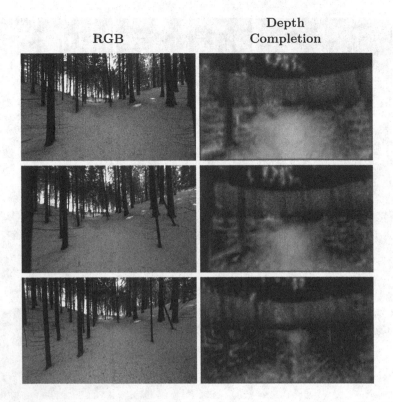

Fig. 8. FuseNet [10] Qualitative Results. The first column shows three different input images and the second column depicts the corresponding depth completion results for every input scene. FuseNet [10] generalizes fairly well, including the areas where no depth information was provided on the sparse depth maps, however some of the fine edges and structures from the objects in the forest could not be recovered.

6   Conclusion

In this paper, we introduced *Finn Woodlands*, a unique dataset for scene understanding in forest environments. *Finn Woodlands* provides manual GT annotations for instance segmentation, semantic segmentation, and panoptic segmentation in addition to sparse depth maps, which are necessary for holistic scene representation. Our dataset contains unstructured objects commonly found in forest scenarios and focuses on detecting and segmenting tree trunks from three different tree species, namely "Spruce", "Birch", and "Pine" tree trunks. We collected data with a relatively simple data-collection setup which can easily be replicated to produce similar data and extend *Finn Woodlands* dataset. We also provided an initial benchmark by testing our data with three deep neural network architectures, Mask R-CNN [18] for instance segmentation, EfficientPS [24] for panoptic segmentation, and FuseNet [10] for depth completion. The results reveal some of the challenges of computer vision when models are deployed in very unstructured scenarios such as forests, thus highlighting opportunities for improvement in similar scenarios. Our dataset can potentially impact the development of forestry applications and research in the field of computer vision in forest-like scenarios.

References

1. Adelson, E.H.: On seeing stuff: the perception of materials by humans and machines. In: IS&T/SPIE Electronic Imaging (2001)
2. Bac, C.W., Van Henten, E., Hemming, J., Edan, Y.: Harvesting robots for high-value crops: state-of-the-art review and challenges ahead. J. Field Robot. **31** (2014). https://doi.org/10.1002/rob.21525
3. Bechar, A., Vigneault, C.: Agricultural robots for field operations: concepts and components. Biosyst. Eng. **149**, 94–111 (2016)
4. Behley, J., et al.: A dataset for semantic segmentation of point cloud sequences. CoRR abs/1904.01416 (2019). https://arxiv.org/abs/1904.01416
5. Brostow, G.J., Fauqueur, J., Cipolla, R.: Semantic object classes in video: a high-definition ground truth database. Pattern Recogn. Lett. (2008)
6. Bulò, S.R., Porzi, L., Kontschieder, P.: In-place activated batchnorm for memory-optimized training of DNNs. CoRR abs/1712.02616 (2017). https://arxiv.org/abs/1712.02616
7. Cabon, Y., Murray, N., Humenberger, M.: Virtual KITTI 2. CoRR abs/2001.10773 (2020). https://arxiv.org/abs/2001.10773
8. Caesar, H., et al.: nuscenes: A multimodal dataset for autonomous driving. CoRR abs/1903.11027 (2019). https://arxiv.org/abs/1903.11027
9. Che, Z., et al.: D2-city: a large-scale dashcam video dataset of diverse traffic scenarios. arXiv abs/1904.01975 (2019)
10. Chen, Y., Yang, B., Liang, M., Urtasun, R.: Learning joint 2D-3D representations for depth completion. CoRR abs/2012.12402 (2020). https://arxiv.org/abs/2012.12402
11. Cordts, M., et al.: The cityscapes dataset for semantic urban scene understanding. CoRR abs/1604.01685 (2016). https://arxiv.org/abs/1604.01685
12. Deng, J., Dong, W., Socher, R., Li, L.J., Li, K., Fei-Fei, L.: ImageNet: a large-scale hierarchical image database. In: 2009 IEEE Conference on Computer Vision and Pattern Recognition, pp. 248–255 (2009). https://doi.org/10.1109/CVPR.2009.5206848
13. Everingham, M., Van Gool, L., Williams, C.K.I., Winn, J., Zisserman, A.: The PASCAL visual object classes challenge 2012 (VOC2012) results (2012). https://www.pascal-network.org/challenges/VOC/voc2012/workshop/index.html
14. Fortin, J.M., Gamache, O., Grondin, V., Pomerleau, F., Giguère, P.: Instance segmentation for autonomous log grasping in forestry operations (2022). https://doi.org/10.48550/ARXIV.2203.01902, https://arxiv.org/abs/2203.01902
15. Geiger, A., Lenz, P., Urtasun, R.: Are we ready for autonomous driving? The KITTI vision benchmark suite. In: Conference on Computer Vision and Pattern Recognition (CVPR) (2012)
16. Geyer, J., et al.: A2D2: Audi autonomous driving dataset. CoRR abs/2004.06320 (2020). https://arxiv.org/abs/2004.06320
17. Grondin, V., Fortin, J.M., Pomerleau, F., Giguère, P.: Tree detection and diameter estimation based on deep learning. Forestry: Int. J. Forest Res. (2022)

18. He, K., Gkioxari, G., Dollár, P., Girshick, R.B.: Mask R-CNN. CoRR abs/1703.06870 (2017). https://arxiv.org/abs/1703.06870
19. Ioffe, S., Szegedy, C.: Batch normalization: accelerating deep network training by reducing internal covariate shift. CoRR abs/1502.03167 (2015). https://arxiv.org/abs/1502.03167
20. Jiang, P., Osteen, P.R., Wigness, M.B., Saripalli, S.: RELLIS-3D dataset: data, benchmarks and analysis. CoRR abs/2011.12954 (2020). https://arxiv.org/abs/2011.12954
21. Juliani, A., Berges, V., Vckay, E., Gao, Y., Henry, H., Mattar, M., Lange, D.: Unity: a general platform for intelligent agents. CoRR abs/1809.02627 (2018). https://arxiv.org/abs/1809.02627
22. Lin, T., et al.: Microsoft COCO: common objects in context. CoRR abs/1405.0312 (2014). https://arxiv.org/abs/1405.0312
23. Mitra, A., et al.: Everything you wanted to know about smart agriculture. CoRR abs/2201.04754 (2022). https://arxiv.org/abs/2201.04754
24. Mohan, R., Valada, A.: EfficientPS: efficient panoptic segmentation. CoRR abs/2004.02307 (2020). https://arxiv.org/abs/2004.02307
25. Silberman, N., Hoiem, D., Kohli, P., Fergus, R.: Indoor segmentation and support inference from RGBD images. In: Fitzgibbon, A., Lazebnik, S., Perona, P., Sato, Y., Schmid, C. (eds.) ECCV 2012. LNCS, vol. 7576, pp. 746–760. Springer, Heidelberg (2012). https://doi.org/10.1007/978-3-642-33715-4_54
26. Neuhold, G., Ollmann, T., Bulà, S.R., Kontschieder, P.: The mapillary vistas dataset for semantic understanding of street scenes. In: 2017 IEEE International Conference on Computer Vision (ICCV),. pp. 5000–5009 (2017). https://doi.org/10.1109/ICCV.2017.534
27. Ringdahl, O.: Automation in forestry: development of unmanned forwarders. Ph.D. thesis, Umeå University, May 2011
28. Russakovsky, O., et al.: Imagenet large scale visual recognition challenge. CoRR abs/1409.0575 (2014). https://arxiv.org/abs/1409.0575
29. Sekachev, B., et al.: OpenCV/CVAT: v1.1.0, August 2020. https://doi.org/10.5281/zenodo.4009388
30. Shamshiri, R., et al.: Research and development in agricultural robotics: a perspective of digital farming. Int. J. Agric. Biol. Eng. 11, 1–14 (2018). https://doi.org/10.25165/j.ijabe.20181104.4278
31. da Silva, D.Q., dos Santos, F.N.: ForTrunkDet - forest dataset of visible and thermal annotated images for object detection. J. Imaging (2021)
32. Sinha, R.K., Pandey, R., Pattnaik, R.: Deep learning for computer vision tasks: a review. CoRR abs/1804.03928 (2018). https://arxiv.org/abs/1804.03928
33. Tan, M., Le, Q.V.: Efficientnet: rethinking model scaling for convolutional neural networks. CoRR abs/1905.11946 (2019). https://arxiv.org/abs/1905.11946
34. Wang, W., et al.: Tartanair: a dataset to push the limits of visual SLAM. CoRR abs/2003.14338 (2020). https://arxiv.org/abs/2003.14338
35. Wigness, M., Eum, S., Rogers, J.G., Han, D., Kwon, H.: A RUGD dataset for autonomous navigation and visual perception in unstructured outdoor environments. In: International Conference on Intelligent Robots and Systems (IROS) (2019)
36. Yu, F., Xian, W., Chen, Y., Liu, F., Liao, M., Madhavan, V., Darrell, T.: BDD100K: a diverse driving video database with scalable annotation tooling. CoRR abs/1805.04687 (2018). https://arxiv.org/abs/1805.04687

Re-identification of Saimaa Ringed Seals from Image Sequences

Ekaterina Nepovinnykh[✉][iD], Antti Vilkman, Tuomas Eerola[iD],
and Heikki Kälviäinen[iD]

Computer Vision and Pattern Recognition Laboratory,
Department of Computational Engineering,
Lappeenranta-Lahti University of Technology LUT, Lappeenranta, Finland
{ekaterina.nepovinnykh,tuomas.eerola,heikki.kalviainen}@lut.fi,
antti.vilkman@hotmail.com

Abstract. Automatic game cameras are commonly used for monitoring wildlife as they allow to document of the activity of animals in a non-invasive manner. By utilizing a large number of cameras and identifying individual animals from the images, it is possible to, for example, estimate the population size and study the migration patterns of the animals. Large image volumes produced by the cameras call for automated methods for the analysis. Re-identification of animals has commonly been implemented through one-to-one matching, where images are processed individually and the best match is searched from the database of known individuals one by one. Game cameras can be configured to produce a sequence of images that allows capturing the animal from multiple angles potentially improving the re-identification accuracy. In this work, the re-identification of the endangered Saimaa ringed seal (*pusa hispida saimensis*) from image sequences is studied. The individual identification is realized through Saimaa ringed seal's unique pelage pattern. The proposed one-to-many and many-to-many matching methods aggregate the pelage pattern features over the whole sequence providing better embeddings for the re-identification tasks. We show that the proposed aggregation method outperforms traditional one-to-one matching based re-identification by a large margin.

Keywords: saimaa ringed seal · computer vision · image processing · re-identification · one-to-many · many-to-many

1 Introduction

Animal population monitoring including estimating the changes in population size over time and tracking the migration patterns of individual animals is important for conservation efforts of the endangered species [31]. Observing animals from image data is a non-invasive method and can be done by utilizing automatic camera traps or image material obtained using citizen science projects. Due to large amounts of data generated through these methods, automatic computer

R. Gade et al. (Eds.): SCIA 2023, LNCS 13885, pp. 111–125, 2023.
https://doi.org/10.1007/978-3-031-31435-3_8

vision methods are needed. These include detecting animals in images and identifying or re-identifying the individual animals based on, for example, unique fur or skin patterns.

While general-purpose object detection and segmentation methods have matured to the point where they can be relatively easily applied to new application areas including different animal species, the fully automated animal individual re-identification remains largely unsolved. The reasons for this include subtle differences in fur pattern, large variations in animal pose, appearance, and illumination, low image quality, and often extreme dataset bias issues [24].

This paper focuses on the Saimaa ringed seal (*pusa hispida saimensis*), an endangered fresh-water seal species only found in Lake Saimaa, Finland. With a population size of only 430–440 individuals, it is among the most endangered pinnipeds in the world [15]. Wildlife photo-identification has been applied in conservation efforts to monitor the seal population as it is a non-invasive alternative to methods that require catching the seal, thus, reducing the amount of stress that is caused to the animals. Saimaa ringed seals have a ring pattern on their fur which is both permanent and unique for each individual making it possible to re-identify individuals based on the pattern, as shown in Fig. 1. Identification from images has traditionally been done manually [12], but recent progress in computer vision methods has enabled automatic solutions capable of reducing the amount of manual work [24].

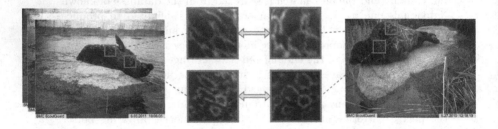

Fig. 1. Re-identification of Saimaa ringed seals based on pelage pattern.

The existing methods for animal re-identification focus on one-to-one matching, i.e. query images are analyzed individually and the matching individual is searched from the database of known individuals via one-by-one comparisons. Automatic camera traps typically capture multiple images per individual with slightly different poses, appearance,s and varying illumination. Utilizing multiple images simultaneously has evident benefits as it allows capturing a larger portion of the fur pattern and reduces the negative effects of low image quality. The contributions of the paper are the following: 1) improvement to an existing re-identification method [24] by introducing a pattern feature aggregation approach that allows aggregating over an image sequence and 2) a new dataset (SealID_seq) consisting of more than 32 000 images of Saimaa ringed seals organized into sequences. We show that the proposed aggregated pattern features produce superior re-identification compared to one-to-one matching on a challenging Saimaa ringed seal data. The SealID_seq dataset has been made publicly available at https://doi.org/10.23729/a84bf6be-1f2a-4164-b7a0-ef6103152f6f. The codes for the described experiments are available at https://github.com/kwadraterry/Norppa.

1.1 Related Work

Various methods for automatic animal re-identification [7,30] exist in the literature, many achieving reasonably high accuracies on certain animal species. Often these methods make use of identifiable features in animal fur patterns (such as stripes of zebras or point patterns of whale sharks). The use of Scale Invariant Feature Transform (SIFT) features to find and describe keypoints in images [19] has been a popular method in animal re-identification due to their invariance to scale and orientation. For example, in the Wild-ID algorithm [3], the SIFT features are extracted from images of giraffes and the identification is performed by finding the database image with the most similar features, with a modified Random Sample Consensus (RANSAC) algorithm [9] being used to ensure the geometric consistency of the matched features.

The HotSpotter algorithm [7] is built upon a similar approach, using Root-SIFT [1] for feature descriptors and searching database images with similar descriptors. HotSpotter uses RANSAC to ensure the consistency of the descriptors in the found matches and computes the combined scores for each label in the database to identify the query image instead of simply using the highest scoring image. HotSpotter has been demonstrated on multiple patterned species including jaguars, giraffes, zebras, and lionfish [7].

Modern approaches to animal re-identification are based on deep learning, particularly convolutional neural networks (CNNs) [30,31]. CNNs are applied to different biological traits such as primate faces [4,8], Amur tiger stripes [16–18] or cattle muzzle pattern [14].

Saimaa ringed seal offers a challenging task in automatic re-identification, with large variance in poses, appearances, and low contrast between the ring pattern and the rest of the fur. These challenges render the re-identification problem considerably harder than those presented in the earlier studies and cause the existing re-identification methods to produce subpar accuracy on ringed seals (see e.g. [24]). Multiple methods [5,6,24,27,28,33] have been proposed for automatic re-identification of Saimaa ringed seals, matching features found on the pelage patterns. All proposed methods start with seal segmentation to detect the seal and remove the background followed by pelage pattern feature extraction. Various approaches to encode the pattern features have been proposed [6,24,27,33]. The most successful methods employ pattern extraction step [24,27] to construct a binary representation of the pelage pattern and metric learning-based pattern encoding. The extracted features are aggregated to form a descriptor of the seal, and the descriptor is then compared against descriptors from known identified seals from the database to find the most similar matches.

The matching of Saimaa ringed seals has been previously done in a one-to-one manner, comparing images from the database to the query image one by one. Aggregating information from multiple images of an individual has the potential to provide more versatile and enhanced features by utilizing multiple angles. Comparing descriptors built from single or multiple query images to descriptors built from multiple database images, the use of one-to-many, many-to-one, and many-to-many matching has the ability to improve speed [7] and accuracy [25] of re-identification algorithms. While image sequences and multiple views have been used to improve methods in human re-identification [21,29], only a few works [2,18,25] utilizing such techniques exist in the realm of automatic animal re-identification.

2   Proposed Method

2.1   Pipeline

The proposed method for Saimaa ringed seal re-identification is based on the method proposed in [24]. The seal is first segmented and the pelage pattern is extracted from the segmented image. Regions of interest are searched from the pattern images and the corresponding pattern image patches are extracted and encoded. Finally, encodings are aggregated to a single Fisher vector of a fixed length, and the most similar vector is searched from the database of known individuals by computing distances between query and database vectors. In [24] features from each image were aggregated to Fisher Vectors separately. With the proposed one-to-many and many-to-many matching methods, the Fisher Vectors are instead formed over all database images of individual seals or all images in a query sequence. The updated pipeline is presented in Fig. 2.

2.2   Data Preprocessing

Preprocessing of the data consists of three steps: tonemapping, segmentation, and pattern extraction. All dataset images are obtained using the outdoor camera traps under different weather and light conditions, which resulted in images of varying quality. To address this problem, we utilize tone mapping techniques to balance the contrast levels between the dark and bright areas of the image. Saimaa ringed seals exhibit strong site-fidelity meaning that they tend to stay in the same regions. Because of this the images of certain individual are often captured with the same game camera and have similar background. To prevent the risk of the re-identification algorithm learning to recognize the background instead of the seal itself, the background should be removed completely. That is why mask segmentation is preferred over bounding box extraction. Then, in order to emphasize the pelage pattern as the basis of the re-identification and discard irrelevant information the pattern is extracted.

Tonemapping is performed using the method described in [20] which is able to produce natural tonemapped images without introducing any visible defects.

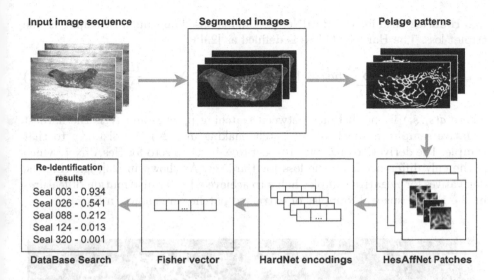

Fig. 2. The proposed Saimaa ringed seal re-identification pipeline.

Instance segmentation of seals and feature extraction follow the same process that is presented in [24]. Mask R-CNN [10] is used to detect the seal and remove the background from images. The U-net-based encoder-decoder model is then used to extract the pelage pattern of the seal. As the images vary in resolution, the pattern images are resized to equalize the thickness of the pattern lines. Examples of the pattern images are shown in Fig. 3.

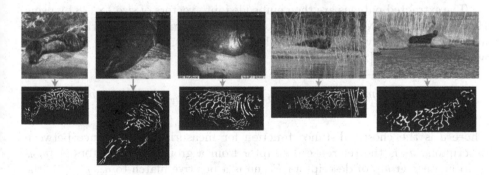

Fig. 3. Examples of pattern images extracted by U-net CNN.

With the pattern extracted, affine covariant regions are found and extracted from the pattern image by using HesAffNet [23]. HesAffNet extracts local regions from images and transforms them according to the estimated local affine transformation, generating affine-invariant patches of pelage patterns. HesAffNet is trained by using the HardNegC loss function [23]. The loss is a variation of a loss HardNeg used for the HardNet [22] and is based on similar principles as the triplet loss. The HardNegC loss is defined as [23]

$$L = \frac{1}{n} \sum_{i=1,n} max(0, 1 + d(s_i, \dot{s}_i) - d(s_i, N)), \quad \frac{\partial L}{\partial N} := 0, \quad (1)$$

where $d(s_i, \dot{s}_i)$ is the distance between matching patches and N is the hardest negative sample in the training batch, making $d(s_i, N)$ the distance to that sample. The derivative of L with respect to N is set to zero for HesAffNet, which is the only difference from the loss for HardNet. As shown in [23], ignoring the derivatives for negative samples helps to achieve a better distribution of features in the feature space. Examples of extracted patches are shown in Fig. 4.

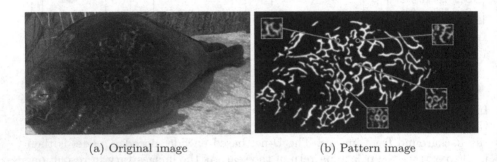

(a) Original image   (b) Pattern image

Fig. 4. Patches extracted by HesAffNet and their locations in the pattern image.

The extracted patches are then embedded into vectors of size 1×128 by using HardNet [22]. HardNet is trained to correctly embed and match descriptors while avoiding false positives from similar descriptors by using the triplet margin loss which is defined as [22]

$$L = \frac{1}{n} \sum_{i=1,n} max(0, 1 + d(a_i, p_i) - min(d(a_i, p_{j_{min}}), d(a_{k_{min}}, p_i))), \quad (2)$$

where d is the chosen distance function for measuring the distance between descriptors, a_i is the reference descriptor from a group of descriptors A, p_i is from another group of descriptors P and is a positive match to a_i, $p_{j_{min}}$ is the closest negative match to a_i from P and $a_{k_{min}}$ is the closest negative match to p_i from A. Choosing the minimum distance between the positive and negative matches in the loss function effectively always picks the most difficult sample into the triplet, improving the model's ability to avoid false positives. After the HardNet embedding, PCA is applied to the features to decorrelate them and reduce dimensionality.

2.3   Feature Aggregation

In [24], a Fisher vector is generated for each image. However, since every image of the same individual should, in theory, contain a similar and possibly complementing set of pattern features, it is beneficial to perform an aggregation over a sequence of images assuming that we can be sure that each image in the sequence contains the same individual. This makes it possible to aggregate over all database images of a known individual to obtain database embeddings describing the whole pelage pattern (e.g. both sides of the animal). Furthermore, as game cameras can be configured to capture multiple images while an animal is present, an image sequence with pose and appearance variation can be obtained, and better query embeddings can be extracted by aggregating the whole sequence.

To compute the Fisher vector, a visual vocabulary is constructed using Gaussian Mixture Model (GMM) on the database features. This vocabulary is used to create Fisher Vectors for each image sequence. The resulting descriptors are then L2 and Power normalized, as proposed in [11], since those normalizations are shown to increase the accuracy.

2.4   Re-Identification

The final re-identification is performed by measuring the distance between the query descriptor and each of the database descriptors. This is achieved by computing the cosine distance between the descriptors. The cosine distance is calculated as [32]

$$D_{cos} = 1 - \frac{u \cdot v}{||u||_2||v||_2} \tag{3}$$

where u and v are the seal descriptor Fisher Vectors.

Once the distances between the query descriptor and all database descriptors have been computed, the class of the database descriptor with the shortest distance to the query descriptor is chosen as the predicted class.

Three different methods of utilizing the aggregated features were implemented with respect to whether the query features, the database features, or both features were aggregated from multiple images. The implemented methods were one-to-many where the database features are aggregated, many-to-one where the query features are aggregated, and many-to-many where both the query and database features from multiple images are aggregated to a single descriptor. The different methods are illustrated in Fig. 5.

3   Experiments

3.1   Data

To evaluate the proposed method, a new dataset (SealID_seq) consisting of image sequences of Saimaa ringed seals was collected. The data was gathered on

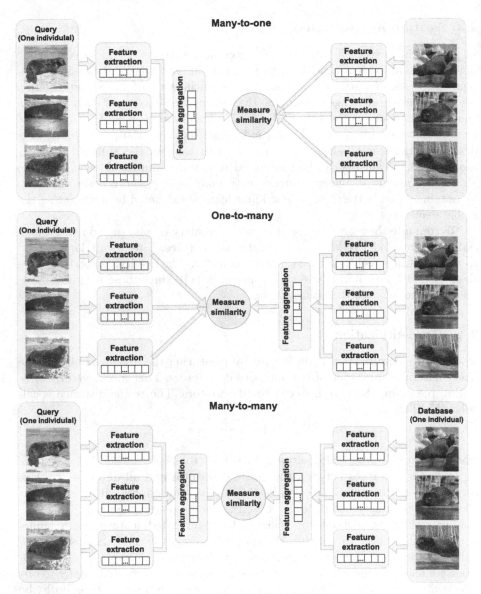

Fig. 5. The different manners of grouping are visualized, showing when the features from the query, the database or both are aggregated.

a yearly basis during the molting season of the Saimaa ringed seal (mid-April–mid-June) from 2010 to 2017 by game camera traps (Scout Guard SG550, Scout Guard SG560, and Uovision UV785). The game cameras were set in motion sensitivity (2 photos over a 0.5–2 min time span) or time-laps (2 photos every 10 min) and were installed in locations previously found during the boat survey [12,13]. All seal images were identified based on the unique fur patterns of each individual.

The dataset is based on SealID [26], an earlier Saimaa ringed seal re-identification dataset. The same database of known individuals is used in both datasets. It consists of a minimal amount of high-quality images for each known individual to cover the full view of a seal body (see Fig. 6). The SealID_seq dataset contains a larger query set consisting of camera trap images. Individual query images in The SealID_seq are more challenging due to lower image quality and worse pelage pattern extraction performance, but unlike in the SealID dataset, the dataset is divided into image sequences with the same individual allowing to use of multiple images simultaneously. Data collection was carried out in Lake Saimaa, Finland (61° 05'-62° 36'N, 27°15'-30° 00' E) under permits by the local environmental authorities (ELY-centre, Metsähallitus). The Photo-ID data were collected annually during the Saimaa ringed seal molting season (mid-April–mid-June) from the year 2010 to 2019 by game camera traps.

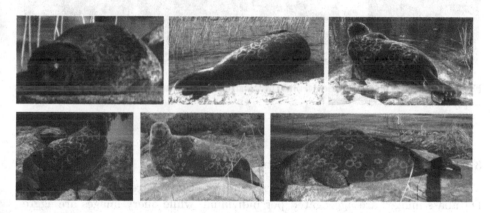

Fig. 6. Example images of various individuals from the database of known individuals.

Image sequences in SealID_seq were separated by the location and time period when they were captured. All sequences were checked by the experts to verify that each contains only one individual over the whole time period. The algorithm is intended to be used in a semi-automatic manner, and verifying that the individual does not change in a sequence is a relatively simple task compared to the final goal of re-identification (finding the match in the database). After segmenting and extracting patterns from the raw images, images with less than 10% of the area containing the pattern or images with no patches found were discarded. After this preprocessing step, the dataset used for the re-identification experiments includes images of 30 individuals, consisting of 32077 images in 120 separate sequences. The mean sequence length is 267 images, with the shortest sequence containing 5 images while the longest contains 1564 images.

Saimaa ringed seals tend to stay on the same haul-out rocks for long periods of time allowing them to obtain long image sequences with varying pose and even illumination. It allows for acquiring a better representation of the entire pelage pattern. The lighting conditions can also change within a sequence. For example, the seal is initially illuminated by the sun, but as time goes on, the images become darker or even captured in the night vision mode of the game camera. An example sequence from SealID_seq is shown in Fig. 7 where a gradual change in the pose and illumination can be seen throughout the sequence.

Fig. 7. Example of sequences from the SealID_seq dataset. Each row represents a sequence.

3.2 Description of Experiments

Various matching grouping strategies were tested for the re-identification task. Database images are aggregated per individual, while query images are aggregated per sequence. In a real-world scenario, it is reasonable to assume that the seal captured on consecutive shots of an automatic camera does not change, meaning that the aggregation can be performed for sequences of images from game cameras. Manual verification might be required, but this is generally a safe assumption, especially for short sequences. The effects of sequence lengths on matching are also tested by limiting the number of aggregated images from each sequence. Many-to-one matching is performed by taking a maximum of k first images from each sequence before aggregating them.

Since the images in SealID_seq dataset are generally of low quality, the effects of the tonemapping preprocessing step are also tested on that dataset to evaluate the usefulness of that step to the whole re-identification pipeline. For the tone-mapping of images, the algorithm proposed in [20] was used.

3.3 Evaluation Criteria

To evaluate the re-identification performance of the proposed methods, the top-1 accuracy and the top-5 accuracy were used as the metrics to determine the accuracy of the predictions. When using Top-1 accuracy, a correct prediction is one where the nearest matching image predicted by the model is of the correct class. When using Top-5 accuracy a correct prediction is such that an image of the correct class is found within the five nearest images predicted by the model. The accuracy is then calculated as

$$accuracy = \frac{Number\ of\ correct\ predictions}{Number\ of\ query\ samples} \tag{4}$$

3.4 Results

Results of the experiments compared to vsMany version of the HotSpotter algorithm are presented in Table 1. For all the experiments, NORPPA method performed better on TOP-5 than HotSpotter. For different modifications of NORPPA, aggregating query features yields a large improvement to the accuracy while aggregating database features slightly reduces accuracy. Such a large jump in accuracy in the case of the many-to-one and many-to-many matching can be explained by considering the low-quality images or images where only a small part of the pattern is recognizable. The aggregation of features from multiple query images helps to correctly classify even images where there is not enough visual information to recognize the individual by complementing that with information from other images.

On the other hand, the drop in accuracy of the one-to-many and many-to-many approaches compared to their to-one counterparts can be explained by the fact that the database only contains a minimal amount of images covering the full body and full pelage pattern cannot fit on one image. For example, if the query image contains information about the left side of a seal, the distance to the database image of the left side would naturally be lower than to the image of the right side. However, aggregating both database images of the left and right sides

Table 1. Re-identification results on the SealID_seq dataset.

Method	Tonemapped	Top-1 accuracy	Top-5 accuracy
HotSpotter (vsMany) [7]	No	44.15%	46.09%
	Yes	54.26%	56.54%
One-to-one	No	37.99%	51.88%
	Yes	48.07%	61.53%
One-to-many	No	27.32%	43.71%
	Yes	39.17%	57.18%
Many-to-one	No	77.93%	89.27%
	Yes	92.92%	98.28%
Many-to-many	No	65.19%	87.40%
	Yes	91.46%	95.85%

might actually move the final descriptor further from the query descriptor of the left side, resulting in misclassification. It is possible that instead of aggregating all database images of an individual into a descriptor, it would be better to aggregate them into several descriptors that would correspond to different parts or views of the pattern.

Additionally, the results clearly indicate that the tonemapping preprocessing step consistently increases the accuracy of re-identification, making it essential to the final re-identification pipeline.

The results from testing the effect of the query sequence length on the accuracy are presented in Fig. 8. Generally, both top-1 and top-5 re-identification accuracies increase with the length of a sequence. The longer the sequence, the higher the chance that more representative features are extracted and aggregated, resulting in more robust descriptors. However, the increase in accuracy is steepest with smaller sequences, meaning that even very short sequences can substantially increase the re-identification accuracy of the algorithm.

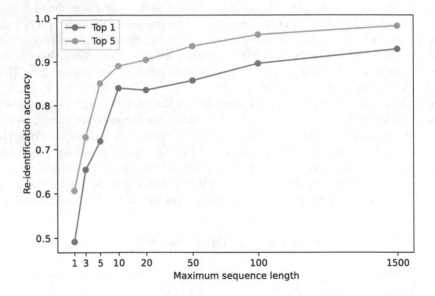

Fig. 8. Dependency of the re-identification accuracy on the length of sequences.

4   Conclusion

In this work, one-to-many, many-to-one, and many-to-many matching of the Saimaa ringed seals are considered. The idea is to use information from multiple images or multiple views to obtain a more comprehensive descriptor for the re-identification task. Individual features from multiple images are aggregated into a single Fisher Vector descriptor, containing information about a seal individual.

In the experiments, three different ways of aggregating the features were considered: 1) one-to-many where database features for a seal were aggregated from multiple images, 2) many-to-one where query features for a seal were aggregated from multiple images, and 3) many-to-many where both query and database features for a seal were aggregated from multiple images. The results indicate that query feature aggregation contributes to the significant increase in re-identification accuracy, with a jump of 45% in the top-1 accuracy and 37% in the top-5 accuracy compared to the standard one-to-one matching approach. On the other hand, aggregating database features decreases accuracy which suggests that a different aggregation method might be necessary for the database.

Acknowledgements. The authors would like to thank Raija ja Ossi Tuuliaisen Säätiö Foundation and the CoExist project (Project ID: KS1549) for funding the research. In addition, the authors would like to thank Vincent Biard, Piia Mutka, Marja Niemi, and Mervi Kunnasranta from the Department of Environmental and Biological Sciences at the University of Eastern Finland (UEF) for providing the data of the Saimaa ringed seals and their expert knowledge of identifying each individual.

References

1. Arandjelović, R., Zisserman, A.: Three things everyone should know to improve object retrieval. In: IEEE Conference on Computer Vision and Pattern Recognition, pp. 2911–2918 (2012)
2. Bergamini, L., et al.: Multi-views embedding for cattle re-identification. In: International Conference on Signal-Image Technology & Internet-Based Systems (SITIS), pp. 184–191 (2018). https://doi.org/10.1109/SITIS.2018.00036
3. Bolger, D.T., Morrison, T.A., Vance, B., Lee, D., Farid, H.: A computer-assisted system for photographic mark-recapture analysis. Methods Ecol. Evol. **3**(5), 813–822 (2012)
4. Brust, C.A., at al.: Towards automated visual monitoring of individual gorillas in the wild. In: International Conference on Computer Vision Workshop (ICCVW) (2017). https://doi.org/10.1109/iccvw.2017.333
5. Chehrsimin, T., et al.: Automatic individual identification of Saimaa ringed seals. IET Comput. Vision **12**(2), 146–152 (2018)
6. Chelak, I., Nepovinnykh, E., Eerola, T., Kalviainen, H., Belykh, I.: EDEN: deep feature distribution pooling for Saimaa ringed seals pattern matching. arXiv preprint arXiv:2105.13979 (2021)
7. Crall, J.P., Stewart, C.V., Berger-Wolf, T.Y., Rubenstein, D.I., Sundaresan, S.R.: Hotspotter-patterned species instance recognition. In: IEEE Workshop on Applications of Computer Vision, pp. 230–237 (2013)
8. Deb, D., et al.:: Face recognition: primates in the wild. In: IEEE International Conference on Biometrics Theory, Applications and Systems, pp. 1–10 (2018)
9. Fischler, M.A., Bolles, R.C.: Random sample consensus: a paradigm for model fitting with applications to image analysis and automated cartography. Commun. ACM **24**(6), 381–395 (1981)
10. He, K., Gkioxari, G., Dollár, P., Girshick, R.: Mask R-CNN. In: IEEE International Conference on Computer Vision, pp. 2961–2969 (2017)

11. Perronnin, F., Sánchez, J., Mensink, T.: Improving the fisher kernel for large-scale image classification. In: Daniilidis, K., Maragos, P., Paragios, N. (eds.) ECCV 2010. LNCS, vol. 6314, pp. 143–156. Springer, Heidelberg (2010). https://doi.org/10.1007/978-3-642-15561-1_11

12. Koivuniemi, M., Auttila, M., Niemi, M., Levänen, R., Kunnasranta, M.: Photo-id as a tool for studying and monitoring the endangered Saimaa ringed seal. Endang. Spec. Res. **30**, 29–36 (2016)

13. Koivuniemi, M., Kurkilahti, M., Niemi, M., Auttila, M., Kunnasranta, M.: A mark-recapture approach for estimating population size of the endangered ringed seal (Phoca hispida saimensis). PLoS ONE **14**, 214–269 (2019). https://doi.org/10.1371/journal.pone.0214269

14. Kumar, S., et al.: Deep learning framework for recognition of cattle using muzzle point image pattern. Measurement **116**, 1–17 (2018). https://doi.org/10.1016/j.measurement.2017.10.064

15. Kunnasranta, M., Niemi, M., Auttila, M., Valtonen, M., Kammonen, J., Nyman, T.: Sealed in a lake-biology and conservation of the endangered Saimaa ringed seal: a review. Biol. Cons. **253**, 108908 (2021)

16. Li, S., Li, J., Tang, H., Qian, R., Lin, W.: ATRW: a benchmark for amur tiger re-identification in the wild. In: ACM International Conference on Multimedia (2020). https://doi.org/10.1145/3394171.3413569

17. Liu, C., Zhang, R., Guo, L.: Part-pose guided amur tiger re-identification. In: International Conference on Computer Vision Workshop (ICCVW) (2019). https://doi.org/10.1109/ICCVW.2019.00042

18. Liu, N., Zhao, Q., Zhang, N., Cheng, X., Zhu, J.: pose-guided complementary features learning for amur tiger re-identification. In: International Conference on Computer Vision Workshop (ICCVW) (2019). https://doi.org/10.1109/ICCVW.2019.00038

19. Lowe, D.G.: Distinctive image features from scale-invariant keypoints. Int. J. Comput. Vision **60**(2), 91–110 (2004)

20. Mantiuk, R., Myszkowski, K., Seidel, H.P.: A perceptual framework for contrast processing of high dynamic range images. ACM Trans. Appl. Percept. **3**, 286–308 (2006). https://doi.org/10.1145/1166087.1166095

21. McLaughlin, N., Del Rincon, J.M., Miller, P.: Recurrent convolutional network for video-based person re-identification. In: IEEE Conference on Computer Vision and Pattern Recognition, pp. 1325–1334 (2016)

22. Mishchuk, A., Mishkin, D., Radenovic, F., Matas, J.: Working hard to know your neighbor's margins: Local descriptor learning loss. arXiv preprint arXiv:1705.10872 (2017)

23. Mishkin, D., Radenović, F., Matas, J.: Repeatability is not enough: learning affine regions via discriminability. In: Ferrari, V., Hebert, M., Sminchisescu, C., Weiss, Y. (eds.) ECCV 2018. LNCS, vol. 11213, pp. 287–304. Springer, Cham (2018). https://doi.org/10.1007/978-3-030-01240-3_18

24. Nepovinnykh, E., Chelak, I., Eerola, T., Kälviäinen, H.: NORPPA: novel ringed seal re-identification by pelage pattern aggregation. arXiv preprint arXiv:2206.02498 (2022)

25. Nepovinnykh, E., Chelak, I., Lushpanov, A., Eerola, T., Kälviäinen, H., Chirkova, O.: Matching individual ladoga ringed seals across short-term image sequences. Mamm. Biol. **102**, 1–16 (2022). https://doi.org/10.1007/s42991-022-00229-3

26. Nepovinnykh, E., et al.: SealID: Saimaa ringed seal re-identification database. Sensors **22**, 7602 (2022)

27. Nepovinnykh, E., Eerola, T., Kalviainen, H.: Siamese network based pelage pattern matching for ringed seal re-identification. In: IEEE/CVF Winter Conference on Applications of Computer Vision Workshops, pp. 25–34 (2020)
28. Nepovinnykh, E., Eerola, T., Kälviäinen, H., Radchenko, G.: Identification of Saimaa ringed seal individuals using transfer learning. In: International Conference on Advanced Concepts for Intelligent Vision Systems, pp. 211–222 (2018)
29. Parkhi, O.M., Simonyan, K., Vedaldi, A., Zisserman, A.: A compact and discriminative face track descriptor. In: IEEE Conference on Computer Vision and Pattern Recognition, pp. 1693–1700 (2014)
30. Schneider, S., Taylor, G.W., Kremer, S.C.: Similarity learning networks for animal individual re-identification-beyond the capabilities of a human observer. In: IEEE/CVF Winter Conference on Applications of Computer Vision Workshops, pp. 44–52 (2020)
31. Schneider, S., Taylor, G.W., Linquist, S., Kremer, S.C.: Past, present and future approaches using computer vision for animal re-identification from camera trap data. Methods Ecol. Evol. **10**(4), 461–470 (2019)
32. SciPy API reference, distance computations, cosine: the SciPy community. https://docs.scipy.org/doc/scipy/reference/generated/scipy.spatial.distance.cosine.html (2022). Accessed 30 May 2022
33. Zhelezniakov, A., et al.: Segmentation of Saimaa ringed seals for identification purposes. In: International Symposium on Visual Computing, pp. 227–236 (2015)

Action and Behaviour Recognition

Attention-guided Boundary Refinement on Anchor-free Temporal Action Detection

Henglin Shi(ID), Haoyu Chen(ID), and Guoying Zhao(✉)(ID)

Center for Machine Vision and Signal Analysis, University of Oulu, Oulu, Finland
{henglin.shi,haoyu.chen,guoying.zhao}@oulu.fi

Abstract. Modelling temporal dependencies is important for accurate action detection. In this work, we develop a temporal attention unit to mine the global dependencies among features from different temporal locations. Additionally, based on the developed temporal attention unit, we propose an attention-guided boundary refinement module for revising action prediction results. Besides, we integrate the proposed module into a contemporary anchor-free detector for performing temporal action detection. To evaluate the proposed method, experiments are carried out on two large-scale temporal action detection datasets, namely THU-MOS14 and ActivityNet1.3 datasets. Experimental results show that the action detection performance is significantly boosted by the proposed temporal attention module which outperforms several state-of-the-art methods.

Keywords: Action detection · temporal action localization · attention

1 Introduction

Computer vision based temporal action detection is a prospective task which is important for many applications, such as video content understanding, human-machine interactions. However, compared with the widely studied action recognition, temporal action detection is more challenging, which does not only aim to predict the corresponding label of the possible action, but also precisely locate the temporal starting and ending location of the candidate action instance.

Currently, there are three main paradigms for video based temporal action detection, namely anchor-based methods, actionness-based methods, and anchor-free methods. Anchor-based methods perform action detection by firstly generating a set of predefined proposals (i.e., anchors) and then performing

---

This work was supported by the Academy of Finland for Academy Professor project EmotionAI (grants 336116, 345122), project MiGA (grant 316765), the University of Oulu & The Academy of Finland Profi 7 (grant 352788), and Ministry of Education and Culture of Finland for AI forum project. As well, the authors wish to acknowledge CSC - IT Center for Science, Finland, for computational resources.

R. Gade et al. (Eds.): SCIA 2023, LNCS 13885, pp. 129–139, 2023.
https://doi.org/10.1007/978-3-031-31435-3_9

adjustment on these anchors, which is also called top-down paradigm. Similar approaches have achieved extensive success in object detection tasks. Moreover, actionness-based methods perform in a reversed way (bottom-up) which firstly evaluates likeliness of being the starting and ending of an action for each frame, and then heuristically combines each pair of frames to form action proposals based on their boundary likeliness. Lastly, anchor-free paradigm performs in a more concise way that the feature at each temporal location will predict an action location and class. Such paradigm has been widely studied in object detection but is still emerging in temporal action detection.

This work is inspired by a recently proposed anchor-free action detector, AFSD [9], where firstly an anchor-free model performs action detection to produce initial action proposals and classes as coarse predictions, and then a saliency-based refinement module refines these coarse predictions based on the features extracted within the coarse action boundaries. However, the source of these salient boundary features is the Feature Pyramid Network (FPN) feature passed through a temporal convolution, which is insufficient for capturing the temporal dependencies. In order to support performing precise boundary refinement, we utilize the attention mechanism to extract boundary features which captures the temporal interactions between different temporal locations. Experimental results show that the proposed temporal attention feature is effective for improving the boundary refinement process and further enhancing action detection performances.

The contributions of this work are summarized as follows. Firstly, we propose an effective temporal attention unit for capture the dependencies among features from different temporal locations. Additionally, we develop an Attention-guided Refinement Module for revising detected action boundaries and classes under an anchor-free detector, which achieves state-of-the-art performances on THU-MOS14 and ActivityNet1.3 dataset. To our knowledge, this is the first work introduces the temporal attention mechanism in the task of anchor-free temporal action detection. Moreover, to further evaluate the potential of temporal attention mechanism, we explore various strategies for introducing temporal attentions, such as using attention features as the fundamental features for producing the coarse action predictions. Lastly, extensive ablation experiments are conducted to explore the possible instantiations of each encoder of the temporal attention module.

The rest of this paper is organized as follows. Section 2 reviews closely related researches, including recent development of video action analysis, and different paradigms of temporal action detection; Sect. 3 describes the proposed temporal attention module and the adopted framework; Sect. 4 illustrates the experiments and corresponding results on the THUMOS14 and ActivityNet1.3 datasets; Sect. 5 concludes the findings and discusses future research directions.

2 Related Work

2.1 Video Action Detection

Contemporary temporal action detection frameworks can be classified into three main categories, namely anchor-based detectors [5,11,19], actionness-based (or boundary-based) detectors [10,12], and anchor-free detectors [9,20].

Anchor-based action detectors have achieved great success in the task of object detection. Inspired by Faster-RCNN framework, R-C3D was [19] proposed to generate and regress action proposals at each temporal location for action detection. TAL Net [5] further introduced multi-scale architecture that different size of predefined action proposals are managed by network at different scales. These methods perform action detection by firstly predicting action proposals and then performing action recognition on these proposals, which are two isolated stages. SSAD [11] proposed an one-stage framework that with the simultaneous generation of action proposals and classifications.

Actionness-based action detectors localize actions by assigning scores to each temporal location as being the **starting, ending** frame of an action, then these starting and ending candidates are heuristically combined for generating action proposals. Starting and ending frame combinations with high probabilities are treated as action proposal candidates. Then actionness feature are samples from these proposals for performing action detection. Representative actionness-based action detector are Boundary Sensitive Network (BSN) [12] and Boundary Matching Network (BMN) [10].

Anchor-based and actionness-based detectors are commonly known as two stage detectors, where localizing an action and predicting its class are two isolated processes. **Anchor-free detectors** perform generating action proposals and predicting action classes at the same time based on the features at each temporal location. A2Net [20] is one of the earliest works which introduced anchor-free mechanism for action detection, in which the action class score, and the temporal distance from the beginning and ending to the feature's temporal locations are predicted. Regarding the action proposal predicted by the anchor-free detector as the coarse detection results, AFSD [9] proposed a boundary refine module for adjusting the coarse action proposals.

2.2 Feature Extraction in Video Action Analysis

The quality of video action features is extremely important for action detection results. Currently, convolutional neural network (CNN) based backbone (feature extraction) models still dominate video action analysis. For examples, C3D [16], Two-stream Network [15], and I3D [4] have been widely used in various action detection frameworks. Besides, the recent emerging non-local network [18] and attention mechanism [17] are gaining more focus because of their capability from mining global dependencies. In this work, we propose to introduce the attention mechanism to capture the global temporal dependency for refining detected action boundaries.

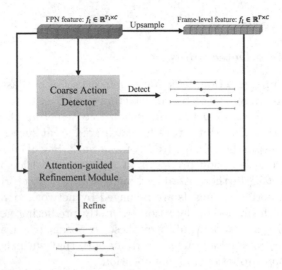

Fig. 1. The architecture of our method, which includes a coarse action detector which is anchor-free and an Attention-guided Refinement Module. The coarse detector receives FPN features and predicts action boundaries and classes (classes features and refinements are eliminated for simplicity). The refinement module takes the FPN feature, frame-level feature, and coarse predictions to produce refined predictions.

3 Proposed Method

As Fig. 1 shows, our method adopts a two-stage, detection-refinement detection paradigm which has been proved effective in the tasks of object detection [14] and action detection [9]. Firstly, the detection module, an anchor-free action detector receives FPN features and performs action detection by predicting the temporal boundary as well as the corresponding class of an action for the feature at each temporal location. The detection results are regarded as coarse predictions.

Moreover, the refinement module revises each coarse detection based on the given FPN feature and frame-level features. In [9], a Saliency-based Refinement module is proposed by pooling the most sensitive features from the input features within the temporal region indicated by coarse predictions. However, in order to perform action boundary refinement more precisely, features are expected to capture more temporal dependencies. In this work, we develop an effective temporal attention model for temporal dependency modelling, and further propose an Attention-guided Refinement module for boundary refinement.

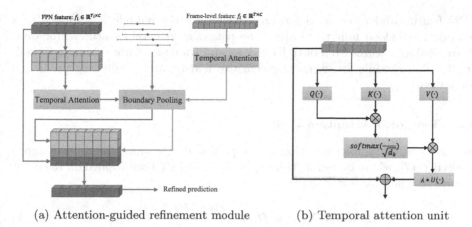

(a) Attention-guided refinement module   (b) Temporal attention unit

Fig. 2. The proposed module and temporal attention unit.

3.1 Anchor-free Action Detection

Temporal action detection aims to localize the **starting** and **ending** time of possible actions within a given video clip, and also predict appropriate classes for these actions. Denoting the video clip as X and $X \in \mathbb{R}^{T \times H \times W \times 3}$, where T, H, W represent the temporal length, frame height, and width of the video, respectively. Then the task of the model is to identify the start and end location of all possible actions, associated with the predict action classes, where each prediction can be encoded into a 3-dimensional vector (t_s^i, t_e^i, c^i).

For most temporal action detectors, the input video will be passed through a commonly used feature extractor such as C3D and I3D, and then all the spatial dimension will be pooled or filtered. AFSD [9] adopts the I3D as the feature extractor and constructs Feature Pyramid Network (FPN) for extracting multi-level features. We suggest readers to the original reference for more details. For simplicity, we take one FPN feature from level l as the example, where the feature $f^l \in \mathbb{R}^{T_l \times C}$, where T_l is the temporal dimension of current level FPN feature and C is the channel number. Then an action prediction $\{t_{i,s}^l, t_{i,e}^l, y_i^l\}$ will be made based on the feature at temporal location i, where $i \in [0, T_l - 1]$. $t_{i,s}^l, t_{i,e}^l, y_i^l$ are predicted starting time, ending time, and class label, respectively. In our two-stage, detection-refinement framework like AFSD [9], these predictions are regarded as coarse predictions.

3.2 Attention-guided Refinement Module

In this work, we introduce the temporal attention unit to encode the global temporal dependencies into a single feature map and further propose the Attention-guided Refinement Module based on the saliency-based Boundary Pooling Module [9]. The architecture of the proposed module is shown in Fig. 2a. Firstly, the

FPN feature and frame-level feature received by the module are applied to a temporal attention unit to produce the temporal attention feature. Moreover, the attention features from both FPN-level and frame-level are fed into a Boundary Pooling Module for extracting sensitive feature and conducting prediction refinement.

3.3   Temporal Attention Unit

In this work, the implementation of the proposed attention module is designed following [17], where three functions $Q(\cdot)$, $K(\cdot)$ and $V(\cdot)$ are applied on the input feature X where $X \in \mathbb{R}^{T \times C}$:

$$Attention(X) = softmax(\frac{Q(X)K(X)^T}{\sqrt{d^k}})V(X), \tag{1}$$

where d^k is the feature dimension of $K(X)$. Besides, the attention feature will be passed through another function $U(\cdot)$ with a scale factor λ and added with original input by a residual path to obtain the final output of the temporal attention module:

$$Y = U(Attention(X)) * \lambda + x \tag{2}$$

Fig. 2b describes the implementation of the proposed attention module, where $Q(\cdot)$, $K(\cdot)$, $V(\cdot)$, and $J(\cdot)$ are implemented by temporal convolutions, and each one is followed by a ReLU activation function and Group Normalization.

4   Experimentation

4.1   Datasets and Settings

Datasets: Extensive experiments are carried out on two large-scale datasets to evaluate the proposed method, namely the THUMOS14 [7] and ActivityNet-1.3 [3]. Thumos14 dataset contains 412 untrimmed videos where 200 of them are assigned as the validation set and 212 videos are assigned as the test set. Each video is annotated into 20 action classes with specific starting and ending times. ActivityNet-1.3 dataset contains 19,949 videos. Similarly, they are also annotated into 200 action classes with starting and ending times. The training/validation/test splitting ratio of ActivityNet-1.3 dataset is 2:1:1.

Implementation Details: The input settings of the experiments are implemented following [9]. For the THUMOS14 dataset, each input video (both RGB and optical flow) are firstly re-sampled at 10 frame-per-second (fps), and then split into clips with equal temporal length T, where $T = 256$. Each pair of neighbouring clips have overlap of 30 frames and 128 frames during the processes of training and testing, respectively. Besides, clips that have less than T frames are padded with zeros. For the ActivityNet1.3 dataset, each video is re-sampled to 768 frames, so no splitting process is needed. The spatial dimension of both datasets is set to 96 × 96 and implemented by cropping. During the training, random cropping and random flipping are used. During the testing, only the center cropping is used.

Since the proposed method is a module relying on the framework of AFSD [9], so the training and optimization are still following the original work. During training, Adam [8] is used as the solver. The learning rate is set to 10^{-5} for THUMOS14 with the loss weight of 10; and the learning rate is set to 10^{-4} for ActivityNet1.3 with the loss weight of 1. For both datasets, the weight decay is set to 10^{-3}. During testing, the outputs of the RGB modality and optical flow modality are averaged to obtain the final result. All final predictions are fed into a Soft-NMS process to reduce the redundant predictions. The tIoU threshold for the Soft-NMS [2] is set to 0.5 for the for THUMOS14 dataset, 0.85 for the ActivityNet1.3 dataset.

Metrics: Mean Average Precisions (mAPs) at different temporal IoU (tIoU) thresholds in following experiments. The thresholds for THUMOS14 are set to [0.3 : 0.1 : 0.7], and the thresholds for ActivityNet1.3 are set to [0.5 : 0.05 : 0.95].

4.2   Ablation Study

To analyze the effectiveness of the temporal attention feature for action detection, we conduct two sets of ablation experiments. Firstly, we experiment to introduce the temporal attention mechanism at different stage of the AFSD action detection framework to analyse the performance difference and explore the best practice of utilizing such temporal attention mechanism for action detection. Secondly, we explore different instantiation choices of these encoders within the temporal attention module to evaluate their necessity. Both sets of experiments are carried out based on the THUMOS14 dataset with RGB modality only.

Temporal Attention Feature at Different Stages: We introduce the temporal attention mechanism at following stages: (1) placed after the backbone model to transform the FPN feature as temporal attention features for generating action proposals and class predictions (**Source**); (2) introduced to transform the boundary pooling feature at current pyramid-level for boundary refinement, but without frame-level boundary feature (**Boundary/self**); (3) introduced to transform the boundary pooling feature at current pyramid-level and frame-level for boundary refinement (**Boundary/all**). Lastly, the baseline model selected is the original AFSD, where the boundary pooling feature at current pyramid-level and frame-level are used, but no temporal attention mechanism introduced (**W/O attention**).

The comparison results are presented in Table 1. As the table shows, the setting of transforming the boundary pooling feature at current pyramid-level and frame-level achieves the best performance (**Boundary/all**). However, when only utilizing the temporal attention feature at current pyramid level without

Table 1. Performance differences when applying temporal attention feature at different stages.

Stage	0.5	0.6	0.7	Avg.
W/O attention [9]	45.9	35.0	23.4	43.5
Source	44.8	34.4	22.8	42.7
Boundary/self	44.7	33.8	21.6	42.3
Boundary/all	**47.3**	**36.0**	**24.2**	**44.6**

frame-level feature for boundary refinement the performances of the model is critically impaired, and even lower than the baseline model, which indicates the importance and necessity of frame-level features (**Boundary/self**). Lastly, from the second row of Table 1 we can see that directly transform the backbone feature to temporal attention feature for action detection does not receive good performance (**Source**). The reason could be that simply applying the temporal attention mechanism on I3D output would harm the semantic structure of the original feature. Thus, in order to successfully utilize temporal attention features for action detection, further experimental explorations are still needed.

Instantiation of Attention Encoders: We explore different choices of instantiating encoding functions within the attention module, namely $K(\cdot)$, $Q(\cdot)$, $V(\cdot)$, and $U(\cdot)$, with following scenarios: (1) K, Q, V, and U are identity functions which will output the same values as the their inputs; (2) K, Q, V are identity functions, and U is a temporal convolution module; (3) K, Q, V, and U are all temporal convolution modules. Besides, we also evaluate two different choices of the hyper-parameter λ. As Table 2 shows, the results are better when all encoders are temporal convolution modules than when they are identity functions. Additionally, compared to $\lambda = 0.5$, $\lambda = 1.0$ achieves a better performance.

Table 2. Performance differences of different instantiation of encoding function in the attention module.

K \ Q \ V	U	λ	0.5	0.6	0.7	Avg.
Identity	Identity	1.0	44.3	34.6	22.1	42.1
Identity	Conv	1.0	45.0	34.2	21.9	42.9
Conv	Conv	1.0	**47.3**	**36.0**	**24.2**	**44.6**
Conv	Conv	0.5	45.5	34.4	22.9	43.2

4.3 Compare with State-of-the-Art

The comparison results with state-of-the-art methods on selected datasets are presented in Table 3. Among these selected comparison methods, R-C3D, SSAD,

Table 3. Comparison with state-of-the-art methods on the THUMOS14 and ActivityNet1.3 datasets. Average mAPs are reported under tIoU thresholds [0.3:0.1:0.7] for THUMOS14 and [0.5:0.05:0.95] for ActivityNet1.3.

Type	Method	back bone	THUMOS14						ActivityNet 1.3			
			0.3	0.4	0.5	0.6	0.7	Avg	0.5	0.75	0.95	Avg
Anchor-based	SSAD [11]	TS	43.0	35.0	24.6	—	—	—	—	—	—	—
	TURN [6]	C3D	44.1	34.9	25.6	—	—	—	—	—	—	—
	R-C3D [19]	C3D	44.8	35.6	28.9	—	—	—	26.8	—	—	—
	TAL [5]	I3D	53.2	48.5	42.8	33.8	20.8	39.8	38.2	18.3	1.3	20.2
	GTAN [13]	I3D	57.8	47.2	38.8	—	—	—	52.6	34.1	8.9	34.3
Actionness-based	SSN [21]	TS	51.0	41.0	29.8	—	—	—	43.2	28.7	5.6	28.3
	BSN [12]	TS	53.5	45.0	36.9	28.4	20.0	36.8	46.5	30.0	8.0	30.0
	BMN [10]	TS	56.0	47.4	38.8	29.7	20.5	38.5	50.1	34.8	8.3	33.9
	BC-GNN [1]	TS	57.1	49.1	40.4	31.2	23.1	40.2	50.6	34.8	**9.4**	34.3
Anchor free	A2Net [20]	I3D	58.6	54.1	45.5	32.5	17.2	41.6	43.6	28.7	3.7	27.8
	AFSD [9]	I3D	67.3	62.4	55.5	43.7	31.1	52.0	52.4	35.3	6.5	34.4
	Proposed	I3D	**68.4**	**63.5**	**56.4**	**45.4**	**32.3**	**53.2**	**52.8**	**35.6**	6.7	**34.8**

and TAL are anchor-based methods, BSN and BMN are actionness-based methods, as well as A2Net and AFSD are anchor-free methods (A2Net combines anchor-based and anchor-free methods). All of these compared methods adopt commonly used methods as the backbone model for feature extraction, namely C3D, TS, and I3D. Our method is developed based on AFSD by placing the temporal attention mechanism to produce attention feature for boundary refinement, so we choose the original AFSD as the baseline model in this comparison.

For **THUMOS14** dataset, the action detection performance of the proposed method outperforms all selected comparison method on all tIoU thresholds. Besides, our method outperforms the baseline model, the original AFSD, with 1.2% of average mAP, and achieves at least a margin of 0.9% on all settings. Especially, on the metric of mAP@0.6, our method outperform the original AFSD for 1.7%.

For **ActivityNet1.3** dataset, our method outperforms these compared methods on the average mAPs. On most tIoU choices, our method also outperforms the original AFSD and other methods, except the mAP@0.95 compared with these actionness-based methods, BSN, BMN, and BC-GNN. Since actionness-based methods heuristically enumerate all combinations of starting and ending frames with high probabilities, so they will generate more proposal for actions, while anchor-free methods only generate one proposal at each temporal location. Thus, our method and other anchor-free methods are more efficient but with the trade-off of sacrificing the performance with high tIoUs.

5   Conclusions

In this work, we introduce the attention mechanism in the task of temporal action detection. Based on the experiment results, we have following findings.

Firstly, the temporal attention mechanism can capture the temporal dependencies among different temporal locations of input, which is helpful for refining predicted action boundaries in temporal action detection models. Moreover, temporal attention mechanism is effective for action detection, but the framework should be carefully designed since simply utilizing the temporal attention feature as the source feature for action detection is proved not feasible by our experiments. In our future work, we will keep investigating the potential of temporal attention feature, and further explore the best practice for temporal action detection task.

References

1. Bai, Y., Wang, Y., Tong, Y., Yang, Y., Liu, Q., Liu, J.: Boundary content graph neural network for temporal action proposal generation. In: Vedaldi, A., Bischof, H., Brox, T., Frahm, J.-M. (eds.) ECCV 2020. LNCS, vol. 12373, pp. 121–137. Springer, Cham (2020). https://doi.org/10.1007/978-3-030-58604-1_8
2. Bodla, N., Singh, B., Chellappa, R., Davis, L.S.: Soft-NMS-improving object detection with one line of code. In: Proceedings of the IEEE International Conference on Computer Vision, pp. 5561–5569 (2017)
3. Caba Heilbron, F., Escorcia, V., Ghanem, B., Carlos Niebles, J.: Activitynet: a large-scale video benchmark for human activity understanding. In: Proceedings of the IEEE Conference on Computer Vision and Pattern Recognition, pp. 961–970 (2015)
4. Carreira, J., Zisserman, A.: Quo vadis, action recognition? a new model and the kinetics dataset. In: Proceedings of the IEEE Conference on Computer Vision and Pattern Recognition, pp. 6299–6308 (2017)
5. Chao, Y.W., Vijayanarasimhan, S., Seybold, B., Ross, D.A., Deng, J., Sukthankar, R.: Rethinking the faster R-CNN architecture for temporal action localization. In: Proceedings of the IEEE Conference on Computer Vision and Pattern Recognition, pp. 1130–1139 (2018)
6. Gao, J., Yang, Z., Chen, K., Sun, C., Nevatia, R.: Turn tap: temporal unit regression network for temporal action proposals. In: Proceedings of the IEEE International Conference on Computer Vision, pp. 3628–3636 (2017)
7. Jiang, Y.G., et al.: Thumos challenge: action recognition with a large number of classes (2014)
8. Kingma, D.P., Ba, J.: Adam: a method for stochastic optimization. arXiv preprint arXiv:1412.6980 (2014)
9. Lin, C., et al.: Learning salient boundary feature for anchor-free temporal action localization. In: Proceedings of the IEEE/CVF Conference on Computer Vision and Pattern Recognition, pp. 3320–3329 (2021)
10. Lin, T., Liu, X., Li, X., Ding, E., Wen, S.: BMN: boundary-matching network for temporal action proposal generation. In: Proceedings of the IEEE/CVF International Conference on Computer Vision, pp. 3889–3898 (2019)
11. Lin, T., Zhao, X., Shou, Z.: Single shot temporal action detection. In: Proceedings of the 25th ACM international conference on Multimedia, pp. 988–996 (2017)
12. Lin, T., Zhao, X., Su, H., Wang, C., Yang, M.: BSN: boundary sensitive network for temporal action proposal generation. In: Ferrari, V., Hebert, M., Sminchisescu, C., Weiss, Y. (eds.) ECCV 2018. LNCS, vol. 11208, pp. 3–21. Springer, Cham (2018). https://doi.org/10.1007/978-3-030-01225-0_1

13. Long, F., Yao, T., Qiu, Z., Tian, X., Luo, J., Mei, T.: Gaussian temporal awareness networks for action localization. In: Proceedings of the IEEE/CVF Conference on Computer Vision and Pattern Recognition, pp. 344–353 (2019)
14. Qiu, H., Ma, Y., Li, Z., Liu, S., Sun, J.: BorderDet: border feature for dense object detection. In: Vedaldi, A., Bischof, H., Brox, T., Frahm, J.-M. (eds.) ECCV 2020. LNCS, vol. 12346, pp. 549–564. Springer, Cham (2020). https://doi.org/10.1007/978-3-030-58452-8_32
15. Simonyan, K., Zisserman, A.: Two-stream convolutional networks for action recognition in videos. arXiv preprint arXiv:1406.2199 (2014)
16. Tran, D., Bourdev, L., Fergus, R., Torresani, L., Paluri, M.: Learning spatiotemporal features with 3d convolutional networks. In: Proceedings of the IEEE International Conference on Computer Vision, pp. 4489–4497 (2015)
17. Vaswani, A., et al.: Attention is all you need. In: Advances in Neural Information Processing Systems, pp. 5998–6008 (2017)
18. Wang, X., Girshick, R., Gupta, A., He, K.: Non-local neural networks. In: Proceedings of the IEEE Conference on Computer Vision and Pattern Recognition, pp. 7794–7803 (2018)
19. Xu, H., Das, A., Saenko, K.: R-c3d: Region Convolutional 3D Network for Temporal Activity detection. In: Proceedings of the IEEE International Conference on Computer Vision, pp. 5783–5792 (2017)
20. Yang, L., Peng, H., Zhang, D., Fu, J., Han, J.: Revisiting anchor mechanisms for temporal action localization. IEEE Trans. Image Process. **29**, 8535–8548 (2020)
21. Zhao, Y., Xiong, Y., Wang, L., Wu, Z., Tang, X., Lin, D.: Temporal action detection with structured segment networks. In: Proceedings of the IEEE International Conference on Computer Vision, pp. 2914–2923 (2017)

Spatio-temporal Attention Graph Convolutions for Skeleton-based Action Recognition

Cuong Le[1,2(✉)] and Xin Liu[1]

[1] Computer Vision and Pattern Recognition Laboratory, School of Engineering Science, Lappeenranta-Lahti University of Technology LUT, Lappeenranta, Finland
cuong.le@liu.se, xin.liu@lut.fi
[2] Computer Vision Laboratory, Department of Electrical Engineering, Linköping University, Linköping, Sweden

Abstract. In skeleton-based action recognition, graph convolutional networks (GCN) have been applied to extract features based on the dynamic of the human body and the method has achieved excellent results recently. However, GCN-based techniques only focus on the spatial correlations between human joints and often overlook the temporal relationships. In an action sequence, the consecutive frames in a neighborhood contain similar poses and using only temporal convolutions for extracting local features limits the flow of useful information into the calculations. In many cases, the discriminative features can present in long-range time steps and it is important to also consider them in the calculations to create stronger representations. We propose an attentional graph convolutional network, which adapts self-attention mechanisms to respectively model the correlations between human joints and between every time steps for skeleton-based action recognition. On two common datasets, the NTU-RGB+D60 and the NTU-RGB+D120, the proposed method achieved competitive classification results compared to state-of-the-art methods. The project's GitHub page: STA-GCN.

Keywords: Computer vision · Action recognition · Graph convolution · Attention mechanism

1 Introduction

Understanding human action plays a crucial role in many applications, such as video surveillance, human-computer interaction, and human behavior analysis [12,20,27,30]. Image-based methods suffer from many difficulties such as illumination changes, sensor noises, and perspective changes. One solution to overcome these problems in human action recognition is to utilize skeleton data.

Earlier skeleton-based action recognition methods extract features using two main approaches: hand-crafted [8,13–15,25] and deep learning. Due to the recent development of large-scale datasets and computing power, deep learning approaches are providing better results in action recognition [27]. Earlier methods that use RNN [11,26] and CNN [5,6] overlook the correlation between joints in the human skeleton, therefore limiting their expressive capability. Recently, graph convolutional networks (GCN) are introduced into skeleton modeling to integrate the natural human topology into the calculations [1–3,9,16,21,28,29], and they are currently state-of-the-art in skeleton-based action recognition. Besides, recent research often adapt the standard [9,21,28] or multi-scale dilated 1D convolution [1,2,16] to extract temporal features. However, in skeleton data, local consecutive frames usually contain similar features which are not useful to effectively discriminate challenging action samples. To overcome this, we could either increase the size of convolution filters or create a very deep model to widen the receptive field, but these methods can be computationally expensive. Another solution is to use the self-attention mechanism [23] that has the ability to pinpoint specific useful frames over a global receptive field, and this is the chosen approach of the paper. Figure 1 illustrates the proposed idea.

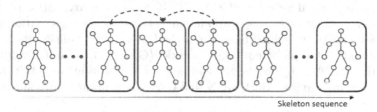

(a) Traditional temporal modeling with convolutions.

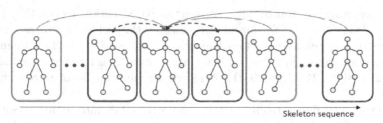

(b) Temporal modeling with added self-attentions.

Fig. 1. Temporal self-attention can pinpoint useful information along the action sequences by assigning weights on each frames. The red color denotes the current frame of calculation and the green frames indicate useful information. For many samples of human action, farther frames along the observed sequence can contain more discriminated features than the local neighborhoods. Aggregating local features is optimal for classifying challenging actions, so it is important to consider long-range dependencies. (Color figure online)

In this paper, a GCN-based attention method for skeleton-based action recognition is presented. In particular, a self-attention mechanism is combined with graph convolutional networks (GCN) for the spatial modeling of skeleton-based human models. In the temporal dimension, self-attention is applied together with multi-scale 1D convolutions to extract time-related features across the skeleton sequences. The proposed model is tested on two large-scale datasets: NTU-RGB+D60 and NTU-RGB+D120, and the classification results proved to be competitive with the current state-of-the-art methods.

2   Related Works

2.1   Deep Learning on Graphs

Standard deep learning toolboxes are optimized for either grid-like data (CNN) or sequences (RNN), thus creating difficulties when applying them to graphs. Recently, GNN was introduced to define the learning task on graph-structure data [4]. Similar to traditional supervised learning methods, each skeleton sequence is represented as a data sample with an associated action label, and the goal is to learn the mapping from data points to labels. One of the GNN variants is graph convolutional networks (GCN) [7]. GCN is an approximation of spectral GNNs and is good at capturing graph features. However, the aggregation weights of GCN are explicitly defined based on node degrees, thus limiting its representation capability. Graph attentional networks (GAT) [24] is later introduced to address this problem by implicitly learning the connection weights through a self-attention mechanism. Both GCN and GAT are utilized in this study.

2.2   Skeleton-based Action Recognition

Earlier deep learning approaches for skeleton-based action recognition include RNN-based [11, 26] and CNN-based [5, 6] consider the skeleton graph as an uncorrelated set of features, and overlook the dynamic connectivity of the human body. GCN-based methods, that integrate the joint connections into their spatial modeling [1, 2, 9, 16–18, 21, 22, 28, 29, 31, 32] record a significant boost in classification accuracy.

Yan et al. [28] first introduced the concept of GCN into action recognition, namely ST-GCN. The skeleton sequence is modeled from two types of edges: spatial edges that express connectivity between human joints, and temporal edges that connect the joints across time steps. Li et al. [9] proposed AS-GCN to capture richer dependencies in the spatial dimension of skeleton data. The method presents a module for capturing action-specific latent dependency between every human joint and extending human topology to represent higher-order dependencies. Shi et al. [21] reasoned that using predetermined and fixed skeleton graph topology for aggregating information is not optimal for diverse samples. Therefore, they proposed 2s-AGCN that captures second-order bone information in addition to joints' dependencies of skeleton data.

Liu et al. [16] proposed a disentangled and unifying graph convolutional network MS-G3D. The disentangling task removes the redundant dependencies between node features when aggregating spatial information and the graph topology is modified to directly obtain information from farther nodes. The combination of the two proposed methods creates a powerful extractor with multi-scale receptive fields across spatial and temporal dimensions. Zhang et al. [31] also consider long-range dependencies by using context-aware graph convolutions (CA-GCN) based on self-attention. In addition to the local modeling of each joint vertex, CA-GCN integrates information from all other vertices within the sequence. Also relying on self-attention, Shi et al. [22] proposed decoupling schemes (DSTA-Net) to model spatio-temporal interactions between joints and frames without knowing their positions or mutual connections.

Chen et al. [1] proposed CTR-GCN that dynamically refine a shared prior topology for each feature channel. CTR-GCN creates multi-channel attention maps to refine the correlation between joints in each skeleton graph. This approach provides an effective modeling scheme from different channels, leading to stronger representation. Zhang et al. [32] propose a Spatial-Temporal Specialized Transformer Encoder (STST) to model the skeleton posture of each frame and capture changes of posture in the temporal dimension, thus providing strong modeling of action sequences. Similarly, various approaches [17,18] also utilizes the transformer architecture to extract spatial and temporal dependencies (ST-TR). Recently, Chi et al. [2] adopted the information bottleneck to derive the objective and the corresponding loss for maximum informative latent representation of skeleton-based actions.

In this paper, we explicitly combine the self-attention modeling of spatial and temporal dependencies from skeleton-based sequences. The proposed method consists of the strong spatial representation from the implicit interaction modeling between joints, and the ability to pinpoint useful time-based information of the temporal attention module.

3   Proposed Method

In this section, we first introduce the related notations on skeleton graphs and graph convolution. Detailed information about our proposed *Spatio-temporal attentional graph convolutions* is presented.

3.1   Preliminaries

A human action can be represented as a sequence of skeleton graphs. A graph $\mathcal{G} = (\mathcal{V}, \mathcal{E})$ is an ordered pair constructed by a set of N vertices (or nodes) $\mathcal{V} = \{v_1, v_2, ..., v_N\}$ and a set of edges $\mathcal{E}$ between these vertices. An edge going from node $u \in \mathcal{V}$ to node $v \in \mathcal{V}$ is noted as $(u, v) \in \mathcal{E}$. A graph can be conveniently formulated by an symmetry adjacency matrix $\mathbf{A} \in \mathbb{R}^{N \times N}$. The strength of edges is presented by the value of matrix's entries a_{ij}. The neighborhood of v_i is the set $\mathcal{N}(v_i) = \{v_j | a_{ij} \neq 0\}$. Action classes, represented as skeleton graph sequences, contain a node feature set $\mathcal{X}$, which can be presented in matrix form $\mathbf{X} \in \mathbb{R}^{T \times N \times C}$. The relationship between nodes within a frame is described by an adjacency matrix $\mathbf{A}$.

The calculation of GCN [7] adapts the idea of symmetric normalization into the node update function for an input skeleton feature $\mathbf{x}$ at layer k as:

$$\mathbf{h}^k = \sigma\left(\tilde{\mathbf{D}}^{-1/2}\tilde{\mathbf{A}}\tilde{\mathbf{D}}^{-1/2}\mathbf{x}^k\mathbf{W}\right) \tag{1}$$

where $\tilde{\mathbf{A}} = \mathbf{A} + \mathbf{I}$ is the adjacency matrix with added self-loop, $\tilde{\mathbf{D}}$ is the degree matrix of $\tilde{\mathbf{A}}$, σ is the activation function, and $\mathbf{W}$ is the weight. Every entry of $\tilde{\mathbf{A}}$ takes the binary form to represent connectivity.

Graph Attentional Networks (GAT) [24] relaxes the entries of the adjacency matrix by adaptively learning it for every pair of connected nodes using the self-attention mechanism. The update function for an input skeleton feature $\mathbf{x}$ at layer k is formulated as:

$$\mathbf{h}^k = \sigma\left(\mathbf{M}\mathbf{x}^k\mathbf{W}\right) \tag{2}$$

where $\mathbf{M}$ is the self-attention score matrix which follows the calculation of [23]. In the original method, [24], the masked version of GAT is applied to only consider connected nodes. However, in this paper, the unmasked version of GAT is used instead to model the interactions between every pair of human joints.

3.2   Model Architecture

Our proposed feature extraction block consists of three modules connected to each other: GCN-GAT-combined spatial self-attention, temporal self-attention, and multi-scale temporal convolution. The proposed STA-GCN block is illustrated in Fig. 2. The architecture includes multiple blocks of STA-GCN, followed by a global average pooling, a fully connected, and a softmax layer.

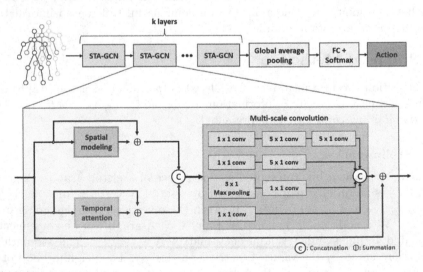

Fig. 2. Model architecture.

3.3   Spatial Modeling

To effectively extract features from the skeleton graphs, we combine the adjacency calculations from GCN and GAT into one update function:

$$\mathbf{C} = \mathbf{M} + \tilde{\mathbf{D}}^{-1/2} \tilde{\mathbf{A}} \tilde{\mathbf{D}}^{-1/2}$$
$$\mathbf{h}^k = \sigma(\mathbf{C}\mathbf{x}^k\mathbf{W}) \tag{3}$$

where $\mathbf{C}$ is the combined adjacency matrix. By combining the two methods, we get the benefit from both. GCN is good for capturing spatial dependency between nodes from the given prior knowledge about human kinetics in the adjacency matrix. Unmasked GAT is good for modeling hidden correlations between human joints that are not visually connected.

The spatial modeling process is illustrated in Fig. 3. First, the input sequence is globally pooled along the temporal dimension. The pooled matrix is used to derive the query and key for computing the attention score. We then multiply the value with matrix $\mathbf{C}$ to get the final embedding tensor of the skeleton sequence. Furthermore, a multi-head version of self-attention is utilized to stabilize the $\mathbf{h}^k$ calculation. The combination of attention map and adjacency matrix also happens head-wised. Therefore, each head has its own combined attention map.

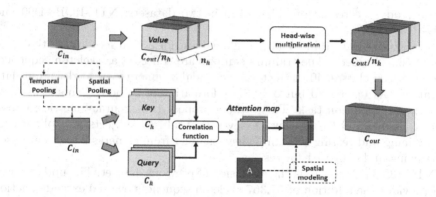

Fig. 3. Spatial and temporal self-attention modeling.

3.4   Temporal Modeling

The main goal of the temporal modeling process is to complement the traditional local extracting methods with a global aggregation approach to capture long-range dependencies within a skeleton sequence. Similar to the spatial modeling process, multi-head self-attention is also applied to temporally model skeleton sequences. First, each sequence is spatially pooled at every skeleton graph. After pooling, the data becomes a classical sequential problem. The attention map between frames is calculated from query and key. The only difference from the spatial modeling is no combination process with the adjacency matrix (Fig. 3). Then, the value is multiplied by the attention map to produce the output.

By using self-attention, we can extract long-range dependencies within one layer. The purpose of temporal self-attention is to pinpoint the most beneficial frames from the skeleton sequence. However, the attention module may put a lot of weight into long-range frames and not consider local neighborhoods. As earlier methods demonstrated, extracting local features is an effective way to ensure a good baseline performance. While self-attentions pinpoint specific useful frames over the whole sequence, temporal convolutions provide direct access to neighborhood features. For this reason, we proposed combining self-attentions with 1D temporal convolutions to complement each other. We adopt the multi-scale dilated convolution module from [1,2,16] with minor changes to extract local temporal features. The dilated convolutions from [16] increase the receptive field while keeping the number of calculations unchanged. However, long-range dependencies are already collected by self-attentions, so standard 1D temporal convolutions are implemented instead to capture richer local dependencies.

4 Experiments

4.1 Datasets

In this study, we tested our algorithm in two datasets: NTU-RGB+D60 and NTU-RGB+D120.

NTU-RGB+D60 [19] is a large-scale dataset for action recognition that consists of 56,578 videos. The training samples are collected as skeleton sequences from 60 action classes, 40 distinct subjects, and 3 camera view angles. Four data modalities were provided but only 3D information of 25 body joints is used for the action recognition task. The authors propose two accuracy metrics: Cross-subject (Xsub) and Cross-view (Xview). In Xsub, 40 subjects are split evenly into training and testing sets. In Xview, samples from cameras 2 and 3 are used for training and camera 1 for testing.

NTU-RGB+D120 [10] is the extension of previous dataset. The updated version provides the addition of 57,367 skeleton sequences over 60 extended action classes. In total, NTU-RGB+D 120 consists of 113,945 training samples over 120 action classes, which are performed by 106 human subjects and captured from 32 different camera setups. The authors propose two evaluation settings: Cross-subject (Xsub) and Cross-setup (Xset). In Xsub, 106 subjects are split evenly into training and testing sets. In Xset, 32 collection setups are split as even-IDs for training and odd-IDs for testing.

4.2 Implementation Details

The experiment results are conducted on two NVIDIA Tesla V100 GPUs from Finland's CSC server with PyTorch deep learning framework. The model is trained with SGD with momentum 0.9, weight decay 0.0004, batch size 64, and an initial learning rate of 0.1 for 65 epochs. The learning rate is scheduled to decay with a rate of 0.1 at epochs 35 and 55. A warm-up strategy is adopted for the first 5 epochs to stabilize the training process. For two datasets NTU-RGB+D 60 and NTU-RGB+D 120, the data pre-processing from [1] is used, and all skeleton sequences are resized to 64 frames each.

Similar to [1,21,29], multiple training with different data modalities are also implemented. In addition to the original data of skeleton joints, bone and velocity modalities are also utilized for training. Thus, there are a total of four different training at each evaluation setting: joint, bone, joint-motion, and bone-motion. The performances of all four modalities are then assembled into a final value of accuracy.

The standard evaluation metric on human action recognition research is accuracy measurement. For a fair comparison, we also measured the classification accuracy on both NTU-RGB+D60 and NTU-RGB+D120. In addition, we also conduct F1 measurement over classes on some related methods that published their model's weights [1,16].

To find the best performing model for STA-GCN, an ablation experiment was conducted. With the same training hyper-parameters, a baseline model consisting of only graph convolutional networks and temporal convolutions was tested. Then, the proposed spatial adaptive attention and temporal attention were added respectively. All ablation study experiments were conducted in the cross-subject setting of the NTU-RGB+D60 dataset.

4.3   Ablation Study

In this section, the proposed spatial-temporal attention graph convolution network is tested on the cross-subject evaluation setting on the NTU-RGB+D60 dataset. An architecture similar to ST-GCN [28] is deployed as the starting baseline. There are three modules that need to be studied: adaptive GCN, additional GAT, and temporal attention modeling. Also, the original 1D convolution in ST-GCN [28] is changed to the multi-scale module for a fair comparison. The experimental results are shown in Table 1.

Table 1. Accuracy comparison for ablation study.

Methods	Params.	Accuracy (%)	Mean last 10 epochs (%)
Baseline	843868	87.50	87.27
Baseline w. adaptive GCN	850118	89.30	89.09
STA-GCN w/o. temp attention	1119422	89.88	89.72
STA-GCN w. temp attention	1174560	89.90	89.87

First, we tested the performance of the original baseline model with normal GCN and multi-scale convolutions. Then, the adaptive characteristic is integrated into the GCN module to observe the improvement. In the third experiment, GAT and self-attentions are fused into the spatial modeling to create the proposed STA-GCN model. Self-attentions along temporal dependencies are separately considered in the fourth experiment. It can be observed that the accuracy of the classifiers increases gradually as more modules are added.

With an overall accuracy of 87.50%, the baseline performs fairly well on the NTU-RGB+D60 dataset. However, the baseline model struggles against many difficult classes, such as: eating snack (action 02), reading (action 11), writing (action 12), taking off shoes (action 17), playing with phone (action 29), sneezing (action 41). These classes differ by small changes in arm movements, thus creating challenges for skeleton-based action recognition. By integrating the proposed modules into the baseline model, an increase in performance is recorded, as shown in Fig. 4.

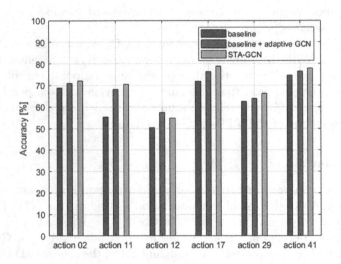

Fig. 4. Comparison of performance on difficult classes.

In addition, we measured the class-wise F1-score for the baseline and STA-GCN. Table 2 illustrates the top five classes with the highest improvement on F1 measurement. The highest improvements was recorded on classes that include small-gesture action samples such as reading (action 11), using a hand-fan (action 49), touching neck (action 47), or putting on glass (action 18). The action reading (action 11) has the highest improvement when STA-GCN was applied with 8% increment in F1-score. This demonstrates the positive impact when applying temporal attention modules to recognize subtle gesture action samples.

Figure 5 shows the last 10 epochs of the training process. In the case of the model without temporal attention, the accuracy oscillates between 89.6% to 89.9%, with a mean accuracy of 89.72%. While the performance of the model with temporal attention is stable at around 89.9% and an average of 89.87%. Therefore, though the final accuracy of the two models is the same, the one with temporal attention proved to be more consistent and superior. Because of these reasons, the STA-GCN module with temporal attention is chosen as the main building block of the final skeleton-based action recognition model.

Table 2. Top five classes with highest F1-score improvement.

Rank	STA-GCN vs baseline		STA-GCN vs adaptive baseline	
	Action	Increment(%)	Ation	Increment(%)
Top 1	action 11	8.05	action 11	5.13
Top 2	action 49	7.47	action 47	3.20
Top 3	action 47	7.12	action 18	2.98
Top 4	action 18	5.73	action 05	2.75
Top 5	action 29	5.66	action 28	2.71

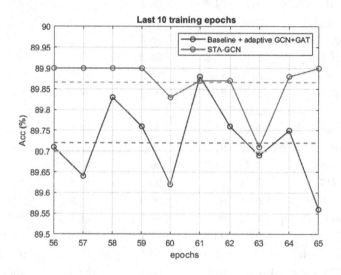

Fig. 5. Comparison between models with and without temporal attention.

4.4 Results

We adopt the same multi-stream fusion framework as [1,21], by fusing four different modalities of data: joint, bone, joint motion, and bone motion. The comparisons in classification accuracy of our approach with other graph-based methods is demonstrated in Table 3.

On NTU-RGB+D60, the final STA-GCN model beats earlier methods [6,11,25], and some recent GCN-based methods such as [9,16,21,28,29], but cannot outperform the current state-of-the-arts [1,2] in Xsub and Xview. However, compared to other methods that also implemented self-attention on temporal domain [17,18,22,31,32], our STA-GCN has a clear advantage, especially on Xsub setting. For NTU-RGB+D 120 dataset, the evaluation quality is the same. STA-GCN still outperforms [17,18,22,31,32] in both Xsub and Xset metrics. When compared to ST-TR [17], the method that is most similar to our approach, STA-GCN achieved closed performance on the NTU-RGB+D60 but outperforms by a large margin on the NTU-RGB+D120. Besides, STA-GCN under-performs CTR-GCN [1] by 0.4% and 0.2% and InfoGCN by 1.3% and 0.8% on Xsub and Xset settings of the NTU-RGB+D120.

Table 3. Classification accuracy compared with state-of-the-art methods on the NTU-RGB+D60 and NTU-RGB+D120 datasets.

Methods	Year	NTU-RGB+D60		NTU-RGB+D120	
		Xsub (%)	Xview (%)	Xsub (%)	Xset (%)
Lie Group [25]	2014	50.1	52.8	–	–
Temporal CNN [6]	2017	74.3	83.1	–	–
Ind-RNN [11]	2018	81.8	88.0	–	–
ST-GCN [28]	2018	81.5	88.3	–	–
AS-GCN [9]	2019	86.8	94.2	–	–
2s-AGCN [21]	2019	88.5	95.1	82.9	84.9
MS-G3D [16]	2020	91.5	96.2	86.9	88.4
CA-GCN [31]	2020	83.5	91.4	–	–
DSTA-Net [22]	2020	91.5	96.4	86.6	89.0
Dynamic GCN [29]	2020	91.5	96.0	87.3	88.6
STST [32]	2021	91.9	96.8	–	–
ST-TR [17]	2021	89.9	96.1	81.9	84.1
CTR-GCN [1]	2021	92.4	96.8	88.9	90.6
Qin's method [18]	2022	90.5	96.1	85.7	86.8
InfoGCN [2]	2022	93.0	97.1	89.8	91.2
STA-GCN (Net)		92.4	96.5	88.5	90.4
STA-GCN (Joint+Bone)		92.0	96.4	88.4	90.0
STA-GCN (Joint)		89.9	94.9	84.7	86.3
STA-GCN (Joint-motion)		87.4	93.3	81.4	83.1
STA-GCN (Bone)		90.3	94.9	86.3	87.8
STA-GCN (Bone-motion)		87.4	91.9	81.2	83.0

As previously mentioned, F1-score measurement is also carried out for some related methods that published their model weights, specifically MS-G3D [16] and CTR-GCN [1]. The measurement is shown in Table 4. Because NTU-RGB+D60 and NTU-GRB+D120 is fairly balanced datasets, the F1-scores did not vary much from the accuracy measurements.

Table 4. F1-score measurement results

Methods	NTU-RGB+D60		NTU-RGB+D120	
	Xsub-joint (%)	Xsub-bone (%)	Xsub-joint (%)	Xsub-bone (%)
MS-G3D [16]	89.4	90.1	83.8	86.0
CTR-GCN [1]	89.9	90.5	85.2	85.7
STA-GCN	89.9	90.2	84.9	86.4

5 Conclusion

In this study, a spatio-temporal attentional graph convolution network for skeleton-based action recognition is presented. As the results attest, the combination of GCN, self-attentions, and temporal convolutions can effectively learn features and joint relationships from the action sequences. Compared to the state-of-the-arts on two common datasets NTU-RGB+D60 and NTU-RGB+D120, the proposed model STA-GCN achieved competitive classification performance. In the future, we believe it is worthwhile to put more focus on the modeling of micro-movements in human limbs, in order to increase the classification performance of the model in those challenging situations.

References

1. Chen, Y., Zhang, Z., Yuan, C., Li, B., Deng, Y., Hu, W.: Channel-wise topology refinement graph convolution for skeleton-based action recognition. In: IEEE/CVF International Conference on Computer Vision (ICCV), pp. 13359–13368 (2021)
2. Chi, H.G., Ha, M.H., Chi, S., Lee, S.W., Huang, Q., Ramani, K.: INFOGCN: representation learning for human skeleton-based action recognition. In: IEEE/CVF Conference on Computer Vision and Pattern Recognition (CVPR), pp. 20186–20196 (2022)
3. Gao, R., Liu, X., Yang, J., Yue, H.: CDCLR: llip-driven contrastive learning for skeleton-based action recognition. In: 2022 IEEE International Conference on Visual Communications and Image Processing (VCIP), pp. 1–5. IEEE (2022)
4. Hamilton, W.L.: Graph representation learning. Synthesis Lectures on Artificial Intelligence and Machine Learning 14(3), 1–159 (2020)
5. Ke, Q., Bennamoun, M., An, S., Sohel, F., Boussaid, F.: A new representation of skeleton sequences for 3d action recognition. In: IEEE Conference on Computer Vision and Pattern Recognition (CVPR), pp. 4570–4579 (2017)
6. Kim, T.S., Reiter, A.: Interpretable 3d human action analysis with temporal convolutional networks. In: IEEE Conference on Computer Vision and Pattern Recognition Workshops (CVPRW), pp. 1623–1631 (2017)
7. Kipf, T.N., Welling, M.: Semi-supervised classification with graph convolutional networks. In: International Conference on Learning Representations (ICLR) (2017)
8. Koniusz, P., Cherian, A., Porikli, F.: Tensor representations via kernel linearization for action recognition from 3D skeletons. In: Leibe, B., Matas, J., Sebe, N., Welling, M. (eds.) ECCV 2016. LNCS, vol. 9908, pp. 37–53. Springer, Cham (2016). https://doi.org/10.1007/978-3-319-46493-0_3

9. Li, M., Chen, S., Chen, X., Zhang, Y., Wang, Y., Tian, Q.: Actional-structural graph convolutional networks for skeleton-based action recognition. In: IEEE/CVF Conference on Computer Vision and Pattern Recognition (CVPR), pp. 3590–3598 (2019)

10. Liu, J., Shahroudy, A., Perez, M., Wang, G., Duan, L.Y., Kot, A.C.: Ntu rgb+d 120: a large-scale benchmark for 3d human activity understanding. IEEE Trans. Pattern Anal. Mach. Intell. (PAMI) **42**(10), 2684–2701 (2020)

11. Liu, J., Shahroudy, A., Xu, D., Kot, A.C., Wang, G.: Skeleton-based action recognition using spatio-temporal LSTM network with trust gates. IEEE Trans. Pattern Anal. Mach. Intell. (PAMI) **40**(12), 3007–3021 (2018)

12. Liu, X., Shi, H., Chen, H., Yu, Z., Li, X., Zhao, G.: imigue: an identity-free video dataset for micro-gesture understanding and emotion analysis. In: IEEE/CVF Conference on Computer Vision and Pattern Recognition (CVPR), pp. 10631–10642 (2021)

13. Liu, X., Shi, H., Hong, X., Chen, H., Tao, D., Zhao, G.: Hidden states exploration for 3d skeleton-based gesture recognition. In: 2019 IEEE Winter Conference on Applications of Computer Vision (WACV), pp. 1846–1855 (2019)

14. Liu, X., Shi, H., Hong, X., Chen, H., Tao, D., Zhao, G.: 3d skeletal gesture recognition via hidden states exploration. IEEE Trans. Image Process. **29**, 4583–4597 (2020)

15. Liu, X., Zhao, G.: 3d skeletal gesture recognition via discriminative coding on time-warping invariant riemannian trajectories. IEEE Trans. Multimedia **23**, 1841–1854 (2021)

16. Liu, Z., Zhang, H., Chen, Z., Wang, Z., Ouyang, W.: Disentangling and unifying graph convolutions for skeleton-based action recognition. In: IEEE/CVF Conference on Computer Vision and Pattern Recognition (CVPR), pp. 140–149 (2020)

17. Plizzari, C., Cannici, M., Matteucci, M.: Spatial temporal transformer network for skeleton-based action recognition. In: Pattern Recognition. ICPR International Workshops and Challenges: Virtual Event, Part III, pp. 694–701 (2021)

18. Qin, X., Cai, R., Yu, J., He, C., Zhang, X.: An efficient self-attention network for skeleton-based action recognition. Sci. Rep. **12**, 2045–2322 (2022)

19. Shahroudy, A., Liu, J., Ng, T.T., Wang, G.: Ntu rgb+d: a large scale dataset for 3d human activity analysis. In: IEEE Conference on Computer Vision and Pattern Recognition (CVPR), pp. 1010–1019 (2016)

20. Shi, H., Peng, W., Chen, H., Liu, X., Zhao, G.: Multiscale 3d-shift graph convolution network for emotion recognition from human actions. IEEE Intell. Syst. **37**(4), 103–110 (2022)

21. Shi, L., Zhang, Y., Cheng, J., Lu, H.: Two-stream adaptive graph convolutional networks for skeleton-based action recognition. In: IEEE/CVF Conference on Computer Vision and Pattern Recognition (CVPR), pp. 12018–12027 (2019)

22. Shi, L., Zhang, Y., Cheng, J., Lu, H.: Decoupled spatial-temporal attention network for skeleton-based action-gesture recognition. In: Asian Conference on Computer Vision (ACCV), pp. 38–53 (2020)

23. Vaswani, A., et al.: Attention is all you need. In: International Conference on Neural Information Processing Systems (NIPS), pp. 6000–6010 (2017)

24. Veličković, P., Cucurull, G., Casanova, A., Romero, A., Liò, P., Bengio, Y.: Graph attention networks. In: International Conference on Learning Representations (ICLR) (2018)

25. Vemulapalli, R., Arrate, F., Chellappa, R.: Human action recognition by representing 3d skeletons as points in a lie group. In: IEEE Conference on Computer Vision and Pattern Recognition (CVPR), pp. 588–595 (2014)

26. Wang, H., Wang, L.: Modeling temporal dynamics and spatial configurations of actions using two-stream recurrent neural networks. In: IEEE Conference on Computer Vision and Pattern Recognition (CVPR), pp. 3633–3642 (2017)
27. Wang, L., Huynh, D.Q., Koniusz, P.: A comparative review of recent kinect-based action recognition algorithms. IEEE Trans. Image Process. **29**, 15–28 (2020)
28. Yan, S., Xiong, Y., Lin, D.: Spatial temporal graph convolutional networks for skeleton-based action recognition. In: AAAI Conference on Artificial Intelligence, pp. 3482–3489 (2018)
29. Ye, F., Pu, S., Zhong, Q., Li, C., Xie, D., Tang, H.: Dynamic GCN: context-enriched topology learning for skeleton-based action recognition. In: ACM International Conference on Multimedia, pp. 55–63 (2020)
30. Yu, Z., et al.: Searching multi-rate and multi-modal temporal enhanced networks for gesture recognition. IEEE Trans. Image Process. **30**, 5626–5640 (2021)
31. Zhang, X., Xu, C., Tao, D.: Context aware graph convolution for skeleton-based action recognition. In: IEEE/CVF Conference on Computer Vision and Pattern Recognition (CVPR), pp. 14321–14330 (2020)
32. Zhang, Y., Wu, B., Li, W., Duan, L., Gan, C.: STST: Spatial-temporal specialized transformer for skeleton-based action recognition. In: ACM International Conference on Multimedia, pp. 3229–3237 (2021)

Image and Video Processing, Analysis, and Understanding

RELIEF: Joint Low-Light Image Enhancement and Super-Resolution with Transformers

Andreas Aakerberg[1]([✉]), Kamal Nasrollahi[1,2], and Thomas B. Moeslund[1]

[1] Visual Analysis and Perception Laboratory, Aalborg University, Aalborg, Denmark
{anaa,kn,tbm}@create.aau.dk
[2] Research Department, Milestone Systems A/S, Brondby, Denmark

Abstract. The goal of Single-Image Super-Resolution (SISR) is to reconstruct a High-Resolution (HR) version of a degraded Low-Resolution (LR) image. Existing Super-Resolution (SR) methods mostly assume that the LR image is a result of blurring and downsampling the HR image, while in reality LR images are often degraded by additional factors such as low-light, low-contrast, noise, and color distortion. Due to this, current State-of-the-Art (SoTA) SR methods cannot reconstruct real low-light low-resolution images, and a straightforward strategy is, therefore, to first perform Low-Light Enhancement (LLE), followed by SR, using dedicated methods for each task. Unfortunately, this approach leads to poor performance, which motivates us to propose a method for joint LLE and SR. However, since LLE and SR are both ill-posed and ill-conditioned inverse problems, the joint reconstruction task becomes highly challenging, which calls for efficient ways to leverage as much as possible of the available information in the degraded image during reconstruction. In this paper, we propose REsolution and LIght Enhancement transFormer (RELIEF), a novel Transformer-based multi-scale hierarchical encoder-decoder network with efficient cross-shaped attention mechanisms that can extract informative features from large training patches due to its strong long-range dependency modeling capabilities. This in turn leads to significant improvements in reconstruction performance on real Low-Light Low-Resolution (LLLR) images. We evaluate our method on two publicly available datasets and present SoTA results on both.

Keywords: Transformers · super-resolution · low-light enhancement · image restoration

1 Introduction

SISR aims at increasing the spatial resolution and produce HR details given a LR input image. Due to the many practical applications of enhancing details in images, SR has been an active research field for decades. However, current SoTA SR methods are trained on well-illuminated images and they are therefore not

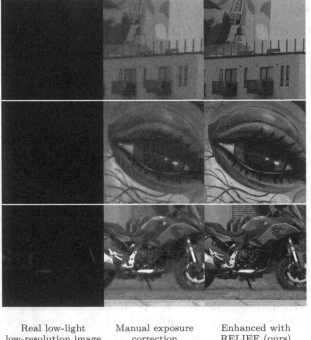

Real low-light   Manual exposure   Enhanced with
low-resolution image   correction   RELIEF (ours)

Fig. 1. Our proposed Resolution and Light Enhancement Transformer (RELIEF) can produce high-quality images from real low-light low-resolution inputs with severe noise and color distortions.

suitable for reconstruction of real LR images captured in poor lighting conditions, e.g., by surveillance or remote sensing cameras. The conventional strategy is therefore to correct the exposure level with dedicated LLE algorithms before super-resolving the image. However, this sequential processing scheme leads to poor reconstruction accuracy mainly due to error accumulation. On the other hand, it has been shown that joint processing e.g. joint SR and denoising [64], SR and demosaicing [25], and SR and deblurring [29] leads to superior performance, compared to sequential processing. This motivates us to jointly handle the LLE and SR reconstruction problem.

Current SoTA SR methods are based on Convolutional Neural Networks (CNNs) which are typically trained on LR patches with a dimension of 64×64 pixels and their corresponding HR patch, typically of $\times 2$, $\times 3$, or $\times 4$ times larger scale. As reconstruction of HR details are mostly a local problem, i.e. distant neighbor pixels provides little information regarding the reconstruction of the local pixel, SR models do not benefit much from using larger training patches [39,48]. However, for the problem of LLE, the use of more global contextual information can provide valuable cues about the light enhancement level of specific pixels, see Fig. 2. Yet this has not been explored in the literature, which can possibly be explained by the ineffective long-range dependency modeling capabilities of CNNs, which limits their ability to benefit from more global contextual information.

In this paper, we propose to use Transformers to effectively utilize additional global contextual information for reconstruction of LLLR images, as Transformers have recently shown impressive performance on both high- and low-level vision tasks due to their high capability in modeling long-range dependencies. A key component in Transformers is the self-attention mechanism, but due to high computational cost, and memory requirements, it is not feasible to apply full self-attention on larger images. Attempts to mitigate this problem have been made by either limiting the attention to fine-grained local self-attention [31,58] or coarse-grained global self-attention [46,51]. However, most approaches hinder the modeling capability of the original self-attention mechanism. Another challenge with vanilla Transformers is the lack of locality mechanisms which is essential for vision tasks. To this end, we propose a novel efficient Transformer block, Enhanced Cross-Shaped Window (ECSWin) that utilizes cross-shaped attention windows [14] together with locality enhancement in the positional encoding and feed-forward network to effectively capture long-range pixel dependencies while also leveraging local context. Our ECSWin Transformer block is used in a novel multi-scale hierarchical network, RELIEF, to perform reconstruction of real LLLR images. RELIEF can benefit from large training patch sizes due to its efficient local and global self-attention mechanism, which is applied in multiple encoder-decoders with skip-connections at different scales to aid the reconstruction process. As a result, our RELIEF is capable of achieving better visual quality and more accurate reconstructions that can help reveal information previously hidden in the LLLR images (See Fig. 1). We conduct experiments on the REL-LISUR [1] and SICE [5] datasets and our empirical results show that RELIEF brings significant performance improvements over existing methods.

The contributions of our work are twofold:

- We propose a novel Transformer-based multi-scale hierarchical encoder-decoder network with an efficient cross-shaped attention mechanism for accurate reconstruction of real low-light low-resolution images. To our knowledge, RELIEF is the first method for joint LLE and SR of real LLLR images.
- We demonstrate that increased use of global information, obtained by efficient global self-attention and large training patches results in significant performance improvements on two benchmark datasets.

2   Background

2.1   Low-Light Image Enhancement

Low-light image enhancement has been an active research topic in the past several years resulting in a large number of methods for enhancing the light level of images. Early attempts at LLE relied on histogram equalization [11,42],

Fig. 2. Example of different training patch sizes on an image from [1] (Blue: 64 × 64 pixels, red: 256 × 256 pixels). With its long-range modelling capabilities, RELIEF are able to utilize the information available in the larger patch, which leads to more accurate reconstructions. (Color figure online)

illumination map estimation [17], and Retinex theory [15,45] to correct the image illumination. However, as these methods fail to consider the inherent noise in the Low-Light (LL) images, the reconstruction results are often unsatisfactory. Recently, deep-learning has been utilized to learn an end-to-end mapping between LL and Normal-Light (NL) images [32]. The Retinex theory was further explored in combination with deep learning in [50,60], where a CNNs were used to learn decomposition and illumination enhancement, and most recently, a self-reinforced Retinex projection model was proposed in [35]. Furthermore, Generative Adversarial Networks (GANs) [8,20] have also been applied to the LL image enhancement problem. Nonetheless, LLE methods do not increase the spatial resolution of the images, but mainly aim at correcting the brightness level. As such, these methods only recover limited additional details in the image.

2.2   Image Super-Resolution

Like LLE, image super-resolution is one of the fundamental low-level computer vision problems [36]. From the first CNN based SR network [13], researchers have improved the reconstruction performance of the SR models by extending the network depth [22], utilizing residual learning [26,30], applying dense connections [48,62], and attention mechanisms [61]. Research has also been focusing on improving the perceptual quality, and not only the reconstruction accuracy, by the use of feature losses [21,59] and GANs [26,34,48]. However, most approaches assume that the LR images are created by an ideal bicubic downsampling kernel, which is an oversimplification of the real-world situation [2]. Furthermore, real-world images are often degraded by additional factors besides just downsampling, e.g. blur, low-contrast, color-distortion, noise, and low-light to name a few. To remedy this, a research direction focused on SR methods that can handle more diverse degradations has emerged. These methods often improve upon classical SR methods by extending the degradation model to include more diverse degradations e.g. Gaussian noise, blur, and compression artifacts in the

LR training images [33,47,57]. Yet, only very few works in the literature consider LR images degraded by low-light. Some of the most closely related works to our goal of SR of real natural LLLR RGB images are [16,18,54], which address the problem within different image-specific domains. In [16], a GAN-based method for reconstruction of synthetic LLLR face images is presented. In [18] a dedicated method for SR of LL Near-Infrared (NIR) images is presented, while a method for SR of LL images captured by intensified charge-coupled devices is presented in [54]. Guo et al. [16] experiments on synthetic LLLR face images created by gamma correction and downsampling which is another oversimplification of the complex degradation in real LLLR images. Furthermore, the method is only applicable to face images with a fixed size of 32×32 pixels. Likewise, since the method proposed by Han et al. [18] relies on paired NIR and visible images to enhance the NIR images, the method does not apply to SR or RGB images. The latter also applies to the method proposed by Ying et al. [54] since it only applies to image sensors with a proximity-focused image intensifier and requires a photon image. Therefore, as discussed above, no existing SR model has been developed for reconstructing real LLLR RGB images. Hence, with RELIEF we provide the first method to enhance the visibility, quality, and details of such images.

2.3   Vision Transformer

The Transformer was initially developed for natural language processing [44], but recently Transformers has also achieved great success in high-level vision tasks such as object detection [7,31,66], human pose estimation [6,27] and semantic segmentation [31,52,63]. Different from CNNs, most vision Transformers decompose an image into a sequence of patches and learn long-range dependencies between each patch. Due to their promising performance, Transformers have also been studied for different low-level vision problems [23,28,37]. However, their potential in joint LLE and SR has not been explored in the literature. As such, we design a novel Transformer based network that proves to be highly effective for the task of reconstructing real LLLR images by joint LLE and SR.

3   Method

In this section, we describe the proposed RELIEF for joint LLE and SR starting with an overview of the overall pipeline, followed by descriptions of the individual components. Figure 3 shows the architecture of RELIEF which is designed as a U-shaped [38] multi-scale hierarchical Transformer network.

3.1   Overall Pipeline

Given an LLLR image $I_{LLLR} \in \mathbb{R}^{H \times W \times 3}$, where W and H are the width and height, respectively, our goal is to restore its Normal-Light High-Resolution (NLHR) version I_{NLHR}. To accomplish this, RELIEF first extracts low-level

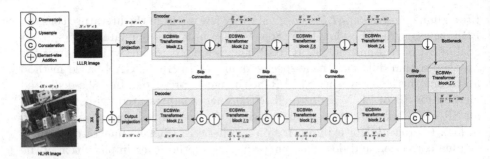

Fig. 3. The architecture of our RELIEF for joint LLE and SR. RELIEF consists of multiple ECSWin Transformer blocks organized in a U-shaped multi-scale hierarchical network with skip-connections.

features $F_0 \in \mathbb{R}^{H \times W \times C}$, where C is the number of channels, from I_{LLLR}. F_0 is obtained by a 3×3 convolutional layer with LeakyReLU. Next, deep features F_d are extracted from the low-level features F_0 in K symmetrical encoder-decoder levels. Each level contains multiple ECSWin Transformer blocks with large attention areas to capture long-range dependencies. After each encoder level, the features are reshaped to 2D feature maps and downsampled, while the number of channels is increased. We perform this operation using a 4×4 convolutional operation with stride 2. We use $K = 4$ encoder levels and as such the latent feature output at the last encoder stage is $F_l \in \mathbb{R}^{\frac{H}{8} \times \frac{W}{8} \times 8C}$ given an input feature map $F_0 \in \mathbb{R}^{H \times W \times C}$. Next, to capture even longer dependencies, we incorporate a bottleneck stage between the encoder and decoder at the lowest level. The output from the bottleneck stage is processed by a 2×2 transposed convolution operation with stride 2 to upsample the size of the latent features and reduce the channel number before entering the first decoder level. To improve the reconstruction process, skip connections are used to concatenate encoder and decoder features resulting in feature maps with twice the amount of channels. After each decoder Transformer block, the features are upsampled with a transposed convolution operation similar to the one used after the bottleneck stage. Then, at the last decoder level, the deep features F_d is reshaped using a 3×3 convolutional layer to obtain a residual image $I_R \in \mathbb{R}^{H \times W \times 3}$. Finally, the reconstructed HR and light-enhanced image is obtained as $\hat{I}_{NLHR} = (I_{LLLR} + I_R) \uparrow_s$, where s is the scaling factor of the upsampling operation. The latter is performed with pixel-shuffle [41] and 3×3 convolutional operations. We optimize RELIEF with L_1 pixel loss.

3.2   ECSWin Self-attention Transformer Block

The computational complexity of the original full self-attention mechanism grows quadratically with the input size and is therefore not feasible to use in combination with large training image patches. Several works have tried to reduce the computational complexity by shifted [31], halo [43], and focal [53] windows to perform self-attention. However, for most methods, the effective receptive field grows slowly, which hinders the long-range modeling capability. To reduce the computational burden, while maintaining strong long-range modeling capability, we use a Cross-Shaped Window (CSWin) attention mechanism [14]. With CSWin, self-attention is calculated in horizontal and vertical stripes by splitting the multi-heads into parallel groups to achieve efficient global self-attention. We gradually increase the widths of the stripes throughout the depth of the network to further enlarge the attention area and limit the computational cost. To improve the use of local contextual information we combine the CSWin self-attention mechanism, with Locally-enhanced Feed-Forward (LeFF) and Locally-Enhanced Positional Encoding (LePE) and form our ECSWin Transformer block. The different components will be described in detail in the following sections.

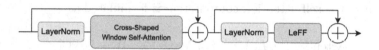

Fig. 4. Illustration of our ECSWin Self-Attention Transformer block.

As illustrated in Fig. 4, each ECSWin Transformer block is composed of layer normalization (LN) layers [3], a CSWin self-attention module, residual connections and the LeFF layer. More formally, the ECSWin Transformer block can be defined as:

$$\hat{X}^l = \text{CSWin-Attention}\left(\text{LN}\left(X^{l-1}\right)\right) + X^{l-1},$$
$$X^l = \text{LeFF}\left(\text{LN}\left(\hat{X}^l\right)\right) + \hat{X}^l, \tag{1}$$

where LN represents the layer normalization [3], and $\hat{X}^l$ and X^l are the outputs of the CSWin and LeFF modules, respectively. We design our RELIEF architecture to contain multiple CSWin Transformer blocks at each encoder-decoder level. Next, we describe the locally-enhanced feed-forward network and positional encoding in ECSWin.

3.3   Locally-Enhanced Feed-Forward Network

To better utilize local context, which is essential in image restoration, we exchange the Multi-Layer Perceptron (MLP) based feed-forward network used in the vanilla Transformer block with a LeFF layer [55]. In the LeFF layer, the feature dimension of the tokens is increased with a linear projection layer and hereafter reshaped to 2D feature maps. Next, a 3 × 3 depth-wise convolutional operation is applied to the reshaped feature maps. Lastly, the feature maps are flattened to tokens, and the channels are reduced with a linear layer such that the dimension of the enhanced tokens matches the dimension of the input. A Gaussian Error Linear Unit (GELU) [19] activation function is used after each linear and convolutional layer.

3.4   Locally-Enhanced Positional Encoding

As the self-attention mechanism inherently ignores positional information in the 2D image space, we use positional encoding to add such information back. Different from the typical encoding mechanisms Absolute Positional Encoding (APE) [44], Relative Positional Encoding (RPE) [40], and Conditional Positional Encoding (CPE) [10] that adds positional information into the input tokens before the Transformer Blocks, we use LePE [14], implemented with a depth-wise convolution operator [9], to incorporate positional information within each Transformer block. Hence, the self-attention computation is formulated as:

$$\text{Attention}(Q, K, V) = \text{SoftMax}(\frac{QK^T}{\sqrt{d_k}})V + \text{DWC}(V) \tag{2}$$

where d_k is the dimension of the queries and keys and DWC is the depth-wise convolution operator. As seen in Fig. 5, LePE operates in parallel directly on V from the query (Q), key (K), and value (V) pairs obtained by a linear transformation of the input X.

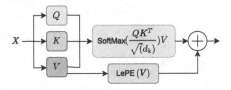

Fig. 5. Illustration of the LePE positional encoding mechanism.

4   Experiments and Analysis

4.1   Datasets

RELLISUR. The recent RELLISUR dataset [1], is the only publicly available dataset of real degraded LLLR images and their high-quality NLHR counterparts. The RELLISUR dataset contains 850 distinct sequences of LLLR images, with five different degrees of under-exposure in each sequence, paired with NLHR images of three different scale levels. In our work, we experiment with ×4 upscaling which is the most challenging scale factor in the dataset. We follow the predefined split from [1], and as such the number of train, val, and test images are 3610, 215, and 425, respectively.

SICE. SICE [5] is a dataset of 589 various scenes captured at different exposure levels, ranging from under to overexposed including a correctly exposed Ground-Truth (GT) image. We follow the train test split defined in [5], resulting in 58 test and 531 train images. We use the GT normal-light images as is, but use only the darkest exposure of each scene as the LL image during both training and testing. We synthetically create degraded LR versions of the LL images to obtain paired degraded LLLR and clean NLHR images. The LL images is degraded by convolving with an 11×11 Gaussian blur kernel with a standard deviation of 1.5 before downsampling with factor $\times 4$. Next, we model sensor noise by adding Gaussian noise with zero mean and a standard deviation of 8. Finally, we store the images in JPEG format with a quality setting of 70 to add compression artifacts. A total of 8 images, which resolutions are less than 256×256 pixels after the downsampling, are discarded from the training set. Evaluation is performed on 256×256 center crops.

4.2 Evaluation Metrics

We use two hand-crafted (PSNR, SSIM [49]) and one learning-based (DISTS [12]) Full-Reference Image Quality Assessment (FR-IQA) metrics for our quantitative comparisons. PSNR is a measure of the peak error between the reconstructed image and the GT, while SSIM is more focused on visible structure and texture differences. However, none of these metrics correlates well with the perceived image quality [4]. To this end, we use DISTS [12] which better captures the perceptual image quality as judged by human observers. Moreover, DISTS is also robust to mild geometric transformations. For all metrics, we report scores computed on the RGB channels.

4.3 Implementation Details

Our RELIEF model is trained from scratch for 5×10^5 iterations with a batch size of 16 using L_1 loss. We use the ADAM optimizer [24] with a learning rate of $2e-4$ which we decrease with a factor 0.5 at 2×10^5, 4×10^5 and 4.5×10^5. For data augmentation, we perform rotation and horizontal and vertical flips. We use 4 encoder-decoder levels in our RELIEF implementation, with two ECSWin Transformer blocks at each level, including the bottleneck. The number of attention heads and dimensions of the stripe widths in the encoder are set to [4,8,16,32] and [1,2,8,8], respectively, which are mirrored in the decoder. In the bottleneck, 32 heads and a stripe width of 8 are used. We use channel dimension $C = 48$ for the first encoder level in all experiments. As such, the resulting number of feature channels from level-1 to level-5 becomes [48,96,192,384,768].

4.4 Comparison with Existing Methods

To the best of our knowledge, no existing method in the literature can handle reconstruction of real LLLR RGB images. To this end, we compare our proposed

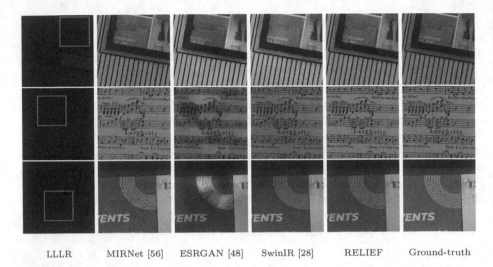

LLLR   MIRNet [56]   ESRGAN [48]   SwinIR [28]   RELIEF   Ground-truth

Fig. 6. Visual comparison for joint LLE and ×4 SR on the RELLISUR [1] dataset. Compared to the other approaches, our RELIEF produces more visually faithful results with less artifacts.

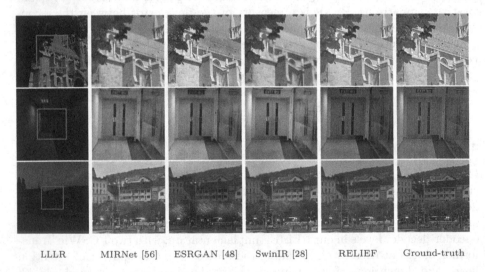

LLLR   MIRNet [56]   ESRGAN [48]   SwinIR [28]   RELIEF   Ground-truth

Fig. 7. Visual comparison for joint LLE and ×4 SR on the SICE [5] dataset. Our RELIEF is better at restoring the correct colors and removing undesirable artifacts.

method against dedicated methods for LLE, SR, and general image restoration. MIRNet [56] and ESRGAN [48] are SoTA methods for LLE and SR, respectively. To enable upsampling together with LLE we append a Pixel-shuffle [41] layer to MIRNet. As the VGG-discriminator in ESRGAN [48] is not compatible with large training patches, we use the patch discriminator from [65] instead. SwinIR [28] is a SoTA Transformer based method for general image restoration e.g. SR,

JPEG compression artifact reduction, and denoising. We use the real-world SR configuration[1] and Pixel-shuffle upsampling for SwinIR. We re-train all competing methods using the same training hyper-parameters used for our RELIEF for a fair comparison. We use a LR training patch size of 256 × 256 pixels for all methods, although the performance of RELIEF can be further improved by using an even larger training patch size as shown in Sect. 4.6. MIRNet and SwinIR are optimized with L1 loss, while ESRGAN is optimized with a combination of L1, perceptual and adversarial loss as proposed by the authors. We emphasize that none of the above-mentioned exiting methods are designed for joint LLE and SR, but once trained on such data they can still serve as baselines against our proposed method.

Table 1. Overview of different models and the number of parameters $\times 10^6$ and GMACs.

Model	Parameters	GMACs
MIRNet [56] w. Pixel-shuffle [41]	31.8	51.0
ESRGAN [48]	23.2	100.0
SwinIR [28]	11.6	47.2
RELIEF	46.3	5.7

As seen in Table 1, our RELIEF has the highest number of parameters, but a significantly lower computational burden than any of the compared methods, e.g. 5.7 vs. 47.2 GMACs for SwinIR [28]. However, as proved by empirical evidence in Sect. 4.6, we can obtain comparable performance with a RELIEF variant with less than half the parameters. The time it takes to process an input image of 256 × 256 pixels with RELIEF using a RTX 3090 GPU is ≈ 59 ms.

4.5   Results

Quantitative Results. As seen in Table 2, sequential processing with MIRNet followed by ESRGAN, performs worse than the jointly trained MIRNet with PixelShuffle upsampling. The best performance is obtained by RELIEF which obtains gains in PSNR of 0.28 and 0.78 dB on the RELLISUR and SICE datasets, respectively. Similarly, our RELIEF also achieves the best perceptual quality, according to the DISTS [12] metric, even though our method is not optimized with perceptual losses like ESRGAN.

---

[1] https://github.com/cszn/KAIR/blob/master/options/swinir/train_swinir_sr_realworld_x4_psnr.json.

Table 2. Quantitative comparison of state-of-the-art methods for joint LLE and ×4 SR on the RELLISUR and SICE datasets. Our RELIEF sets state-of-the-art results on both datasets.

Method	RELLISUR [1]			SICE [5]		
	PSNR ↑	SSIM ↑	DISTS [12] ↓	PSNR ↑	SSIM ↑	DISTS [12] ↓
MIRNet + ESRGAN [48,56]	19.81	0.7100	0.2017	–	–	–
MIRNet [56] w. Pixel-shuffle [41]	21.04	0.7619	0.1609	18.02	0.6760	0.2749
ESRGAN [48]	17.49	0.6724	0.1518	16.44	0.6271	0.2611
SwinIR [28]	18.99	0.7478	0.1705	17.66	0.6867	0.2753
RELIEF	**21.32**	**0.7686**	**0.1364**	**18.80**	**0.6980**	**0.2606**

Qualitative Results. We show visual comparisons of different methods on both the RELLISUR and SICE datasets in Fig. 6 and Fig. 7. As seen, our RELIEF also shows its clear advantages against the other methods, by producing the most visually pleasing reconstructions with the lowest amount of artifacts. In the RELLISUR dataset, there are severe noise and color distortions hiding in the extremely low-light low-resolution images, which methods like MIRNet and ESRGAN struggle to remove. In comparison, SwinIR produces fewer artifacts, but our RELIEF reconstructs images with the most accurate colors and the least artifacts while preserving most of the structural content. This is especially noticeable in Fig. 6 second and third row, where our method is the only one that manages to reconstruct a uniform and clean background as intended, without compromising edges and fine details. The same trend can be observed with the visual results from the SICE dataset, where images produced by MIRNet and ESRGAN contain severe visual defects, while our method is more faithful to the ground-truth. The main difference between SwinIR and our method is that the reconstructions produced by our method appear much sharper and with less color distortions.

4.6   Ablation Studies

In this section, we investigate the effectiveness and necessity of the components in RELIEF. All evaluations are conducted on RELLISUR [1] using a LR training patch size of 64 × 64 and a channel dimension $C = 48$, unless otherwise stated.

Training Patch Size. We study how the size of the LR training patch affects the reconstruction performance. As seen in Table 3, larger patch sizes result in increased reconstruction accuracy. A significant improvement in reconstruction accuracy of 2.24 dB is obtained by using a patch size of 384 × 384 pixels instead of 64 × 64 pixels. The improvement can also be confirmed visually from the results shown in Fig. 8, where it can be seen that a larger patch size contributes to more details in the reconstructions (Fig. 8, top row), while also ensuring that smooth regions appear more uniform and with fewer artifacts (Fig. 8, bottom row). Based on this we conclude that our RELIEF is effective in terms of leveraging more global contextual information for joint LLE and SR.

Impact of Skip Connections and Bottleneck Layer. Table 4 shows thee variants of our network: RELIEF, RELIEF without skip-connections, and RELIEF without the bottleneck layer. From the table, it can be seen that the skip connections and bottleneck layer are both important components as the PSNR drops by 0.64 and 0.58 dB by removal of these network components, respectively.

Table 3. Ablation on training patch sizes.

LR patch size	PSNR ↑	SSIM ↑	DISTS [12] ↓
64 × 64	19.78	0.7430	0.1917
128 × 128	20.25	0.7491	0.1716
256 × 256	21.32	0.7686	0.1364
384 × 384	**22.02**	**0.7790**	**0.1268**

Table 4. Ablation on different network designs.

Design	w/o skip-conn	w/o bottleneck	RELIEF
PSNR ↑	19.14	19.20	**19.78**

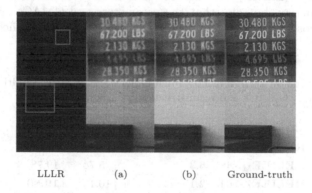

LLLR   (a)   (b)   Ground-truth

Fig. 8. Visual effect of training RELIEF with different LR patch sizes, (a) 64 × 64 pixels and (b) 256 × 256 pixels (samples from the RELLISUR [1] dataset).

Attention and Locality. We compare different multi-headed self-attention mechanisms, feed-forward networks, and positional-encoding mechanisms for the Transformer blocks in RELIEF to show the effect on the reconstruction performance. As seen in Table 5, the best performing configuration with cross-shaped window attention, and enhanced locality in the feed-forward network and positional-embedding yields 0.97 dB improvement over the configuration with shifted-window attention [31], MLP feed-forward network and relative-positional encoding [40] without locality enhancement. Compared to CSWin, our ECSWin block with locality enhanced feed-forward network results in 0.15 dB PSNR gain.

Table 5. Ablation on different multi-headed self-attention mechanisms, feed-forward networks, and positional-encoding mechanisms. † is the result of our ECSWin Transformer block.

MSA	FFN		PE		
	MLP	LeFF [55]	RPE [40]	LePE [14]	PSNR ↑
Swin	✓	–	✓	–	18.81
	✓	–	–	✓	19.49
CSWin	–	✓	–	–	19.63
	–	✓	–	✓	**19.78** †

Model Parameters. We experiment with different amounts of model parameters to find a trade-off between accuracy and complexity by varying the channel number C. As shown in Table 6, we design three variants of RELIEF: $RELIEF_S$, $RELIEF_M$, and $RELIEF_L$. We observe that the PSNR is correlated with the number parameters, but also that the parameters and GMACs grow quadratically. We choose a channel number of 48 to balance performance and model size.

Table 6. Comparison of different channel dimensions and the resulting number of model parameters, GMACs and reconstruction accuracy.

Model	C	Parameters $\times 10^6$	GMACs	PSNR ↑
RELIEF$_S$	32	20.6	2.59	19.68
RELIEF$_M$	48	46.3	5.74	19.78
RELIEF$_L$	64	82.1	10.13	**19.80**

5   Conclusion

In this paper, we introduced RELIEF, a novel U-shaped multi-scale hierarchical Transformer network for joint LLE and SR of real LLLR images. With its efficient ECSWin Transformer blocks, capable of capturing long-range dependencies and utilizing local context, RELIEF can benefit from large training patches which leads to better reconstruction performance. As such, RELIEF is capable of revealing details previously hidden in the dark while also removing undesired artifacts. Experimental results on two benchmark datasets show that RELIEF outperforms the state-of-the-art methods in terms of both reconstruction accuracy and visual quality. In the future, we plan to explore our RELIEF architecture for other image reconstruction tasks.

References

1. Aakerberg, A., Nasrollahi, K., Moeslund, T.B.: RELLISUR: a real low-light image super-resolution dataset. In: NeurIPS (2021)
2. Andreas Lugmayr et al.: Ntire 2020 challenge on real-world image super-resolution: methods and results. In: CVPRW (2020)
3. Ba, L.J., Kiros, J.R., Hinton, G.E.: Layer normalization (2016)
4. Blau, Y., Mechrez, R., Timofte, R., Michaeli, T., Zelnik-Manor, L.: The 2018 PIRM challenge on perceptual image super-resolution. In: Leal-Taixé, L., Roth, S. (eds.) ECCV 2018. LNCS, vol. 11133, pp. 334–355. Springer, Cham (2019). https://doi.org/10.1007/978-3-030-11021-5_21
5. Cai, J., Gu, S., Zhang, L.: Learning a deep single image contrast enhancer from multi-exposure images. TIP (2018)
6. Cai, Y., et al.: Learning delicate local representations for multi-person pose estimation. In: Vedaldi, A., Bischof, H., Brox, T., Frahm, J.-M. (eds.) ECCV 2020. LNCS, vol. 12348, pp. 455–472. Springer, Cham (2020). https://doi.org/10.1007/978-3-030-58580-8_27
7. Carion, N., Massa, F., Synnaeve, G., Usunier, N., Kirillov, A., Zagoruyko, S.: End-to-end object detection with transformers. In: Vedaldi, A., Bischof, H., Brox, T., Frahm, J.-M. (eds.) ECCV 2020. LNCS, vol. 12346, pp. 213–229. Springer, Cham (2020). https://doi.org/10.1007/978-3-030-58452-8_13
8. Chen, Y., Wang, Y., Kao, M., Chuang, Y.: Deep photo enhancer: unpaired learning for image enhancement from photographs with GANs. In: CVPR (2018)
9. Chollet, F.: Xception: Deep learning with depthwise separable convolutions. In: CVPR (2017)
10. Chu, X., et al.: Conditional positional encodings for vision transformers (2021)
11. Coltuc, D., Bolon, P., Chassery, J.: Exact histogram specification. TIP (2006)
12. Ding, K., Ma, K., Wang, S., Simoncelli, E.P.: Image quality assessment: unifying structure and texture similarity (2020)
13. Dong, C., Loy, C., He, K., Tang, X.: Image super-resolution using deep convolutional networks. TPAMI 38, 295–307 (2016)
14. Dong, X., et al.: CSWIN transformer: a general vision transformer backbone with cross-shaped windows (2021)
15. Fu, X., Liao, Y., Zeng, D., Huang, Y., Zhang, X.S., Ding, X.: A probabilistic method for image enhancement with simultaneous illumination and reflectance estimation. TIP 24, 4965–4977 (2015)
16. Guo, K., et al.: Deep illumination-enhanced face super-resolution network for low-light images. In: TOMM (2022)
17. Guo, X., Li, Y., Ling, H.: LIME: low-light image enhancement via illumination map estimation. TIP 26, 982–993 (2017)
18. Han, T.Y., Kim, Y.J., Song, B.C.: Convolutional neural network-based infrared image super resolution under low light environment. In: EUSIPCO (2017)
19. Hendrycks, D., Gimpel, K.: Gaussian error linear units (gelus) (2016)
20. Jiang, Y., et al.: Enlightengan: deep light enhancement without paired supervision. TIP 30, 2340–2349 (2021)
21. Johnson, J., Alahi, A., Fei-Fei, L.: Perceptual losses for real-time style transfer and super-resolution. In: Leibe, B., Matas, J., Sebe, N., Welling, M. (eds.) ECCV 2016. LNCS, vol. 9906, pp. 694–711. Springer, Cham (2016). https://doi.org/10.1007/978-3-319-46475-6_43

22. Kim, J., Lee, J.K., Lee, K.M.: Accurate image super-resolution using very deep convolutional networks. In: CVPR (2016)
23. Kim, T.H., Sajjadi, M.S.M., Hirsch, M., Schölkopf, B.: Spatio-temporal transformer network for video restoration. In: Ferrari, V., Hebert, M., Sminchisescu, C., Weiss, Y. (eds.) ECCV 2018. LNCS, vol. 11207, pp. 111–127. Springer, Cham (2018). https://doi.org/10.1007/978-3-030-01219-9_7
24. Kingma, D.P., Ba, J.: Adam: a method for stochastic optimization (2014)
25. Klatzer, T., Hammernik, K., Knöbelreiter, P., Pock, T.: Learning joint demosaicing and denoising based on sequential energy minimization. In: ICCP (2016)
26. Ledig, C., et al.: Photo-realistic single image super-resolution using a generative adversarial network. In: CVPR (2017)
27. Li, K., Wang, S., Zhang, X., Xu, Y., Xu, W., Tu, Z.: Pose recognition with cascade transformers. In: CVPR (2021)
28. Liang, J., Cao, J., Sun, G., Zhang, K., Van Gool, L., Timofte, R.: Swinir: image restoration using swin transformer. In: ICCVW (2021)
29. Liang, Z., Zhang, D., Shao, J.: Jointly solving deblurring and super-resolution problems with dual supervised network. In: ICME (2019)
30. Lim, B., Son, S., Kim, H., Nah, S., Lee, K.M.: Enhanced deep residual networks for single image super-resolution. In: CVPRW (2017)
31. Liu, Z., et al.: Swin transformer: hierarchical vision transformer using shifted windows (2021)
32. Lore, K.G., Akintayo, A., Sarkar, S.: Llnet: a deep autoencoder approach to natural low-light image enhancement (2017)
33. Luo, Z., Huang, Y., Li, S., Wang, L., Tan, T.: Learning the degradation distribution for blind image super-resolution. In: CVPR (2022)
34. Ma, C., Yan, B., Tan, W., Jiang, X.: Perception-oriented stereo image super-resolution. In: ACM MM (2021)
35. Ma, L., Liu, R., Wang, Y., Fan, X., Luo, Z.: Low-light image enhancement via self-reinforced retinex projection model. IEEE Trans. Multimedia (2022)
36. Nasrollahi, K., Moeslund, T.B.: Super-resolution: A comprehensive survey. In: Mach. Vision Appl. (2014)
37. Qin, Q., Yan, J., Wang, Q., Wang, X., Li, M., Wang, Y.: Etdnet: An efficient transformer deraining model. In: IEEE Access (2021)
38. Ronneberger, O., Fischer, P., Brox, T.: U-net: convolutional networks for biomedical image segmentation. In: Navab, N., Hornegger, J., Wells, W.M., Frangi, A.F. (eds.) MICCAI 2015. LNCS, vol. 9351, pp. 234–241. Springer, Cham (2015). https://doi.org/10.1007/978-3-319-24574-4_28
39. Sajjadi, M.S.M., Schölkopf, B., Hirsch, M.: EnhanceNet: single image super-resolution through automated texture synthesis. In: ICCV (2017)
40. Shaw, P., Uszkoreit, J., Vaswani, A.: Self-attention with relative position representations. In: NAACL-HLT (2018)
41. Shi, W., et al.: Real-time single image and video super-resolution using an efficient sub-pixel convolutional neural network. In: CVPR (2016)
42. Stark, J.A.: Adaptive image contrast enhancement using generalizations of histogram equalization. TIP 9, 889–896 (2000)
43. Vaswani, A., Ramachandran, P., Srinivas, A., Parmar, N., Hechtman, B.A., Shlens, J.: Scaling local self-attention for parameter efficient visual backbones. In: CVPR (2021)
44. Vaswani, A., et al.: In: NeurIPS (2017)
45. Wang, S., Zheng, J., Hu, H., Li, B.: Naturalness preserved enhancement algorithm for non-uniform illumination images. TIP 22, 3538–3548 (2013)

46. Wang, W., et al.: Pyramid vision transformer: a versatile backbone for dense prediction without convolutions (2021)
47. Wang, X., Xie, L., Dong, C., Shan, Y.: Real-esrgan: Training real-world blind super-resolution with pure synthetic data (2021)
48. Wang, X., Yu, K., Wu, S., Gu, J., Liu, Y., Dong, C., Qiao, Y., Loy, C.C.: ESRGAN: enhanced super-resolution generative adversarial networks. In: ECCVW (2019)
49. Wang, Z., Bovik, A.C., Sheikh, H.R.: Image quality assessment: from error visibility to structural similarity. TIP **13**, 600–612 (2004)
50. Wei, C., Wang, W., Yang, W., Liu, J.: Deep retinex decomposition for low-light enhancement. In: BMVC (2018)
51. Wu, H., et al.: CVT: introducing convolutions to vision transformers (2021)
52. Xie, E., Wang, W., Yu, Z., Anandkumar, A., Alvarez, J.M., Luo, P.: Segformer: simple and efficient design for semantic segmentation with transformers (2021)
53. Yang, J., et al.: Focal self-attention for local-global interactions in vision transformers (2021)
54. Ying, C., Zhao, P., Li, Y.: Low-light-level image super-resolution reconstruction based on iterative projection photon localization algorithm. J. Electron. Imaging **27**, 013026 (2018)
55. Yuan, K., Guo, S., Liu, Z., Zhou, A., Yu, F., Wu, W.: Incorporating convolution designs into visual transformers (2021)
56. Zamir, S.W., et al.: Learning enriched features for real image restoration and enhancement. In: Vedaldi, A., Bischof, H., Brox, T., Frahm, J.-M. (eds.) ECCV 2020. LNCS, vol. 12370, pp. 492–511. Springer, Cham (2020). https://doi.org/10.1007/978-3-030-58595-2_30
57. Zhang, K., Liang, J., Gool, L.V., Timofte, R.: Designing a practical degradation model for deep blind image super-resolution (2021)
58. Zhang, P., et al.: Multi-scale vision longformer: a new vision transformer for high-resolution image encoding (2021)
59. Zhang, R., Isola, P., Efros, A.A., Shechtman, E., Wang, O.: The unreasonable effectiveness of deep features as a perceptual metric. In: CVPR (2018)
60. Zhang, Y., Zhang, J., Guo, X.: Kindling the darkness: a practical low-light image enhancer. In: ACM MM (2019)
61. Zhang, Y., Li, K., Li, K., Wang, L., Zhong, B., Fu, Y.: Image super-resolution using very deep residual channel attention networks. In: Ferrari, V., Hebert, M., Sminchisescu, C., Weiss, Y. (eds.) ECCV 2018. LNCS, vol. 11211, pp. 294–310. Springer, Cham (2018). https://doi.org/10.1007/978-3-030-01234-2_18
62. Zhang, Y., Tian, Y., Kong, Y., Zhong, B., Fu, Y.: Residual dense network for image super-resolution. In: CVPR (2018)
63. Zheng, S., et al.: Rethinking semantic segmentation from a sequence-to-sequence perspective with transformers. In: CVPR (2021)
64. Zhou, R., El Helou, M., Sage, D., Laroche, T., Seitz, A., Süsstrunk, S.: W2S: microscopy data with joint denoising and super-resolution for widefield to SIM mapping. In: Bartoli, A., Fusiello, A. (eds.) ECCV 2020. LNCS, vol. 12535, pp. 474–491. Springer, Cham (2020). https://doi.org/10.1007/978-3-030-66415-2_31
65. Zhu, J., Park, T., Isola, P., Efros, A.A.: Unpaired image-to-image translation using cycle-consistent adversarial networks. In: ICCV (2017)
66. Zhu, X., Su, W., Lu, L., Li, B., Wang, X., Dai, J.: Deformable DETR: deformable transformers for end-to-end object detection. In: ICLR (2021)

To Quantify an Image Relevance Relative to a Target 3D Object

Marie Pelissier-Combescure[(✉)], Géraldine Morin, and Sylvie Chambon

University of Toulouse, IRIT, Toulouse INP, Toulouse, France
{marie.pelissier-combescure,geraldine.morin,
sylvie.chambon}@irit.fr

Abstract. Given a 3D object, our purpose is to find, among numerous 2D images within databases, which ones represent this object the best. Selected images should be both informative and offer a relevant view of the object, *i.e.* a pose that presents the essential characteristic information about the 3D object. To estimate the quality of the view, we propose to rely on repeatable, second order features, extracted with a curvilinear saliency detector, in order to both compute the pose within the image and to build a *relevance score*, independent of colors and textures. Based on this score, and given a set of images containing the same object, we are able to rank images from the one that best showcases the object to the worst one. Neural networks dedicated to detection and classification are able to recognise the object with a *confidence score*. So, we also develop an automatic approach based on a *confidence score* extracted from Convolutional Neural Networks. For evaluating and comparing the deterministic and the learning based methods, we use an objective image ranking based on gradual simulated degradations. We also provide visual qualitative results on a real dataset. The results demonstrate the efficiency of the approaches, the robustness of the deterministic method and help understanding the behavior of the methods based on *confidence score*.

Keywords: 2D/3D · saliency · image relevance ranking · learning based method · deterministic method

1 Introduction

Nowadays, massive visual data are available since they are so easily generated. These large volumes of data, such as images, depth maps, videos, 3D models, dynamic or not, temporal or not, contain more or less relevant and important information. In particular, databases offer thousands of images, such as `Imagenet` [4], or with tens of hours of videos, such as `ToCaDa` [17]. `Pascal3D+` [30] dataset contains a set of images, with multiple annotations, and also 3D models of the objects present in the images. These objects can appear in the image foreground or rather be part of the image background. Furthermore, objects may be truncated, occluded or not showcased in the image.

R. Gade et al. (Eds.): SCIA 2023, LNCS 13885, pp. 174–189, 2023.
https://doi.org/10.1007/978-3-031-31435-3_12

In many applications, from medicine to autonomous cars, seeking the pertinent images to envision an object of interest is crucial. For example, in augmented reality medical applications, the images can be registered in real-time with respect to the 3D model of a given organ in a tracking-by-detection paradigm [3]. Once the examination has been performed, the physician needs to extract relevant data to document the situation in order to establish the appropriate treatment protocol. For this, he or she might need to select the viewpoints that best present the structure of the organ and the pathology in question.

An automatic procedure to extract these relevant views is of significant help for the physician. Extracting most relevant views among a large dataset, like images or frames generated by a video is cumbersome and time consuming. This is a crucial need, however, and whereas related work exists, none directly and automatically quantifies the relevance of images of a target object independently of texture or appearance. Consequently, the problematic of this work is to order images from the most to the least relevant one, as illustrated in Fig. 1. Note that the last images least showcase the object because it is truncated or badly positioned. On the other hand, the images without environment or with a dominant object with very little occlusions are always placed at the top of the ranking.

Fig. 1. Automatic ranking of a set of real images according to their capacity to showcase an object of interest (here a sofa): from the image which best showcases the object (left) to the worst one (right).

Our proposition can be related to image quality assessment. Indeed, triage[1] [33] or image summarization[2] [23] find the most relevant images in a set of images. However, these approaches work on images having significant redundancies, while we have to deal with weakly correlated images representing a similar object in various environments, with by definition, low overlap. Assessing this image quality is

---

[1] Triage consists in selecting images with the best quality from a collection of correlated images, like frames from a video.

[2] Image summarization consists in extracting from an image collection a smaller subset of images that most attractively resumes this set.

also important for restoration applications [2]. This task can be done using subjective metrics, based on user studies [1], and objective metrics [18]. Other methods are based on low-level image features as score, such as color, lighting and texture [28,32]. High-level image features, such as aesthetics [16] are also used. All these approaches only rely on image, and thus the quality is not relative to a target object but globally assesses a score for the image. As our goal is to evaluate and compare image relevance for showcasing an object of interest, we refer to the 3D object for detection. In particular, we target and compare images independently of the object texture. Moreover, the object of interest may be viewed from different sides and we expect the ranking to assess the view point relevance.

Consequently, in this paper, our main contribution is to rank the most relevant views of a given object. Two possibilities are studied: based on *relevance score*, our proposition, or based on *confidence score*, computed by neural networks, to which we compare. After introducing the proposed pipeline in Sect. 2, the paper focuses on the method for estimating the image relevance. The following two sections present two different scores: Sect. 3 the deterministic *relevance score* and Sect. 4 the *confidence score* output by neural networks. Finally, the validation protocol is presented in Sect. 5 before showing and analysing the results and the comparison between them in Sect. 6.

2   Deterministic Method for Relevance Evaluation

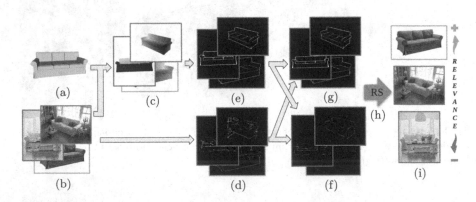

Fig. 2. Pipeline for ranking images according to a 3D object relevance: (a) the target 3D model and (b) the set of images containing the same object. Thanks to the pose estimation, for each image, we compute the corresponding depth map in (c). Then, we estimate the multi-scale curvilinear saliency, MCS, maps associated to each image in (d), as well as the curvilinear saliency, CS, maps associated to each depth map, in (e). Then, we take the intersection (f) and the union (g) between (d) and (e). We only keep salient points belonging to the object. The *relevance score RS* (h) of each image leads to the image ranking in (i). The computation of this score is detailed in the Sect. 3.

We are interested in quantifying the capacity of an image to showcase a 3D object, *i.e.* from a set of images containing an identified 3D object, quantify the capacity of each image to highlight its characteristic elements, as previously illustrated, in the introduction, in Fig. 1. Intuitively, an image is very relevant if the object is upfront, that is, not occluded or truncated, and it appears in the foreground, in a simple scene. We want the relevance to be computed for an object given its geometry, independently of its appearance, and considering the most informative viewpoint. To determine the quality of the viewpoint, we propose to use the salient information extracted in each type of data, namely the 2D images and the 3D model.

Figure 2 sums up our proposed pipeline for the deterministic approach and consists of the following steps:

1. Determine the common representation between 2D and 3D data, step (c). In the literature, a classical choice is to generate the depth maps related to the point of view. In consequence, for each image containing the target object, we compute the associated depth map with a classical rendering pipeline [8].
2. Extract salient points in each modality, see (e) and (d). This important step generate salient point maps and it will be described in the Sect. 2.1.
3. Compare salient point maps between 2D and 3D in order to evaluate the number of salient points that are present in the image. More precisely, we take into account the intersection and the union between the two maps, see (f) and (g). More details are given in the Sect. 2.2.
4. Compute the *relevance score*, see the blue arrow (h). Section 3 is dedicated to the details of this contribution.
5. Rank the set of images according to this score, see Fig. 2.(i).

2.1 Salient Points in 3D and in 2D

In 2D, many detectors rely on first order image derivatives, as the most classical among them [9]. Another category of detectors uses regions [24] where the interest of the point depends on the shape and the significance of the region it belongs to. A last family of detectors is based on second order derivatives and, more precisely, curvature [6]. More recently, multi-scale analysis has been introduced for each of these detector types and one of the most famous approaches is SIFT, *Scale Invariant Features Transform* [15]. In 3D, we find the same kinds of approaches, namely first order, second order and curvature based methods. Most of the techniques rely on the generalization of detectors from 2D to 3D, in particular, by extending SIFT detector to 3D [7].

In this work, we rely on a detector that can recover repeatable salient features, both in 2D and 3D. The main difficulty lies in the nature of the data manipulated. Indeed, a 2D image presents a texture, a background and a lighting that we do not find in the non-textured 3D model. Our detector [20] is based on the notion of curvature while diminishing the effect of texture (thanks to a multi-scale analysis and the consideration of the concept of focus maps). The results presented in [20] show that this detector provides the best repeatability of the extracted points on the different modalities in comparison with state-of-the-art detectors such as SIFT, and allows to recover the object pose. Thus, in this paper, we rely on this detector to characterize and compare the object in both modalities.

For each depth map, we compute the curvilinear saliency (CS) map, see Fig. 2.(c) and (e), estimating, for all points p of the smoothed associated depth map, by:

$$CS(p) = \lambda_1(p) - \lambda_2(p) \tag{1}$$

with $\lambda_2(p) \leq \lambda_1(p)$, the two principal curvatures of the depth map at point p. This expression is efficiently estimated using a simple expression of the Hessian matrix of the depth function, since we are in a functional setting. Then, we filter the noise that we consider linked to low values of CS.

Images, unlike the generated depth maps, contain different shapes and textures. Therefore, as advocated in [20], in order to filter texture information, we apply multi-scale curvilinear saliency, MCS. A multi-scale Gaussian pyramid of the images and their associated curvilinear saliency maps are computed. We keep the points of interest which appear in consecutive scales, see Fig. 2.(d).

After the estimation of these CS and MCS maps, we proposed a method to compare them in the next paragraph.

2.2   Intersection and Union Maps

Preliminary, as the pose estimation can be slightly inaccurate, there may exist a small offset between the two maps. In order to correct this error, we perform a global re-alignment by simply consider an exhaustive set of translations and we select the one that maximizes the number of salient points in common between MCS and CS maps.

The depth map provides characteristic information relative to the object geometry, whereas the image salient features may also occur because of texture or context. Thus, taking the intersection between the CS and MCS maps sorts out the salient points belonging to the object. Furthermore, computing the union helps to penalize coarse location of the salient points in the images, *i.e.*blurry effects on the detection. Unfortunately, when computing the intersection or the union, some image salient points can be due only to texture and not to the object geometry. In consequence, we try to take into account these erroneous salient points by considering the neighborhood. More precisely, we assume that two points are homologous if their neighborhoods are similar, *i.e.*contain the same distribution of salient points. Thus, we compare the two neighborhoods of the two corresponding points between the image and the depth map. Then, we further eliminate all the salient points that have less than 70% similarity with the neighborhood of their homologous points. There are different metrics in the literature to measure the similarity between two neighborhoods and we have chosen the Hausdorff distance which is the most efficient [19]. Finally, we select as many salient points as in the CS map, with 10% margin.

At the end of the filtering step, we suppose that the intersection and union maps contain only the salient points belonging to the studied object, that is, due to geometry and not texture.

3 Relevance Score

Photographers have a set of recommendations to obtain high quality photos and, in computer vision, these guidelines have been used, for example, to identify areas of attention in order to obtain saliency maps [12]. The deterministic *relevance score* depends on three concepts inspired, to some extent, by these rules.

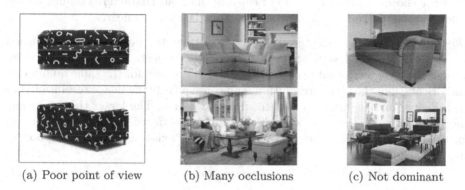

(a) Poor point of view   (b) Many occlusions   (c) Not dominant

Fig. 3. Example of three possible defaults. In this figure we illustrate three aspects we want to penalize with the proposed relevance score.

Characteristic Information, ψ – When taking photographs, it is often recommended to favor a view that shows as much details as possible on the object, while avoiding scene elements that can attract the eye and interfere with the attention given to the main subject. Consequently, this first term consists in determining if the environment and the object pose within a given image allows to extract as much object characteristic information as possible, see Fig. 3.(a). Moreover, the whole studied object should be visible, *i.e.* the object should neither be cropped nor occluded, see Fig. 3.(b). We quantify these aspects by measuring the number of visible salient points due to the object geometry in the image: the more salient points are visible, the more advantageous is the image. To calculate the term ψ, we take the ratio between the number of salient points present in the filtered intersection, cf. Fig. 2.(f), and the number of salient points belonging to the filtered union map, cf. Fig. 2.(g):

$$\psi = \frac{\#Filtered\ Intersection}{\#Filtered\ Union}, \tag{2}$$

where *Filtered Intersection* (respectively *Filtered Union*) is the set of salient points, belonging to the object, present in both (respectively at least one of) CS and MCS maps.

Dominance, β – In photography, another fairly obvious rule is to use a close framing, this means that the object is dominant in the image. Indeed, an object can show a lot of details but be insignificant in a very large image, see Fig. 3.(c).

So, the second criterion, the dominance parameter, β, considers the ratio between the diagonal length of the object axis-aligned bounding box, of size (w, h), and the diagonal length of the whole image, of size (W, H):

$$\beta = \frac{\sqrt{w^2 + h^2}}{\sqrt{W^2 + H^2}}. \tag{3}$$

We have chosen a linear measure of the size since our characteristic information follows contours, that is, linear patterns and thus varies linearly.

Object Resolution, γ – Naturally, in photography, the best possible resolution is expected and the last criterion focuses on this aspect. An object, seen in an image, can have a relevant pose, but be small. Moreover, for the same dominance and orientation, the object resolution, and consequently the image resolution, has an influence on the number of salient points detected. The larger the resolution, the more salient points are available. Similarly to β, we consider a linear size for the object, that is, γ the diagonal length of its axis-aligned bounding box, of size (h, w):

$$\gamma = \sqrt{w^2 + h^2}. \tag{4}$$

Values of γ are mapped to range between 0 and 1.

Relevance Score – The final *relevance score* corresponds to the product of this three parameters:

$$RS = \psi \times \beta \times \gamma. \tag{5}$$

4   Confidence Score

In this section, we propose a learning based alternative to the *relevance score*, the *confidence score*: both scores allow to rank images according to their capacity to showcase a given object. The *confidence score* is computed by a neural network.

Some networks are trained to assess the aesthetics of an image. This task can be considered as a binary classification problem. In that case, the prediction of the aesthetic quality is based on both global and local information [16]. Another possibility is to introduce an attention mechanism [29], or to consider a regression task using different losses [11]. Finally, a recent approach uses a support vector method to estimate the score distribution [31]. This problem is similar to ours since we also study the image quality. However, we are interested in the quality relative to the relevance of an object rather than in the global image quality. More classical networks solve classification problems, *i.e.*have as output a class label for each input image with a *confidence score*, such as the **Inception** [26] network. In 2015, the **ResNet** [10] network won the **Imagenet** challenge after introducing the notion of residual block. Many other networks followed, and today, one of the most famous is the **EfficientNetv2** [27] network. Some studies have shown that neural networks can mimic a human interpretation. For example, we can indicate networks that can be used for predicting the memorability[3] of objects in images [5]. Other works rely on the concept of typicality, *i.e.*the object characteristic that is most remarked by humans to demonstrate the link between the results obtained by a neural network and those obtained by a user experience [13]. However, it is necessary to balance the importance of imitating human perception since it can sometimes be influenced when it is corrupted by an ideal vision of the target object. For example, in [13], the representation automatically identified as the most typical of a banana is a yellow one with some brown spots whereas the human ideal representation is a perfectly yellow banana without any defaults. Moreover, in this paper, as the study of the relevance of an image containing a target object can be seen as a problem closed to classification or regression problem, we focus on this kind of networks. The goal of this kind of neural networks is to identify with limited ambiguity an object or an object class. Consequently, some of these networks provide the object class but also a recognition confidence level. It is therefore legitimate to consider this *confidence score* as an estimation of a measure close to the *relevance score* that we propose and thus, to compare these two scores.

We have chosen to consider three neural networks: a traditional object detection network **Faster** R-CNN [22] and the state of art **YOLOv5** [21], both pre-trained on the COCOdatabase [14]. In fact, the results of **Faster** were not competitive enough to be kept and for reason of space, we do not present the corresponding results. For the classification networks, we use the state of the art **EfficientNetv2** [27] pre-trained on the Imagenet database [4].

5 Validation Protocol

Image Databases with Reference Rankings – As far as we are concerned, no dataset is available for evaluating approaches presented in this paper and we have chosen to create our own reference image ranking. To create a reference ranking, we choose an initial image to which we apply three different kinds of degradations successively, see Fig. 4: augmenting the background size without changing the object resolution, see Fig. 4.(b), adding occlusions, see Fig. 4.(c), scaling of the image, see Fig. 4.(d). We iterate these three degradations as long as the image resolution remains acceptable: after three iterations of these degradations, image sides are reduced by 40%. More precisely, during one iteration, the image width (respectively, the image height) is augmenting by $aug\_w\%$ (respectively, $aug\_h\%$), then the targeted object is occluded on $add\_o\%$ of its visible surface, finally, the image resolution is scaling with a scale sc. The chosen values for these parameters are indicated in Sect. 6.

Once these degradations have been performed, we obtain a set of images ordered from the least degraded (initial image) to the most degraded. As we sequentially apply degradations, the order of the images is objectively determined. The goal is to recover the objective order of these images.

---

[3] Memorability is related to an importance score attributed to each segmented regions of the scene.

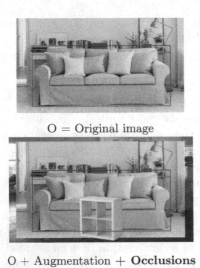

O = Original image

O + **Augmentation**

O + Augmentation + **Occlusions**

O + Augmentation + Occlusions +
Scaling

Fig. 4. Example of simulated degradations. We apply successively three kinds of degradations and it allows us to have a ground truth ranking from the top right to the bottom left image.

Significant Rankings – Given a reference image ranking, in order to evaluate the proposed rankings, see Fig. 5.(c) and (d), based on *relevance scores* or on *confidence scores* compared to this reference ordering, see Fig. 5.(e), we compute the Spearman Rank Order Correlation Coefficient, SROCC, like in [13], see Fig. 5.(f) and (g). This correlation uses the Pearson Linear Correlation Coefficient, PLCC, with the rank variables, which is adapted to compare rankings:

$$SROCC(X, Y) = \frac{cov(R(X), R(Y))}{\rho_{R(X)} \cdot \rho_{R(Y)}}, \tag{6}$$

with:

$R(X)$ the rank variable,

$cov(R(X), R(Y))$ the covariance of the rank variables and

$\rho_{R(X)}$ and $\rho_{R(Y)}$, the standard deviations of the rank variables.

Values of SROCC range from -1 to 1. The closer these coefficients are to 1, the more effective the method is.

Moreover, each correlation coefficient is provided with a value called p-value: the smaller this value, the more significant the correlation coefficient. In the literature, the p-value is compared to a threshold named α, usually equals to 5.10^{-2}, or 1.10^{-2}. Accordingly, we keep only proposed rankings with a significant correlation.

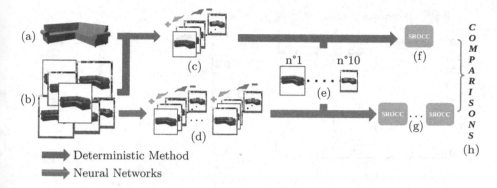

Deterministic Method

Neural Networks

Fig. 5. Evaluation of ranking methods. (a), object 3D mesh, (b) images represented this object, (c) and (d) rankings established respectively with the *relevance* and *confidence scores* (from the highest to the lowest score), (f) and (g) Spearman correlation coefficient computed between the reference ranking in (e), and each ranking obtained in (c) and (d), (h), correlation comparisons.

Finally, in order to compare the three selected methods, for a given reference ranking, if one of the tested approaches provides a ranking with an insignificant correlation coefficient, then that image set and its associated reference ranking will not be considered for comparison between the approaches. Consequently, the larger the number of competing methods in an experiment, the smaller the number of significant correlation coefficients.

6   Results on Reference Rankings

Parameter Settings – To compute the CS map, first, the associated depth map is smoothed with a Gaussian filter; noise is filtered out by removing all points of the CS map with a curvilinear saliency value lower than 1% of the maximum curvilinear saliency value. Second, to generate the MCS map based on a Gaussian pyramid, we set $\sigma = 1.4$ and $nb\_scale = 4$. We only keep points that appear in at least 3 consecutive scales. Third, in order to remove matching errors, a registration of the CS map with respect to the MCS map is estimated in order to maximize the number of salient points that lie within the binary intersection. More precisely, overlaps between the CS map and the MCS map translated with vector (i, j) are computed $\forall i, j \in [-5, 5]$. Finally, to generate the reference rankings, the parameters for each degradation are applied alternatively, in three iterations, using the following values: $aug\_w = \{5\%, 5\%, 10\%\}$, $aug\_h = \{5\%, 5\%, 10\%\}$, $add\_o = \{10\%, 10\%, 15\%\}$, $sc = \{0.9, 0.8, 0.7\}$.

Performance of Each Method – The statistical summaries of correlation coefficients are shown in Fig. 6.(a), by displaying box-and-whisker diagrams. As we mentioned, from each method, only significant proposed rankings are kept. Among our 163 generated reference rankings (of 10 images each), **YOLOv5**,

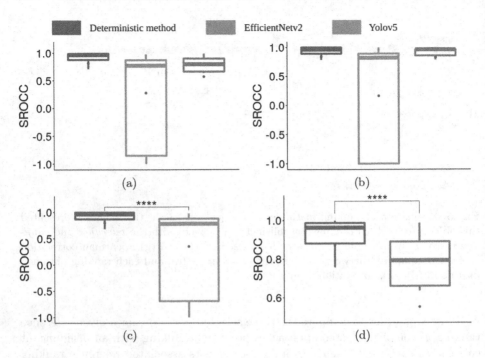

Fig. 6. Performance of each method by comparison with reference rankings (a); comparison with rankings of images containing occlusions (b); comparisons between the deterministic method and each neural networks: Yolov5 (c) and EfficientNetv2 (d). Experiments are made with a p-value threshold $= 5.10^{-2}$. Red dots represent the mean.

(respectively **EfficientNetv2**), provides 96 (respectively 103) significant proposed rankings, and the deterministic method, 136. According to Fig. 6.(a), 75% of correlation coefficients computed by the deterministic method are larger than 0.875, and the associated mean is 0.811. Moreover, **EfficientNetv2** obtains good results, that is, 75% of correlation coefficients are larger than 0.66, whereas the mean of correlation coefficients computed by **YOLOv5** is very low, around 0.274.

To extend our study, we lower the p-value threshold to 1.10^{-2}. Thus the experimental conditions are stricter. The deterministic method is not much affected by this new parameter; among its 136 significant proposed rankings, 89% of them are still significant. Whereas for **YOLOv5** (respectively **EfficientNetv2**), 19% (respectively 35%) of these significant proposed rankings are not significant. According to these results, the deterministic method based on *relevance scores* is more robust to rank from the most to the least relevant image. **EfficientNetv2** also provides satisfying results and provides a good learning based solution. However, **YOLOv5** does not produce such good results.

To better understand the behavior of **YOLOv5**, we focus on learning proposed rankings themselves, see Fig. 5.(d). Among the 10 images from each significant proposed ranking, on average, **YOLOv5** managed to detect the studied object and provide a *confidence score* on only 7 images whereas **EfficientNetv2** succeed in classifying about 9 images. Images that did not receive a *confidence score* are either too low resolution or contain too many occlusions. This explains the poor performance of **YOLOv5** on some of the ground truth rankings. Keeping only the rankings where confidence scores can be computed would clearly favor the learning based approach.

In addition, for *confidence scores*, two observations can be done: scores are not always decreasing and they are affected by the presence of occlusions. Indeed, these networks are trained to be robust to scaling, both of the object or/and the context, by performing data augmentation during the learning phase. Thus, the computed *confidence score* is robust to these two types of degradations (actual and relative size of the object) but is affected by occlusions. To confirm this behavior, we have generated another dataset of reference rankings containing only occlusions. There are 441 rankings with images with occlusions among which 323 (deterministic method), 186 **YOLOv5**, 302 **EfficientNetv2** are significant. On these rankings, the performance based on the *confidence score* of **EfficientNetv2** is greatly improved. Indeed, 75% of these significant proposed rankings based on **EfficientNetv2** have a correlation coefficient higher than 0.86, see Fig. 6.(b). The deterministic method has always the best statistical results, see Fig. 6.(b) whereas, **YOLOv5** is still affected. Almost all rankings find the perfect score up to one of two images swapping, as long as the detection or classification of the object is possible.

Comparison of the Deterministic and Learning Based Approaches – We now compare the performance of the different methods on common significant rankings, see Fig. 6.(c) and (d), where the asterisks represent the level of significant differences between the deterministic mean and each learning method mean. These information come from a Tukey's HSD (Honestly Significant Difference) test. More precisely, the deterministic method and **YOLOv5** (respectively **EfficientNetv2**) are assessed on 780 (respectively 900) images sorted 10 by 10.

Generally, the deterministic method, based on the *relevance score*, has a higher median and mean (white dot) than both learning methods based on the *confidence score*. Second, the Inter-Quartile Range, IQR, of the deterministic method is narrower than the others. In both comparisons, the deterministic method obtains the best median, around 0.96, and the best mean, around 0.84. Moreover, compared to the object detection network or classification network, 75% of correlation coefficients based on deterministic method are greater than around 0.87, see in Fig. 6.(c) and (d). However, **EfficientNetv2** also obtains correct results: 50% of correlation coefficients are higher than 0.79. The overall performance of **YOLOv5** is rather low. Consequently, they are not as good as the ones based on *relevance scores* but also less predictable. We have also seen that **YOLOv5** lacks reliability on providing a confidence score.

Additionally, we have also compared the three methods on the same set of significant rankings (results are not shown here to avoid redundancy – since the comparison of the two learning based methods are not significant): 75% of correlation coefficients based on deterministic method are greater from 0.91 whereas correlation coefficients based on neural network methods are more scattered. Finally, we have also reproduced these two significant comparisons by decreasing the threshold of the p-value to 1.10^{-2}. Under these conditions, the results of the method based on *relevance scores* are still the most satisfying.

7   Results on Real Images

We need images which contain an object to be studied, as well as the 3D model of this object, so we have chosen the Pix3D [25]. The previous section has shown the capacity of the proposed methods to retrieve the correct ordering of degraded images, but also the better performance of the method based on *relevance score* relative to the one based on the confidence score. Now, we can apply the deterministic approach to a set of raw images containing the same object to order

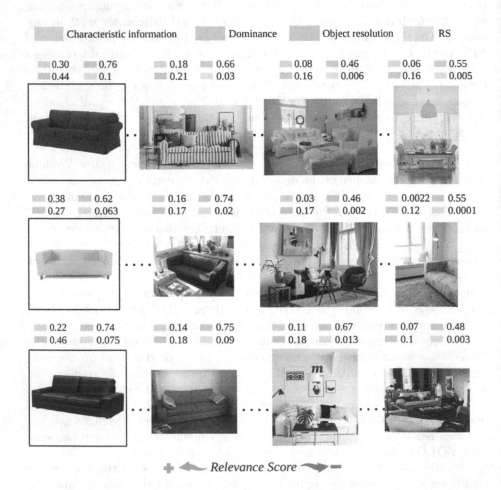

Fig. 7. Visual quality rankings with the relevance score. From left to right the best to the worst score. Note the robustness to the change in texture and appearance.

them relative to their capacity of best showcasing the studied object. We have tested our approach on 9 models of the category *Sofa*, that is, 1048 images. Ranking examples are shown in Fig. 1 and Fig. 7. Note that the last images least showcase the object because it is truncated or badly positioned. More precisely, the last images have the lowest *relevance score* due to occlusions, weak characteristic information and low dominance. On the other hand, the images without environment or with a dominant object with very little occlusions are always placed at the top of the ranking. These examples illustrate the robustness of the deterministic method.

8 Conclusion

We have proposed and validated different efficient methods to measure the subjective quality of an image by quantifying its ability to showcase a given object. We have shown that the deterministic approach, based on a deterministic *relevance score* performs better than the ones based on the *confidence score* of a network, and is also the most robust. The proposed choice of the most relevant image is thus automatic, and does not require training on a large dataset.

One perspective is to rely on networks capacity to learn human judgment, such as in [31]. Another perspective of this work is to consider salient primitives no longer in a depth map (2D projection of the 3D model from different viewpoints) but directly in the 3D model. In the longer term, we wish to generalize this work to temporal data.

References

1. BT, R.I.R.: Methodology for the subjective assessment of the quality of television pictures. International Telecommunication Union (2002)
2. Chambah, M., Rizzi, A., Saint Jean, C.: Image quality and automatic color equalization. In: SPIE Electronic Imaging (2007)
3. Collins, T., et al.: Augmented reality guided laparoscopic surgery of the uterus. IEEE Trans. Med. Imaging **40**(1), 371–380 (2020)
4. Deng, J., Dong, W., Socher, R., Li, L.J., Li, K., Fei-Fei, L.: Imagenet: a large-scale hierarchical image database. In: IEEE Conference on Computer Vision Pattern Recognition (2009)
5. Dubey, R., Peterson, J., Khosla, A., Yang, M., Ghanem, B.: What makes an object memorable? In: IEEE Conference on Computer Vision Pattern Recognition (2015)
6. Fischer, P., Brox, T.: Image descriptors based on curvature histograms. In: German Conference on Pattern Recognition (2014)
7. Flitton, G., Breckon, T., Megherbi Bouallagu, N.: Object recognition using 3D SIFT in complex CT. In: British Machine Vision Conference (2010)
8. Foley, J., Feiner, S., Hughes, J.: Computer Graphics: Principle and Practice. Addion Wesley, Boston (1994)
9. Harris, C., Stephens, M.: A combined corner and edge detector. In: Alvey Vision Conference (1988)
10. He, K., Zhang, X., Ren, S., Sun, J.: Deep residual learning for image recognition. In: IEEE Conference on Computer Vision Pattern Recognition (2016)

11. Kong, S., Shen, X., Lin, Z., Mech, R., Fowlkes, C.: Photo aesthetics ranking network with attributes and content adaptation. In: Leibe, B., Matas, J., Sebe, N., Welling, M. (eds.) ECCV 2016. LNCS, vol. 9905, pp. 662–679. Springer, Cham (2016). https://doi.org/10.1007/978-3-319-46448-0_40

12. Kozegar, E.: Rule of photography in image saliency detection. In: Conference on Knowledge-Based Engineering and Innovation (2016)

13. Lake, B., Zaremba, W., Fergus, R., Gureckis, T.: Deep Neural Networks Predict Category Typicality Ratings for Images. In: Cognitive Science (2015)

14. Lin, T., et al.: Microsoft COCO: common objects in context. CoRR (arXiv) abs/1405.0312 (2014)

15. Lowe, D.: Distinctive image features from scale-invariant keypoints. Int. J. Comput. Vision (2004)

16. Lu, X., Lin, Z., Shen, X., Mech, R., Wang, J.: Deep multi-patch aggregation network for image style, aesthetics, and quality estimation. In: International Conference on Computer Vision, pp. 990–998 (2015)

17. Malon, T., et al.: Toulouse campus surveillance dataset: scenarios, soundtracks, synchronized videos with overlapping and disjoint views. In: ACM Multimedia Systems Conference (2018)

18. Ouni, S., Chambah, M., Herbin, M., Zagrouba, E.: Are existing procedures enough? Image and video quality assessment: review of subjective and objective metrics. In: SPIE Electronic Imaging. Image Quality and System Performance (2008)

19. Pelissier Combescure, M., Morin, G., Chambon, S.: Extraction et comparaison d'information saillante: pose favorable et image 2d révélatrice d'un objet 3d. In: ORASIS (2021). (in French)

20. previous publication, O.: 2D/3D primitive extraction and matching for pose estimation method. IEEE Trans. Image Process. (2019)

21. Redmon, J., Divvala, S., Girshick, R., Farhadi, A.: You only look once: unified, real-time object detection. In: IEEE Conference on Computer Vision Pattern Recognition (2016). https://github.com/ultralytics/yolov5.git

22. Ren, S., He, K., Girshick, R., Sun, J.: Faster R-CNN: towards real-time object detection with region proposal networks. In: Advances in Neural Information Processing Systems, vol. 28 (2015)

23. Riahi Samani, Z., Ebrahimi Moghaddam, M.: Image collection summarization method based on semantic hierarchies. Artif. Intell. (2) (2020)

24. Rosten, E., Drummond, T.: Machine Learning for High-Speed Corner Detection

25. Sun, X., et al.: Pix3D: dataset and methods for single-image 3D shape modeling. In: IEEE Conference on Computer Vision Pattern Recognition (2018)

26. Szegedy, C., et al.: Going deeper with convolutions. In: IEEE Conference on Computer Vision Pattern Recognition (2015)

27. Tan, M., Le, Q.: Efficientnetv2: smaller models and faster training. arXiv 2021. CoRR (arXiv)

28. Tang, H., Joshi, N., Kapoor, A.: Learning a blind measure of perceptual image quality. In: IEEE Conference on Computer Vision Pattern Recognition (2011)

29. Wang, F., et al.: Residual attention network for image classification. In: IEEE Conference on Computer Vision Pattern Recognition, pp. 3156–3164 (2017)

30. Xiang, Y., Mottaghi, R., Savarese, S.: Beyond PASCAL: a benchmark for 3D object detection in the wild. In: IEEE Winter Conference on Applications of Computer Vision (2014)

31. Xu, M., Zhong, J., Ren, Y., Liu, S., Li, G.: Context-aware attention network for predicting image aesthetic subjectivity. In: ACM International Conference on Multimedia (2020)

32. Yuan, L., Sun, J.: Automatic exposure correction of consumer photographs. In: Fitzgibbon, A., Lazebnik, S., Perona, P., Sato, Y., Schmid, C. (eds.) ECCV 2012. LNCS, vol. 7575, pp. 771–785. Springer, Heidelberg (2012). https://doi.org/10. 1007/978-3-642-33765-9_55
33. Zhu, J., Agarwala, A., Efros, A., Shechtman, E., Wang, J.: Mirror mirror: crowd-sourcing better portraits. ACM Trans. Graph. **33** (2014)

Deep Active Learning for Glioblastoma Quantification

Subhashis Banerjee[(✉)][iD] and Robin Strand[iD]

Department of Information Technology, Uppsala University, Uppsala, Sweden
{subhashis.banerjee,robin.strand}@it.uu.se

Abstract. Generating pixel or voxel-wise annotations of radiological images to train deep learning-based segmentation models is a time consuming and expensive job involving precious time and effort of radiologists. Other challenges include obtaining diverse annotated training data that covers the entire spectrum of potential situations. In this paper, we propose an Active Learning (AL) based segmentation strategy involving a human annotator or "Oracle" to annotate interactively. The deep learning-based segmentation model learns in parallel by training in iterations with the annotated samples. A publicly available MRI dataset of brain tumors (Glioma) is used for the experimental studies. The efficiency of the proposed AL-based segmentation model is demonstrated in terms of annotation time requirement compared with the conventional Passive Learning (PL) based strategies. Experimentally it is also demonstrated that the proposed AL-based segmentation strategy achieves comparable or enhanced segmentation performance with much fewer annotations through quantitative and qualitative evaluations of the segmentation results.

Keywords: Active Learning · Convolutional Neural Networks · Segmentation · Brain tumor · Glioblastoma · Magnetic Resonance Imaging

1 Introduction

Cancer is becoming a major cause of death worldwide as people live longer lives, resulting in an aging population. Gliomas are the most common type of primary brain tumor that arises from glial cells in the Central Nervous System (CNS). Gliomas account for 40% of all CNS tumors and 80% of malignant brain tumors. In adults, high-grade gliomas are the most common type of primary brain tumors; among these, the majority are Grade IV gliomas, i.e., Glioblastoma [19]. Surgery is the most common treatment for removing the tumor from the brain. Radiation and chemotherapy are added to improve survival time [12]. Magnetic Resonance Imaging (MRI) [18] plays a major role in the detection, diagnosis, and management of brain cancers in a non-invasive manner. Fast, accurate, precise, and reproducible segmentation and volumetric quantification of Glioblastoma from pre and post-operative MR scans are crucial for volumetric visualization and quantification required for treatment planning and monitoring the therapy effect.

Manual segmentation of Gliomas from MRI is time-consuming and subjected to high inter-and intra-observer variation [4]. The enormous amount of information contained in a typical MRI brain volume scan, as well as difficulties such as partial volume effects, noise, artifacts, etc., puts very high demands on radiologists. Machine learning has recently advanced dramatically, and there is a growing interest in applying cutting-edge machine learning methodology in medicine for applications such as computer-aided diagnosis, with an overall goal of guiding physicians in their clinical work through artificial intelligence [5,15]. Deep learning is a family of machine learning methods that have the potential to find and use patterns in massive amounts of data to perform some of the radiologist's repetitive tasks with high accuracy. Deep learning has achieved accuracy comparable to human experts in some tasks, for example, segmenting a specific tissue/organ or pathologies [3].

Recently Convolutional Neural Networks (CNNs) have been shown to perform impressively on brain tumor segmentation from multimodal MR images and achieved accuracy comparable to human experts. Convolutional Neural Networks proposed and developed for brain tumor segmentation mostly follow an encoder-decoder-based architecture with 2D or 3D convolution operations and solve it by classifying each pixel (2D) or voxel (3D) into tumor or background [2,6,7]. Some of the popular CNN models that were proposed and achieved state-of-the-art performance in Glioma segmentation are "DeepMedic" [11], "Fully Convolutional Network (FCN)" [3], and "U-Net" [16]. Although CNN has achieved state-of-the-art performance, training in a deep CNN-based segmentation model requires a large amount of correctly annotated samples known as ground truth. Generating pixel/voxel-wise annotation for medical image segmentation is a time-consuming and costly task involving the precious time and effort of radiologists. Other difficulties lie in obtaining diverse annotated training data that covers the entire spectrum of potential situations. This problem is typically tackled in regular training by utilizing a large amount of training data, which is very hard to get in the case of medical images.

Active Learning (AL) [17] is proposed as a way to circumvent this problem. Active learning is a semi-supervised machine learning framework in which models learn by training in iterations. In each iteration of active learning, a human annotator or Oracle annotates a new sample from the unlabeled pool and submits it to the labeled pool. Then in parallel, the samples from the labeled pool are used for training in a supervised learning framework. In the beginning, the labeled pool is empty so the parallel training can be initiated after the Oracle annotate at least one or few samples. The continuous learning is carried out in parallel to the annotation process which eventually results in better learning and performance of the models over time. A number of recent studies have explored the use of active learning for medical image segmentation [8,9,20,21]. A weakly-supervised deep active learning framework called COVID-AL is proposed in [20] to diagnose

COVID-19 with CT scans. An AL-based framework for progressively integrating pixel-level annotations during training from the given training data with global image-level labels is proposed in [8] for annotating histology GlaS data of colon cancer. To reduce annotation efforts the authors in [21] integrate active learning and transfer learning (fine-tuning) into a single framework and evaluate the method on three distinct medical imaging applications.

Considering the advantages and applicability of active learning in medical image segmentation, in this paper we propose a deep active learning-based strategy for computerized segmentation of Glioblastoma from Flair MRI. However, the unique challenge posed by the scarcity of annotated training data is managed through the proposed novel zero-shot active learning framework. An interactive semi-automatic region growing-based algorithm is used to generate the tumor regions' initial volumetric segmentation from MR images. After Oracle creates the first volumetric segmentation, the training of an automatic segmentation model based on the deep CNN is initiated. Both the semi-automatic annotation by the Oracle and the training process continue in parallel. When the Oracle submits a new annotation in the labeled pool the training process is re-initiated with the previous training weights of the model. Our contributions are as follows.

- We have proposed a novel Active Learning (AL)-based Glioblastoma segmentation method from MRI.
- Unlike the state-of-the-art methods, our proposed technique involves zero-shot AL, achieved through continual training in parallel with semi-automatic annotation by the Oracle.
- Experimental studies have been conducted on a public MRI dataset to demonstrate the superiority of our proposed Al-based model compared to other conventional PL-based approaches.
- We experimentally demonstrated that the proposed AL-based annotation required around five times shorter time to annotate the same number of MRI volumes and achieved comparable or superior segmentation accuracy than the conventional passive learning-based methods.

The rest of the paper is organized as follows. Sect. 2 discusses the proposed deep active learning-based strategy for segmentation and quantification of Glioblastoma from MR images along with the details of the deep CNN architecture. Sect. 3 describes the experimental setup, the dataset used, and the quantitative and qualitative results. Sect. 4 concludes the paper.

2   Deep Active Learning for Glioma Segmentation

This section describes the proposed deep active learning model for Glioma segmentation along with the CNN architecture. Figure 1 shows different phases of deep active learning-based Glioma segmentation. It starts with the selection of

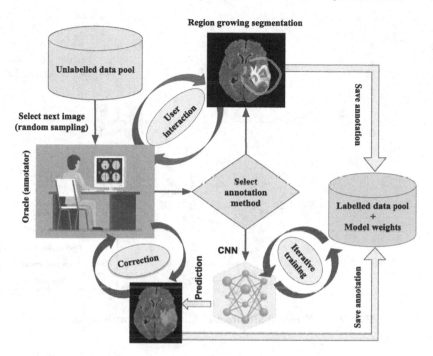

Fig. 1. Deep active learning model for Glioma segmentation.

an MRI image from the unlabelled data pool. Then the annotator or the Oracle selects the suitable method for segmentation. For a few initial annotations
the Oracle has to use the semi-automated region growing-based segmentation
tool [10]. As soon as the Oracle pushes the initial ground truth in the labeled
data pool the training of the CNN model is initiated in parallel and it is re-
initiated with savings of each new ground truth label. After completion of a few
annotations, the Oracle can use the trained model instead of using the region-
growing segmentation to get the initial segmentation which may be not proper
but the Oracle can correct the predictions if required manually and push it into
the labeled data pool. The prediction from the CNN gets better with time and
with more annotations in the label data pool. We experimentally saw that the
CNN model learns very quickly and for Glioma segmentation it approximately
requires 10 to 15 labeled data to train a good-performing model which gives
around 60–70% accuracy in segmenting a new sample from the Unlabelled data
pool.

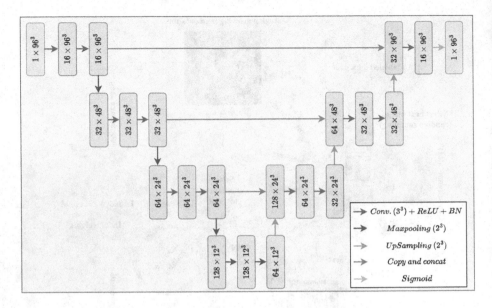

Fig. 2. CNN model architecture, used for segmentation.

The CNN used to learn Glioma segmentation from MRI follows the standard encoder decoder-based architecture like U-Net [16]. The network has four encoding and four decoding blocks each having 16, 32, 64, and 128 convolution filters. Each encoder block is made up of two consecutive 3D convolution layers with ReLU nonlinearity and batch normalization followed by a max-pooling layer to reduce the spatial dimension of the response map in half. The decoder blocks are made up of one upsampling layer to un-project the response map into twice the dimensions of the input. Residual and skip connections are used within encoder and decoder blocks, as well as from the encoder to the decoder to preserve small anatomical structures and ensure gradient flow. Convolution kernels of size $3 \times 3 \times 3$ are used throughout the network, and $2 \times 2 \times 2$ projection windows are used for the max-pooling and upsampling layers. The architecture of the CNN is illustrated in Fig. 2.

3   Experimental Setup and Results

In this section, we discuss the experimental setup, the dataset used for the experiments, and quantitative and qualitative results.

For the experiments, we have collected 3D Flair MR images from 60 Glioma patients (pre-operative multi-institutional scans) from The Cancer Imaging Archive (TCIA) Glioblastoma Multiforme (TCGA-GBM) data collection [1]. The dataset is publicly available and also contains the manual segmentation mask or the ground truth labels of the tumor. For experimental studies, we randomly selected 10 samples as the test set to measure the model performance, and the remaining 50 samples are used for active learning-based training of the CNN model. To better understand the effect of the proposed active learning-based training strategy we also train the same CNN model without involving the active learning strategy using the same 50 MRI samples and test it on the test set of 10 samples separately. A patch-based training of the CNN models is pursued due to the limited size of the dataset and hardware resource restrictions. During inferencing, non-overlapping patches covering the whole MRI volume are used. The CNN models were developed and trained using PyTorch 1.11.0 in Python 3.9. Experiments were performed on a workstation having NVIDIA Geforce RTX 3090 GPU with 24 GB of memory, 128 GB of RAM, and Intel 11th Gen Core i9 CPU. The CNN model was trained on the patches of size $96 \times 96 \times 96$ randomly extracted from the whole MRI volume. Since deep CNNs entail a large number of free trainable parameters, the effective number of training samples was artificially enhanced using real-time data augmentation in the form of linear transformation like random rotation, horizontal and vertical shifts, and horizontal and vertical flips, etc. The hyperparameters, employed through all the experiments, are provided below.

Hyperparameters	Value
weights & bias	Xavier [14]
optimizer	ADAM [13]
epochs	25
batch_size	16
learning rate	1e−4

Figure 3 shows the prediction from the proposed deep active learning-based segmentation model during different stages of active training. It shows the CNN-based segmentation model's prediction after training with manually annotated 5, 10, 15, 20, and 25 randomly selected sample MRI volumes from the unlabelled data pool. It can be observed qualitatively and quantitatively from Figs. 3 and 4 that the model's prediction improves over time and with more annotated data in the labeled data pool. With the saving of a new annotation, the retraining of the CNN model is invoked and at the end, we measure the model performance on the test set using the Dice score [7]. Dice score measures the voxel-wise overlap of the segmented region(s) of an image with its corresponding ground truth. It is expressed as $Dice(Y, T) = \frac{2 \times |Y_1 \cap T_1|}{|Y_1| + |T_1|}$. Here $Y \in \{0, 1\}$ represents the predicted segmentation mask (binary image) obtained from the CNN, and $T \in \{0, 1\}$ be the corresponding expert-level consensus ground truth image (mask). Here T_1/T_0 and Y_1/Y_0 correspond to the set of voxels where $T = 1/0$ and $Y = 1/0$, respectively.

First, we perform the active learning-based annotation of the samples randomly selected from the unlabelled data pool and save the annotated sample in the labeled data pool. Next, training of the CNN with available annotated data in the labeled data pool is initiated. After the training has been done for a fixed number of epochs we measure the model's performance on the test set with the Dice score. We have reported the mean and standard deviation of the Dice score with respect to the numbers of annotated samples available for training in Fig. 4. It can be observed from the figure that the performance of the proposed active learning-based segmentation model improves gradually with more annotated samples in the labeled data pool, and it achieved a very good segmentation performance in terms of a Dice score of around 0.88 with only 20 annotated samples in the labeled pool. We experimentally found that if we train the CNN model without active learning, the model overfits the training data and we are getting inferior segmentation accuracy compared to the active learning-based training strategy as reported in Table 1. During active learning, every time the CNN model was trained for 50 epochs, when not using active earning the CNN was trained with early stopping to avoid overfitting.

Table 1. CNN segmentation performance with and without active learning with partial and full training data.

Model	Dice (mean $\pm$ sd)	# training samples
CNN with AL	**0.68 $\pm$ 0.04**	0–10*
CNN without AL	0.59 $\pm$ 0.07	10
CNN with AL	**0.89 $\pm$ 0.03**	0–20*
CNN without AL	0.87 $\pm$ 0.02	20
CNN with AL	**0.93 $\pm$ 0.02**	0–50*
CNN without AL	0.90 $\pm$ 0.04	50

* Represents that AL starts with zero annotation and then gradually learns from the partial and refined annotations with user interaction in parallel with model training.

We also compared the total annotation time for the active learning-based approach with the conventional or passive learning-based approach. Figure 5 reports the annotation time required by the proposed active learning-based segmentation method from 1 to 50 samples. For the initial 10 samples, we need to follow the same annotation protocol for both of the approaches and the annotation is done by the semi-automatic interactive region growing-based approach. Then in the AL-based scenario, we can use the prediction produced by the partially trained CNN as the initial segmentation and refine it with some interaction. So after the 10th sample, there is zero interactive annotation time but the manual refinement time is added with the CNN-based segmentation time. It can be observed that the manual refinement time is gradually decreased over time with more annotated examples in the labeled data pool. Whereas in the conventional annotation approach the amount of interactive annotation time is required for all the unlabelled samples in the unlabelled data pool. In our experiment, we found the proposed active learning-based annotation required 129.87 min whereas the

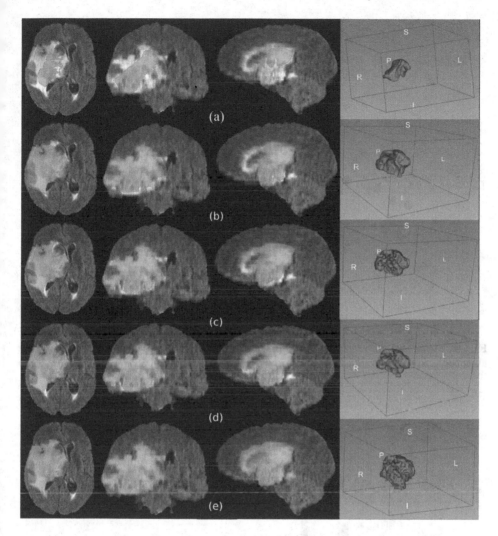

Fig. 3. Active learning-based segmentation of an MRI volume, prediction from a trained model after annotation of (a) 5, (b) 10, (c) 15, (d) 20, and (e) 25 samples from the unlabelled data pool.

conventional annotation method around needed 536 min to annotate the same 50 samples. Along with the annotation time in conventional settings, we also need separate time for training the CNN which is also expensive. Whereas in the AL scenario, the CNN training is done in parallel with the interactive annotation or the refinement steps.

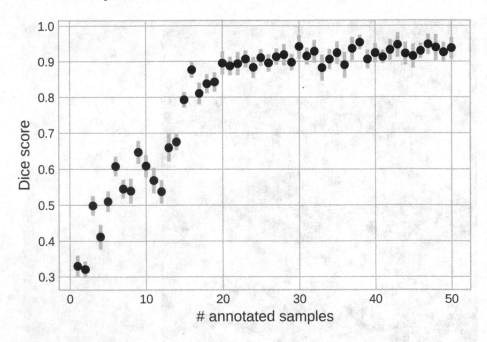

Fig. 4. Mean and standard deviation of the Dice score with respect to the numbers of annotated samples available for training.

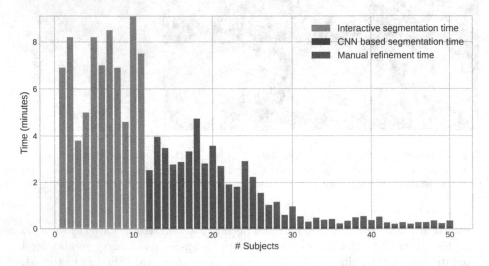

Fig. 5. Subject-wise annotation time by the proposed AL-based segmentation method.

4   Conclusion

In this paper, we proposed an active learning-based segmentation strategy and use it to segment Gliomas from Flair MRI volumes. We found that the proposed AL-based segmentation strategy is a practically viable solution, especially in the case of annotating medical images as it involves precious time and effort of radiologists. We experimentally demonstrated that the proposed AL-based annotation required around five times shorter time to annotate the same number of MRI volumes and achieved comparable or superior segmentation accuracy than the conventional passive learning-based methods. Quantitative and qualitative analysis of the segmentation performance of the proposed AL-based segmentation method is also performed.

References

1. Bakas, S., et al.: Advancing the cancer genome atlas glioma MRI collections with expert segmentation labels and radiomic features. Sci. Data 4(1), 1–13 (2017)
2. Banerjee, S., Dhara, A.K., Wikström, J., Strand, R.: Segmentation of intracranial aneurysm remnant in MRA using dual-attention atrous net. In: 2020 25th International Conference on Pattern Recognition (ICPR), pp. 9265–9272. IEEE (2021)
3. Banerjee, S., Mitra, S.: Novel volumetric sub-region segmentation in brain tumors. Front. Comput. Neurosci. 14, 3 (2020)
4. Banerjee, S., Mitra, S., Shankar, B.U.: Automated 3D segmentation of brain tumor using visual saliency. Inf. Sci. 424, 337–353 (2018)
5. Banerjee, S., Nysjö, F., Toumpanakis, D., Dhara, A.K., Wikström, J., Strand, R.: AI-based solution for improving neuroradiology workflow for cerebrovascular structure monitoring. Research Square (2023)
6. Banerjee, S., Strand, R.: Lifelong learning with dynamic convolutions for glioma segmentation from multi-modal MRI. In: SPIE Medical Imaging 2023 (2023)
7. Banerjee, S., Toumpanakis, D., Dhara, A.K., Wikström, J., Strand, R.: Topology-aware learning for volumetric cerebrovascular segmentation. In: 2022 IEEE 19th International Symposium on Biomedical Imaging (ISBI), pp. 1–4. IEEE (2022)
8. Belharbi, S., Ben Ayed, I., McCaffrey, L., Granger, E.: Deep active learning for joint classification & segmentation with weak annotator. In: Proceedings of the IEEE/CVF Winter Conference on Applications of Computer Vision, pp. 3338–3347 (2021)
9. Budd, S., Robinson, E.C., Kainz, B.: A survey on active learning and human-in-the-loop deep learning for medical image analysis. Med. Image Anal. 71, 102062 (2021)
10. Fedorov, A., et al.: 3D Slicer as an image computing platform for the quantitative imaging network. Magn. Reson. Imaging 30(9), 1323–1341 (2012)
11. Kamnitsas, K., et al.: Deepmedic for brain tumor segmentation. In: International Workshop on Brain Lesion: Glioma, Multiple Sclerosis, Stroke and Traumatic Brain Injuries (2016)
12. Kaur, G., Rana, P.S., Arora, V.: State-of-the-art techniques using pre-operative brain MRI scans for survival prediction of glioblastoma multiforme patients and future research directions. Clin. Transl. Imaging, 1–35 (2022)

13. Kingma, D.P., Ba, J.: Adam: a method for stochastic optimization. arXiv preprint arXiv:1412.6980 (2014)
14. Kumar, S.K.: On weight initialization in deep neural networks. arXiv preprint arXiv:1704.08863 (2017)
15. Malpani, R., Petty, C.W., Bhatt, N., Staib, L.H., Chapiro, J.: Use of artificial intelligence in nononcologic interventional radiology: current state and future directions. Digest. Disease Interv. 5(04), 331–337 (2021)
16. Ronneberger, O., Fischer, P., Brox, T.: U-net: convolutional networks for biomedical image segmentation. In: Navab, N., Hornegger, J., Wells, W.M., Frangi, A.F. (eds.) MICCAI 2015. LNCS, vol. 9351, pp. 234–241. Springer, Cham (2015). https://doi.org/10.1007/978-3-319-24574-4_28
17. Settles, B.: Active learning (2012)
18. Shukla, G., et al.: Advanced magnetic resonance imaging in glioblastoma: a review. Chin. Clin. Oncol. 6(4), 40 (2017)
19. Moliterno Günel, J., Piepmeier, J.M., Baehring, J.M. (eds.): Springer, Cham (2017). https://doi.org/10.1007/978-3-319-49864-5
20. Wu, X., Chen, C., Zhong, M., Wang, J., Shi, J.: Covid-AL: the diagnosis of Covid-19 with deep active learning. Med. Image Anal. 68, 101913 (2021)
21. Zhou, Z., Shin, J.Y., Gurudu, S.R., Gotway, M.B., Liang, J.: Active, continual fine tuning of convolutional neural networks for reducing annotation efforts. Med. Image Anal. 71, 101997 (2021)

Improved Sensitivity of No-Reference Image Visual Quality Metrics to the Presence of Noise

Sheyda Ghanbaralizadeh Bahnemiri, Mykola Ponomarenko[✉], and Karen Egiazarian

Tampere University, Tampere, Finland
{sheyda.ghanbaralizadehbahnemiri,mykola.ponomarenko,
karen.eguiazarian}@tuni.fi

Abstract. A problem of no-reference image visual quality assessment when images are corrupted by noise is considered in this paper. A specialized image set is proposed for the following two tasks: automatic verification of sensitivity of no-reference image visual quality metrics to noise, and analysis of blind noise level estimation methods. As a result, a method to improve the sensitivity of a given no reference quality metric to the presence of noise is proposed by combining this metric with a noise level estimator. The proposed method allows to significantly decrease a probability of wrong quality predictions for noisy images. Efficiency of usage of different noise level estimators in the proposed combined metrics is analyzed.

Keywords: Image visual quality assessment · Blind noise level estimation · Deep neural networks · No-reference image quality metrics

1 Introduction

During last two decades, no-reference image quality assessment (NR-IQA) has been the area of an intensive research [1–4]. NR-IQA metrics are used in many practical applications: automatic selection of image settings during the acquisition process, image visual quality index inside a camera, image indexing in search engines, and as a custom loss function in neural network training.

For training and testing of NR-IQA metrics, several large image databases with mean opinion scores (MOS) are proposed [1, 2, 5–8]. Various quality criteria, such as Spearman Rank Order Correlation Coefficient (SROCC) or Kendall Rank Order Correlation Coefficient, are used to estimate a correspondence between quality predictions of a given NR-IQA metric and MOS of image databases [1].

One of the most informative factors of image visual quality is image sharpness. NR-IQA metrics based only on image sharpness provide high SROCC values with MOS [9, 10]. However, adding noise to an image also may increase image sharpness. Thus, one of the important requirements for the NR-IQA metric is to achieve a decrease in the quality prediction values for noisy images. This is especially important when a NR-IQA metric is used as a loss function for training of deep neural network designed to enhance image visual quality (e.g., deblurring, denoising, etc.) [11–13]. During the training, it is

R. Gade et al. (Eds.): SCIA 2023, LNCS 13885, pp. 201–214, 2023.
https://doi.org/10.1007/978-3-031-31435-3_14

important that an addition of noise to the image will not be interpreted by the metric as an improvement of image quality.

Even though the databases with MOS contain thousands of images, their representativity is still limited [7]. Some distortions that affect image quality are poorly represented in these databases, for example, color distortions [7]. These databases contain a small percentage of noisy images. Usually, these images are obtained in a low-light condition, are blurred, or corrupted by a spatially correlated low-intensity noise. Therefore, it is difficult to use these databases to estimate the sensitivity of a NR-IQA metric to noise of various types and characteristics (e.g., noise level, noise spectrum).

In this paper, we propose a new image dataset called *NoiseSet*, which can be used for an automatic comprehensive assessment of how well NR-IQA metrics respond to noise in images. The dataset includes 1000 images of different visual quality as well as the same 1000 images with added noise of different types and intensities. We propose a criterion of sensitivity of a NR-IQA metric to the presence of noise. It is a probability P of a wrong image quality estimation when the metric predicts for a noisy image better quality than for the same image without noise.

We also propose to combine NR-IQA metrics as a weighted difference of a given NR-IQA metric and a blind noise level estimator (BNLE) [14–16]. It is shown in this paper that P for the combined metric can be decreased almost to zero.

The paper is organized as follows. Section 2 describes the proposed image set NoiseSet and the results of analysis sensitivity to noise of the recent NR-IQA metrics using the set. A comparative analysis of the state-of-the-art BNLE using NoiseSet and databases with MOS is carried out in Sect. 3. The proposed method of combining NR-IQA metrics and BNLE is considered in Sect. 4 as well as a comparative analysis of efficiency of different combinations of NR-IQA metrics and BNLE.

2   Verification of METRIC'S Sensitivity to the Presence of Noise

Figure 1 presents a simplified scheme of a deep convolutional neural network (DCNN) training intended for enhancement of image quality. Examples of this scheme are denoised network [11], deblurring network [12], or end-to-end image restoration network [13].

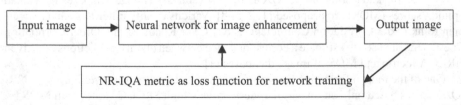

Fig. 1. Deep neural network training with a no-reference image visual quality metric as a loss function

Usually, for neural network training full-reference metrics [17] are used, among which mean square error is used most often. NR-IQA metrics as a loss function can potentially increase quality of output images even in comparison with the ground truth images. In some cases, it will allow to train a network even without ground truth images (recall that NR-IQA metrics do not require any ground truth image). However, there is a specific requirement for the NR-IQA metric, which is a good sensitivity to all factors decreasing image quality, primarily to a noise presence.

During each step of DCNN training, small changes are made to DCNN weights, and a loss function value is calculated. Negative value of selected NR-IQA metric can be used as a such loss function. Then, on a backpropagation stage the gradient of the loss function is computed and used for DCNN weights correction. If noise adding will result in decreasing of loss function value, then the training process can be significantly slowed down or even can go in a wrong direction.

Therefore, to provide a stable training, even adding a small amount of noise to an image by the trained network should result in decrease of the quality prediction of the NR-IQA metric.

In this section, we will propose a simple methodology of verification of sensitivity to noise presence of a given NR-IQA metric (corresponding criterion of the sensitivity will be described below).

2.1 Databases with MOS for Verification of NR-IQA Metrics

The goal of the paper is to show how sensitivity of NR-IQA metrics to noise can be improved. However, at the same time correspondence of the metrics to human perception should not be decreased. Because of this, we need to control both parameters. To control correspondence of metrics to human perception we will use SROCC between MOS of specialized image sets and quality predictions of a given metric.

The following six image databases with MOS values will be used for a verification of NR-IQA metrics correspondence to human perception: *KonIQ-10k* [1], *FLive* [2], *Wild* [5], *NRTID* [6], *HTID* [7] and *SPAQ* [8]. We will calculate SROCC for a given NR-IQA metric separately for MOS of each database.

To calculate an integral SROCC, we merge MOS of these databases using the methodology described in [18]. We created a subset of the merged databases which includes 10000 images: 3000 randomly selected images of *KonIQ-10k*, 1500 images of *FLive*, 1100 images of *Wild*, 500 images of *NRTID*, 1400 images of *HTID* and 2000 images of *SPAQ*. These proportions were chosen considering a quality of MOS and a representativity of each database.

For a comprehensive analysis of the considered task, we selected 14 NR-IQA metrics with different efficiency: modern metrics *HyperIQA* [4], *KonCept512* [1], *IMQNet* [19], *PaQ2PiQ* [2], *DBCNN* [21] based on DCNN, metric *Tres* [3] based on transformers, metrics *FISH* [9] and *SMetric* [22] based on wavelets, and metrics *Desique* [23], *Brisque* [24], *Niqe* [25], *BlurMetric* [10], *CEIQ* [26].

Table 1 shows SROCC values for NR-IQA metrics selected for the analysis. The column "All" shows SROCC values for the selected 10000 images from the merged database.

Table 1. Correlation between values of considered metrics and MOS of six image databases, SROCC

Metric	Kon-IQ 10k	FLive	Wild	NRTID	HTID	SPAQ	All
HyperIQA [4]	0.99	0.46	0.79	0.72	0.72	0.86	0.86
Tres [3]	0.95	0.45	0.81	0.82	0.77	0.86	0.82
KonCept512 [1]	0.95	0.43	0.73	0.72	0.71	0.84	0.79
IMQNet [19]	0.82	0.36	0.73	0.68	0.45	0.82	0.71
PaQ2PiQ [2]	0.72	0.53	0.72	0.75	0.28	0.83	0.70
UIQA [20]	0.62	0.22	0.57	0.59	0.63	0.77	0.62
DBCNN [21]	0.57	0.19	0.43	0.28	0.59	0.62	0.60
FISH [9]	0.61	0.26	0.53	0.60	0.14	0.74	0.48
SMetric [22]	0.61	0.23	0.33	0.71	0.28	0.66	0.42
Desique [23]	0.33	0.09	0.38	0.41	0.20	0.56	0.38
Brisque [24]	0.21	0.08	0.25	0.35	0.35	0.47	0.32
Niqe [25]	0.40	0.03	0.30	-0.07	0.28	0.59	0.32
BlurMetric [10]	0.25	0.13	0.33	0.40	0.02	0.59	0.30
CEIQ [26]	0.34	0.19	0.30	0.24	0.02	0.38	0.16

Some of SROCC values in Table 1 are overestimated. For example, considered *HyperIQA* and *Tres* metrics were pretrained on *KonIQ-10k* images and because of this the quality predictions for *KonIQ-10k* images are abnormally good. However, for goals of the paper, this is not important. Images of *NoiseSet* which is proposed in the paper are not used for training of the considered NR-IQA metrics.

SROCC value in the column "All" smaller than 0.5 means a bad correspondence of the NR-IQA metric to human perception. However, for the purpose of this paper, it is interesting to analyze such metrics too.

2.2 Dataset and Quality Criterion for Verification of Sensitivity of NR-IQA Metrics to the Presence of Noise

Let D be a large image dataset of images, and $M(D_i)$ is quality prediction of image D_i by the metric M. Let us for each image D_i of this dataset create its modification D_i' by adding a noise array η to the image D_i:

$$D_i' = D_i + \eta \tag{1}$$

Let us introduce a criterion of a percentage of wrong quality predictions for noisy images P:

$$P(M) = \frac{100\%}{N} \sum_i^N \delta_i, \; \delta_i = \begin{cases} 1, \; M(D_i) < M(D_i') \\ 0, \; M(D_i) \geq M(D_i') \end{cases} \tag{2}$$

where N is number of images in D.

Here $P(M) = 0$ corresponds to the maximal possible sensitivity of the metric M, and $P(M) = 1$ means that the metric M is treating noise as a quality improving factor, thus, it corresponds to the smallest possible sensitivity.

Proposed criterion P is based on the statement that adding noise to an image degrades its visual quality. Therefore, a good NR-IQA metric M should produce for the image D_i' a smaller value of image quality prediction $M(D_i')$ than $M(D_i)$.

Note that this statement is not always true. Adding a small noise to a blurred image may slightly improve a visual quality of this image by increasing image acutance (subjective perception of image sharpness). Image set *HTID* [7] among other data contains MOS for 48 blurred images and MOS for the same images with added small Gaussian noise to increase image acutance. Figure 2 shows differences between image visual quality for each such pair of images.

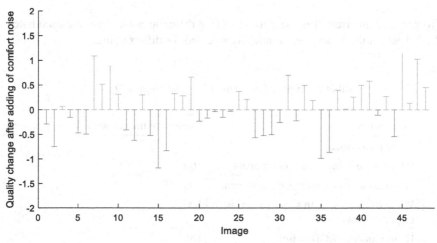

Fig. 2. Changing of image quality due to adding Gaussian noise to blurred images (Color figure online)

Positive differences (marked by green color) mean an increased MOS of the image by adding noise. Negative differences (marked by red color) correspond to a MOS decrease. One can see that even for blurred images adding small noise increase image quality only in half of the cases. Improvement of quality in the cases does not exceed 1.1 (MOS values of HTID are in the range 0...10).

Due to the abovementioned, by decreasing P for a given metric, we increase its sensitivity to noise presence at the cost of reducing its sensitivity to image acutance. However, this is an acceptable price for the possibility to use the metric as a loss function in CNN training.

To create the set D, 1000 color images of *Google Open Images* [28] were manually selected. Histogram of predictions of visual quality of the images by *HyperIQA* metric (the range of predictions of the metric is 0...100) is shown in Fig. 3. One can see that selected images provide good covering of almost whole range of visual quality.

A central part with 1024x768 pixels was extracted from each image and saved losslessly.

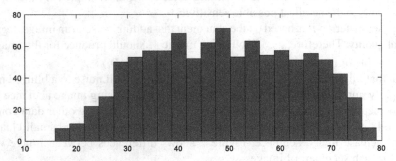

Fig. 3. Predictions of quality of subset D by *HyperIQA* metric

To generate an array of noise η for Eq. (1), 8 different noise types are used listed in Table 2. These noise types were randomly selected for different images.

Table 2. Types of noise used for generation of D_i' set

Type	Description	Number of images with this type of noise
T1	White RGB noise	100
T2	White noise in luminance component	100
T3	White noise in chrominance components	100
T4	White RGB noise with standard deviation bigger in image center	100
T5	High frequency RGB noise	150
T6	High frequency RGB noise with standard deviation larger in image center	150
T7	Spatially correlated RGB noise	150

(continued)

Table 2. (*continued*)

Type	Description	Number of images with this type of noise
T8	Spatially correlated RGB noise with standard deviation larger in image center	150

For all types of noise in Table 2, noise is additive Gaussian. A standard deviation of noise was randomly selected to be in the range 3...30. A size of noise "grain" for spatially correlated noise types has also been selected to be in a wide range.

Noise types *T4*, *T6* and *T8* imitate the situations when noise is added to an image irregularly or when only a part of the image is distorted.

The generated *NoiseSet*, which includes subsets D and D', and the Matlab scripts for the calculation of P for a given metric are available on https://ponomarenko.info/sci a2023.

Figure 4 shows examples of each type of the noise. For better visibility, we have selected the size of noise array of 256×256 pixels for this illustration.

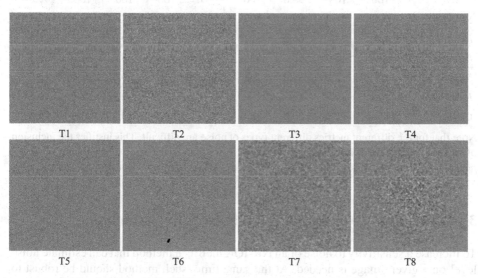

Fig. 4. Examples of different types of noise used in *NoiseSet* design

Table 3 contains the values of P for *NoiseSet* for considered NR-IQA metrics. Smaller P values are better. For better analysis, P values for each type of noise are also shown in Table 3. However, the most informative is the P value for whole *NoiseSet* given in the column "All".

It is well seen that there are no metrics with the maximal sensitivity to a noise presence ($P = 0$).

It is interesting to note that the *FISH* metric, which estimates image sharpness, has almost zero sensitivity to noise. The metric considers any noise added to an image as an increase of sharpness and consequently as an increase of image quality.

Table 3. Wrong quality predictions for noisy images, *P*, %

Metric	Subsets of *TestSet*								All
	T1	*T2*	*T3*	*T4*	*T5*	*T6*	*T7*	*T8*	
hyperIQA	6.0	19.0	4.0	8.0	42.0	27.3	2.7	3.3	15.0
Tres	4.0	8.0	5.0	6.0	36.0	29.3	3.3	3.3	13.1
KonCept512	7.0	10.0	13.0	8.0	28.0	35.3	17.3	17.3	18.5
IMQNet	0.0	26.0	0.0	6.0	2.7	5.3	25.3	31.3	12.9
PaQ2PiQ	27.0	33.0	24.0	24.0	1.3	4.0	65.3	55.3	29.7
UIQA	38.0	41.0	29.0	28.0	40.7	35.3	49.3	41.3	38.6
DBCNN	6.0	14.0	2.0	3.0	8.0	13.3	8.7	8.0	8.2
FISH	100	100	98.0	100	100	100	100	100	99.8
SMetric	0.0	26.0	0.0	99.0	4.0	100	0.0	93.3	42.1
Desique	9.0	2.0	66.0	17.0	13.3	21.3	40.0	53.3	28.6
Brisque	25.0	15.0	62.0	32.0	12.0	18.7	40.0	62.0	33.3
Niqe	1.0	1.0	34.0	1.0	0.0	0.0	30.0	38.0	13.9
BlurMetric	100	100	100	100	100	100	97.3	98.0	99.3
CEIQ	87.0	91.0	20.0	85.0	89.3	86.7	89.3	91.3	81.8

Note that for the different metrics different types of noise are difficult. This justifies the inclusion of various types of noise in *NoiseSet*

3   Comparative Analysis of Blind Noise Level Estimators

To increase a sensitivity to noise of an NR-IQA metric, a method that can estimate noise level on a given image is needed. At the same time, such method should be robust to the presence of sharp texture, contrast edges, and fine details in images (these factors may either degrade or increase image quality). A method also should have very small *P* value, calculated according to (2). At the same time, there should not be a large positive SROCC of the method with MOS of the specialized image databases. It is very desirable for the SROCC to be as close to zero as possible.

Let us have a BNLE *B* that fulfills these requirements. In this case, the following weighted difference of a given NR-IQA metric *M* and *B* can be used to improve a sensitivity to noise of the metric *M*:

$$M' = M - K_{opt}B \tag{3}$$

where K_{opt} is a coefficient which minimizes P with the condition that SROCC remains larger than a selected threshold Tr. Such a combination of M and B decreases P decreasing a resulting sum of δ_i in (2).

For analysis in this paper, we selected the following six BNLE methods.

First, four methods of estimation of standard deviation of additive white Gaussian noise on a given image were chosen: *IEDD* [14], *PCA* [15], *Tanaka* [16] and *Noilap* [29]. These methods are not intended for estimation of level of high frequency or spatially correlated noise. However, for the considered task it is not important.

Second, we included in the analysis the convolutional neural network *SDNet* [30], which estimates a map of noise standard deviations for a given image. To estimate a noise level, median of the map is calculated.

Finally, convolutional neural network *NLNet* [31] was included in the analysis. The network estimates a map of noise levels for the denoiser *DRUNet* [11]. *NLNet* is designed to estimate noise levels in images distorted by Gaussian spatially correlated noise. To estimate a noise level, in this paper, a median of the map is calculated.

Table 4. Wrong quality predictions for noisy images, P, %

Noise level estimator	Subsets of *TestSet*								All
	T1	*T2*	*T3*	*T4*	*T5*	*T6*	*T7*	*T8*	
Noilap [29]	0.0	0.0	2.0	0.0	0.0	0.0	0.0	0.7	0.3
IEDD [14]	0.0	0.0	2.0	0.0	0.0	0.7	0.0	2.0	0.6
SDNet [30]	0.0	0.0	4.0	0.0	2.0	0.0	3.3	6.0	2.1
PCA [15]	0.0	0.0	4.0	0.0	2.7	2.7	2.7	5.3	2.4
NLNet [31]	0.0	0.0	3.0	0.0	17.3	14.0	0.0	0.0	5.0
Tanaka [16]	0.0	0.0	1.0	0.0	27.3	6.7	26.7	17.3	11.8

Table 4 contains values of P for *TestSet* for considered BNLE methods. As one can see, P for considered BNLE methods is, in average, significantly smaller than for the considered NR-IQA metrics (see Table 3). A most sensitive to noise is a simple and fast *Noilap* method. P value for the Tanaka method perhaps is too large for the considered task.

However, a small P is only one of the two requirements for BNLE methods. Table 5 shows values of SROCC between the estimated values of BNLE methods and MOS of image databases.

Table 5. Correlation between values of considered noise level estimators and MOS of six image databases, SROCC

Noise level estimator	Kon-IQ 10k	FLive	Wild	NRTID	HTID	SPAQ	All
Noilap	0.46	0.22	0.47	0.57	0.03	0.67	0.40
IEDD	−0.03	0.09	0.14	0.17	−0.08	0.13	0.13
SDNet	−0.11	0.00	0.01	0.29	0.08	−0.11	0.00
PCA	−0.10	−0.06	−0.10	−0.30	0.04	−0.10	−0.04
NLNet	−0.17	0.00	−0.03	0.25	0.01	0.01	−0.08
Tanaka	−0.10	−0.11	−0.21	−0.39	−0.06	−0.16	−0.18

One can see, that *Noilap* provides too large positive SROCC and, probably, will not be useful in improvement of NR-IQA sensitivity to noise.

The *SDNet* network fits well both requirements: it has a small P and the value of SROCC close to zero.

4   Combining Quality Prediction and Blind Noise Level Estimation

In this Section, we use Eq. (3) to generate the combinations of different NR-IQA metrics and BNLE. K_{opt} is determined for each possible pair of the considered NR-IQA metrics and BNLE.

Figure 5 illustrates the algorithm of obtaining K_{opt} for the pair of *KonCept512* metric and *SDNet* noise estimator.

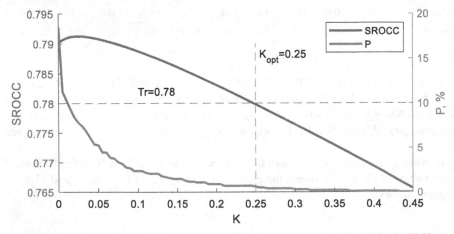

Fig. 5. Selection of K_{opt} for given Tr for the combined metric KonCept512 - $K*$SDNet

KonCept512 metric itself has SROCC 0.79 with MOS of merged databases. Selecting $Tr = 0.78$ one guarantees that a correspondence of the metric to human perception will not be degraded significantly.

As it is seen in the figure, increasing of K results in a small increase of SROCC, and then in a smooth gradual decrease of SROCC. At the same time, P is decreasing rapidly and is close to zero at the end of the curve.

Choosing K_{opt}=0.25 allows to drastically decrease P from 18.5 to 0.5 with SROCC degradation only from 0.79 to 0.78. This compromise is acceptable for a practical use of the combined metric.

The combined metric in this case is calculated as $KonCept512 - 0.25\ SDNet$.

Table 6 contains P values of the optimized weighted differences of NR-IQA metrics and corresponding outputs of BNLE methods.

Table 6. Value of P for optimized sum of a metric and a noise level estimator

Metric (Name, *Tr* value)	Noise level estimator						
	none	*Noilap*	*IEDD*	*SDNet*	*PCA*	*NLNet*	*Tanaka*
HyperIQA, 0.85	15.0	**0.9**	1.1	1.0	3.5	7.1	10.2
Tres, 0.82	13.1	0.6	**0.0**	**0.0**	0.1	**0.0**	**0.0**
KonCept512, 0.78	18.5	6.3	**0.5**	**0.5**	3.3	1.7	1.7
IMQNet, 0.705	12.9	4.7	**0.2**	0.3	3.0	1.6	2.2
PaQ2PiQ, 0.70	29.7	18.3	1.9	**0.7**	7.7	1.3	6.3
UIQA, 0.615	38.6	29.7	3.8	0.3	8.7	**0.0**	9.8
DBCNN, 0.59	8.2	5.5	**0.0**	**0.0**	0.8	**0.0**	0.8
FISH, 0.45	99.8	24.0	23.6	19.2	36.2	**17.6**	56.5
SMetric, 0.39	42.1	40.9	39.5	39.4	40.2	**32.2**	39.8
Desique, 0.37	28.6	23.3	3.0	**0.0**	12.0	**0.0**	12.9
Brisque, 0.31	33.3	28.3	0.3	**0.0**	9.1	**0.0**	9.6
Niqe, 0.31	13.9	12.2	1.8	0.1	6.7	**0.0**	6.5

As one can see, for most of the considered NR-IQA metrics it is possible to provide almost zero P value without any significant degradation of SROCC.

As it is seen from Table 6, different BNLE methods are working best for different NR-IQA metrics. It is possible to recommend for a practical use methods *SDNet*, *IEDD* and *NLNet*, which provide the smallest or near to smallest P values for majority of the considered NR-IQA metrics.

Optimization curves for the pair *PaQ2PiQ* and *SDNet* are interesting to analyze (see Fig. 6).

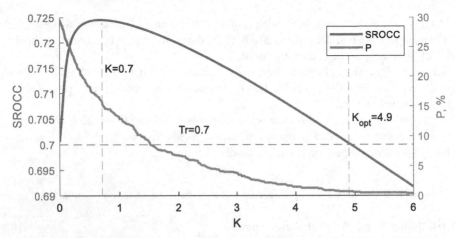

Fig. 6. Selection of K_{opt} for given Tr for the combined metric $PaQ2PiQ - K*SDNet$

For $K = 0.7$, one can see a significant improvement of both parameters. SROCC increases from 0.7 to 0.725, while P decreases from 29.7 to 15.7. This case demonstrates that the proposed methodology sometimes can be used not only for decreasing P value, but also for a simultaneous increase of SROCC.

5   Conclusions

A new methodology of verification and improvement of sensitivity of NR-IQA metrics to noise is proposed in this paper.

It is shown that for a weighted difference of a NR-IQA metric and a BNLE, a value of P can be decreased from 13%...30% to zero without any significant degradation of SROCC with MOS of the specialized databases.

It is shown also that SDNet, NLNet and IEDD are the most efficient BNLE methods to be used in the proposed scheme.

References

1. Hosu, V., Lin, H., Sziranyi, T., Saupe, D.: KonIQ-10k: an ecologically valid database for deep learning of blind image quality assessment. IEEE Trans. Image Process. **29**, 4041–4056 (2020)
2. Ying, Z., Niu, H., Gupta, P., Mahajan, D., Ghadiyaram, D., Bovik, A.: From patches to pictures (PaQ-2-PiQ): Mapping the perceptual space of picture quality. In: Proceedings of the IEEE/CVF Conference on Computer Vision and Pattern Recognition, pp. 3575–3585 (2020)
3. Golestaneh, S.A., Dadsetan, S., Kitani, K.M.: No-reference image quality assessment via transformers, relative ranking, and self-consistency. In: Proceedings of the IEEE/CVF Winter Conference on Applications of Computer Vision, pp. 1220–1230 (2022)
4. Su, S., et al.: Blindly assess image quality in the wild guided by a self-adaptive hyper network. In: Proceedings of the IEEE/CVF Conference on Computer Vision and Pattern Recognition, pp. 3667–3676 (2020)

5. Ghadiyaram, D., Bovik, A.: Massive online crowdsourced study of subjective and objective picture quality. IEEE Trans. Image Process. **25**(1), 372–387 (2015)
6. Ponomarenko, N., Eremeev, O., Egiazarian, K., Lukin, V.: Statistical evaluation of no-reference image visual quality metrics. In: Proceedings of EUVIP, Paris, France, 5p. (2010)
7. Ponomarenko, M., et al.: Color image database HTID for verification of no-reference metrics: peculiarities and preliminary results, EUVIP, 6 p. (2021)
8. Fang, Y., Zhu, H., Zeng, Y., Ma, K., Wang, Z.: Perceptual quality assessment of smartphone photography. In: Proceedings of the IEEE/CVF Conference on Computer Vision and Pattern Recognition, pp. 3677–3686 (2020)
9. Vu, P., Chandler, D.: A fast wavelet-based algorithm for global and local image sharpness estimation. IEEE Signal Processing Letters, pp. 423–426 (2012)
10. Crete, F., Dolmiere, T., Ladret, P., Nicolas, M.: The blur effect: perception and estimation with a new no-reference perceptual blur metric. In: Human vision and electronic imaging XII International Society for Optics and Photonics, vol. 6492 (2007)
11. Zhang, K., Li, Y., Zuo, W., Zhang, L., Van Gool, L., Timofte, R.: Plug-and-play image restoration with deep denoiser prior. IEEE Trans. Pattern Anal. Mach. Intell., 17 p. (2021)
12. Nah, S., Son, S., Lee, S., Timofte, R., Lee, K.M, "NTIRE 2021 challenge on image deblurring", In Proceedings of the IEEE/CVF Conference on Computer Vision and Pattern Recognition, pp. 149–165 (2021)
13. Qian, G., Gu, J., Ren, J., Dong, C., Zhao, F., Lin, J.: Trinity of pixel enhancement: a joint solution for demosaicking, denoising and super-resolution, arXiv preprint arXiv:1905.02538 (2019)
14. Ponomarenko, M., Gapon, N., Voronin, V., Egiazarian, K.: Blind estimation of white gaussian noise variance in highly textured images. Electron. Imaging **2018**(13), 382–391 (2018)
15. Pyatykh, S., Hesser, J., Zheng, L.: Image noise level estimation by principal component analysis. IEEE Trans. Image Process. **22**(2), 687–699 (2012)
16. Liu, X., Tanaka, M., Okutomi, M., "Noise level estimation using weak textured patches of a single noisy image", in 2012 19th IEEE International Conference on Image Processing. IEEE, pp. 665–668 (2012)
17. Sheikh, H.R., Sabir, M.F., Bovik, A.C.: A statistical evaluation of recent full reference image quality assessment algorithms. IEEE Trans. Image Process. **15**, 3440–3451 (2006)
18. Kaipio, A., Ponomarenko, M., Egiazarian, K.: Merging of MOS of large image databases for no-reference image visual quality assessment. IEEE International Workshop on Multimedia Signal Processing, 6 p. (2020)
19. Ponomarenko, M., Ghanbaralizadeh Bahnemiri, S., Egiazarian, K., Transfer learning for no-reference image quality metrics using large temporary image sets. In: Electronic Imaging 2022, Computational Imaging XX, 5 p. (2022)
20. Lu, T., Dooms, A.: Towards content independent no-reference image quality assessment using deep learning. In: IEEE 4th International Conference on Image, Vision and Computing, pp. 276–280 (2019)
21. Zhang, W., Ma, K., Yan, J., Deng, D., Wang, Z.: Blind image quality assessment using a deep bilinear convolutional neural network. IEEE Trans. Circuits Syst. Video Technol., 36–47 (2018)
22. Ponomarenko, N., Lukin, V., Eremeev, O., Egiazarian, K., Astola, J., "Sharpness metric for no-reference image visual quality assessment", Image Processing: Algorithms and Systems X and Parallel Processing for Imaging Applications II, International Society for Optics and Photonics, vol. 8295, 11 p, (2012)
23. Zhang, Y., Chandler, D.: No-reference image quality assessment based on log-derivative statistics of natural scenes. J. Electron. Imaging, vol. 4, 22 p. (2013)

24. Mittal, A., Moorthy, A., Bovik, A.: No-reference image quality assessment in the spatial domain. IEEE Trans. Image Process., 4695–4708 (2012)
25. Mittal, A., Soundararajan, R., Bovik, A.C.: Making a "completely blind" image quality analyzer. IEEE Signal Process. Lett., 209–212 (2012)
26. Yan, J., Li, J., Fu, X.: No-reference quality assessment of contrast-distorted images using contrast enhancement. arXiv preprint arXiv:1904.08879 (2019)
27. Venkatanath, N., Praneeth, D., Chandrasekhar, Bh., Channappayya, S., Medasani, S.: Blind Image Quality Evaluation Using Perception Based Features. In: Proceedings of the 21st National Conference on Communications (NCC), 6 p. (2015)
28. Kuznetsova, A., et al.: The Open Images Dataset V4: Unified image classification, object detection, and visual relationship detection at scale. Int. J. Comput. Vis., 1956–1981 (2020)
29. Immerkr, J.: Fast noise variance estimation. Comput. Vis. Image Underst. **64**(2), 300–302 (1996)
30. Ghanbaralizadeh Bahnemiri, S., Ponomarenko, M., Egiazarian, K.: Learning-based Noise Component Map Estimation for Image Denoising. IEEE Sig. Process. Lett., 5 p. (2022)
31. Ponomarenko, M., Miroshnichenko, O., Lukin, V., Egiazarian, K.: Blind estimation and suppression of additive spatially correlated gaussian noise in images. In: 2021 9th European Workshop on Visual Information Processing (EUVIP), 6 p. (2021)

Rethinking Matching-Based Few-Shot Action Recognition

Juliette Bertrand[1]([✉]), Yannis Kalantidis[2]([✉]), and Giorgos Tolias[1]([✉])

[1] Visual Recognition Group, Faculty of Electrical Engineering,
Czech Technical University in Prague, Prague, Czechia
bertrjul@fel.cvut.cz
[2] NAVER LABS Europe, Grenoble, France
yannis.kalantidis@naverlabs.com

Abstract. Few-shot action recognition, *i.e.* recognizing new action classes given only a few examples, benefits from incorporating temporal information. Prior work either encodes such information in the representation itself and learns classifiers at test time, or obtains frame-level features and performs pairwise temporal matching. We first evaluate a number of matching-based approaches using features from spatio-temporal backbones, a comparison missing from the literature, and show that the gap in performance between simple baselines and more complicated methods is significantly reduced. Inspired by this, we propose Chamfer++, a non-temporal matching function that achieves state-of-the-art results in few-shot action recognition. We show that, when starting from temporal features, our parameter-free and interpretable approach can outperform all other matching based and classifier methods for one-shot action recognition on three common datasets without using temporal information in the matching stage.
Project page: https://jbertrand89.github.io/matching-based-fsar

1 Introduction

Recognizing actions within videos is essential for analyzing trends, enhancing broadcasting experience, or filtering out inappropriate content. However, collecting and annotating enough video examples to train supervised models can be prohibitively time-consuming. It is therefore desirable to recognize new action classes with as few labeled examples as possible. This is the premise behind the task of few-shot learning, where models learn to adapt to a set of unseen classes for which only a few examples are available. In video action recognition, additional challenges arise from the temporal dimension. Recognition methods need to capture the scene's temporal context and temporal dynamics.

One family of approaches is formed by *matching-based* methods [3,21,27,36] where each test example or "query" is compared against all support examples of a class to infer a class confidence score. Most existing matching-based methods use frame-level representations, *i.e.* a 2D convolutional backbone that takes a frame as input, and a feature set is formed by encoding multiple frames. Feature extraction is followed by matching the query feature set Q to the support

R. Gade et al. (Eds.): SCIA 2023, LNCS 13885, pp. 215–236, 2023.
https://doi.org/10.1007/978-3-031-31435-3_15

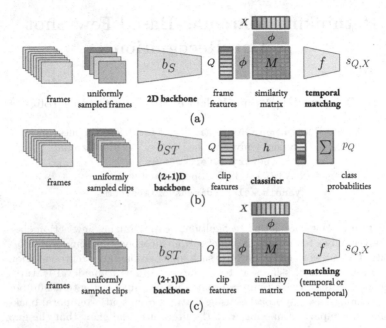

Fig. 1. Different ways of using temporal information in few-shot action recognition. Temporal information is utilized: a) during matching (existing *matching-based* methods) b) in the backbone with a classifier (existing *classifier-based* methods, *e.g.* [33]) c) in the backbone with a matching step (this paper).

example set X, and a similarity $s_{Q,X}$ between the two is computed. This process is depicted in Fig. 1a. Although each feature represents an individual frame and cannot capture temporal information, the feature sets are usually temporal sequences, and the matching process can exploit such information.

Another family of approaches is formed by methods that learn a conventional linear *classifier* at test time, *i.e.* using the handful of examples available. In this case, any temporal context has to be incorporated in the representation, as shown in Fig. 1b. As a representative example, Xian *et al.* [33] adopt the spatio-temporal R(2+1)D architecture [28], where the input is a video *clip*, *i.e.* a sequence of consecutive frames, and convolutions across the temporal dimension enable the features to encode temporal information. Following findings in few-shot learning [5,31], Xian *et al.* further abandon episodic training and instead fine-tune a pre-trained backbone using all training examples of the base classes. Using strong temporal features and by simply learning a linear classifier at test time, they report state-of-the-art results for few-shot action recognition.

Motivated by the two families of approaches presented above, we introduce a new setup depicted in Fig. 1c that aims at answering the following questions:

1. *Do matching-based methods still have something to offer for few-shot action recognition given strong temporal representations?* We level the playing field with respect to representations and evaluate a number of recent

matching-based approaches using strong temporal representations. We find that such approaches perform better than training a classifier at test time.

2. *Is temporal matching necessary when the features capture temporal information?* We show that matching-based methods invariant to the temporal order in the feature sequence (*non-temporal matching*) are performing as good as the ones that do use it (*temporal matching*) on many common benchmarks.

Inspired by the findings above, we further introduce **Chamfer++**, a novel, parameter-free and interpretable matching approach that employs Chamfer matching and is able to achieve a new state-of-the-art for one-shot action recognition on three common benchmarks.

2   Related Work

In the image domain, several methods for few-shot classification are metric-learning-based [2,6,23,26,29] and use a k-nearest-neighbor classifier at test time. Despite the dominance of meta-learning-based methods in the area [9,10,13,18, 22], several recent studies highlight the importance of starting from strong visual representations. It is shown [5,31] that if one learns representations without meta-learning but using instead all available data from all base classes, simple nearest-neighbors [31] or parametric classifier [5] baselines work on-par or better than most methods on common benchmarks. Similar observations are made for few-shot learning in the video domain [33,36,37].

As with the image domain, most methods leverage meta-learning and are grouped into initialization-based [33,37], metric-learning [1,3,21,34–36] and generative-based [8,33] methods. However, unlike the image domain, most methods for few-shot action recognition try to take into account the *temporal* dimension in the visual representations and/or during the matching process. In fact, most methods incorporate temporal matching [1,3,16,17,21,34,35]. The features are either extracted from single frames [3,21,35] or clips [1,33,34] where tempo ral information is already captured in the features.

Some matching methods explicitly aim for temporal alignment and estimate video-to-video similarity via an ordered temporal alignment score. OTAM [3] finds the optimal path on the temporal similarity matrix via a differentiable approximation of the Dynamic Time Warping (DTW) [20] algorithm, a method also used for alignment in other temporal tasks [4,7]. Other methods like ARN [34] use spatio-temporal attention, while TA^{2}N [17] proposes two-stage spatial-temporal alignment. TARN [1] uses temporal attention over clip sequences for alignment. During the training process, the attention parameters, together with the parameters from a subsequent recurrent network, are learned such that features are aligned between the query and a support video from the correct class.

Other methods like TRX [21] or PAL [36] do not explicitly seek alignment but rely on *cross-attention* mechanisms over the query and support features to perform temporal matching. In both cases, class-wise representations are constructed and matching is performed directly in a *video-to-class* manner. For

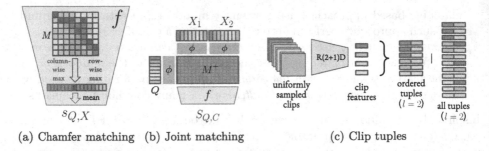

(a) Chamfer matching (b) Joint matching   (c) Clip tuples

Fig. 2. Details of the proposed matching approach. a) The **Chamfer** matching function f_{QS} from Eq.(3). b) Jointly matching multiple examples per class (Chamfer+). c) using clip tuples as representation; one can use only ordered tuples or all tuples; in both cases, the matching part remains *non-temporal*, *i.e.* invariant to the temporal order of features.

TRX, class representations are adapted on-the-fly in a query-conditioned manner, while for PAL the query features are instead adapted to match pre-computed class-wise representations. In our study, we show how such direct video-to-class approaches are highly competitive when more than one videos per class are provided, but that video-to-video methods are superior and more efficient for the one-shot case.

Recently, a number of methods suggest that the meta-learning framework is not optimal for the representation learning phase. Following similar observations in the image classification domain [5,31], some few-shot action recognition methods [33,36,37] learn their representation on the full train set and report higher performance on multiple benchmarks. As discussed in Sect. 3.2, we follow such methods, and use a strong representation learned on the entire train set as the starting point for our study on temporal matching.

3   Method

We first present the video representation and the experimental protocol used in this work in Sect. 3.1. Then, we describe a framework to fairly compare matching-based and classifier-based approaches in Sect. 3.2 and propose a new matching function in Sect. 3.3. In the following, we refer to videos as examples and actions as classes.

3.1   Preliminaries

We describe the video representation that we use and the episodic protocol as formulated in the recent literature of few-shot video action recognition, which we follow in this work.

Video Representation. A clip c_i is a sequence of L consecutive RGB frames in the form of a $L \times H \times W \times 3$ tensor. A deep video backbone b takes a clip c_i as input and maps it to a d-dimensional vector $\mathbf{q}_i = b(c_i) \in \mathbb{R}^d$, named feature. We use the R(2+1)D backbone architecture [28] as in the work of Xian *et al.* [33], which uses efficient and effective separated spatio-temporal convolutions. A video Q is represented by a set of clip features $Q = \{\mathbf{q}_i\}$. The clips are uniformly sampled over the temporal dimension, with possible overlap.

Episodic Protocol. We adopt the commonly used setup [3,21]. The classes of the train and test sets are non-overlapping and usually named base and novel classes, respectively. To simulate the limited annotated data, test episodes are randomly sampled from the test set. Each episode corresponds to a different classification task and comprises query and support examples for a fixed set of classes, where labels of support examples are known, while labels of query examples are unknown. Only k labeled examples per class, also named shots, are available in the support set, with k typically ranging from 1 to 5. The performance is evaluated via classification accuracy on the query examples averaged over all test episodes.

3.2 A Common Setup for Classifier and Matching-Based Approaches

We identify two dominant families of approaches proposed in the recent few-shot action recognition literature: the classifier-based and the matching-based methods[1]. Unfortunately, discrepancies in setup and architecture between existing approaches make it difficult to compare them fairly. We propose to follow the same representation learning strategy and start from a common frozen backbone.

Classifier-Based Approaches. The classifier-based approaches [33,37] train the video representation using a classifier in the form of a linear layer. This is similar to the corresponding work on few-shot learning in the image domain [11].

Classifiers are trained using the whole dataset, but episodic training cannot guarantee to see all the examples because it is intractable to sample all the possible episodes. Xian *et al.* [33] depart from episodic training and propose a *Two-stage Learning (TSL)* process. During the first stage, a R(2+1)D video backbone and a classifier are learned jointly using all the labeled examples of the train set. During the second stage, the backbone remains fixed to avoid overfitting and a newly initialized classifier needs to be trained per test episode using the support examples.

In both stages, a linear classifier with a soft-max function denoted by h : $\mathbb{R}^d \rightarrow \mathbb{R}^C$, where C is the number of classes, is added to the output of the backbone. During the training stage, $C = C_t$ is equal to the number of all classes in the train set. During the second stage, $C = C_f$ is equal to the number of classes per test episode, usually $C_f = 5$. Training is performed by optimizing

---

[1] Prototypical networks can be seen as an extension of matching based methods [21, 27,36]. Hence we group them with the matching-based family.

the class probabilities $h(\mathbf{q}_i)$ for each $\mathbf{q}_i \in Q$ with the cross-entropy loss ($\mathcal{L}_{cls}$), while inference is performed by sum-pooling of the classifier output across clips, i.e. $\sum_{\mathbf{q}_i \in X} h(\mathbf{q}_i)$.

Matching-Based Approaches. The matching-based approaches estimate the similarity between the query and all the support examples of each class to obtain class probabilities. Training is performed on episodes sampled from the train set, which are meant to imitate the episodes of the test set.

Let $Q = \{\mathbf{q}_i\}$ and $X = \{\mathbf{x}_i\}$ be two videos with $|Q| = |X| = n$ and assume that n is constant across videos. We form the *temporal similarity matrix* for the ordered video pair (Q, X) denoted by $M_{Q,X} \in \mathbb{R}^{n \times n}$, or just M for brevity, with elements $m_{ij} = \phi(\mathbf{q}_i)^\top \phi(\mathbf{x}_j)$, where $\phi : \mathbb{R}^d \to \mathbb{R}^D$ is a learnable projection head. The function ϕ consists of a linear layer, a layer normalization, and a ℓ_2 normalization to guarantee bounded similarity values m_{ij}. When no linear projection is used, ϕ is equivalent to the identity mapping with an ℓ_2 normalization. Each element m_{ij} of the matrix M can be seen as a temporal correspondence between clip i of video Q and clip j of video X.

We consider the family of matching approaches that infer a video-to-video similarity $S_{Q,X}$, between video Q and X, solely based on the matrix M, i.e. $S_{Q,X} = f(M)$, with $f : \mathbb{R}^{n \times n} \to \mathbb{R}$, named the *matching function*. The scalar $S_{Q,X}$ should be high if the two videos depict the same action. By definition, the result of function f only depends on the strength and position of the pairwise similarities m_{ij} and does not directly depend on the features themselves. The function f can either be hand-crafted or include learnable parts. A graphical overview of different matching approaches is given in the appendix as long as a more detailed description of them. Some matching functions use temporal information by leveraging the position, either absolute or relative, of the pairwise similarities m_{ij}. The matching functions that use temporal information are called *temporal*, whereas the others are called *non temporal*.

The pairwise video-to-video similarities between query and support examples are averaged per class to obtain class probabilities[2]. During training, the class probabilities are optimized with the cross-entropy loss ($\mathcal{L}_{cls}$). Inference is also performed by estimating the similarity between query and all support examples. This is a form of a k-nearest-neighbor classifier.

A Common Starting Point. We follow the training stage of TSL [33] to learn the R(2+1)D backbone parameters using all the annotated examples. We freeze this backbone, and treat the resulting model as a feature extractor. This is our starting point for both classifier-based and matching-based approaches which enable us to fairly compare the two families of approaches. Specifically, matching-based methods only learn the feature projection function ϕ and the

---

[2] In prototypical networks [21,27,36], the pairwise clip-to-clip similarities are used as weights to compute a class prototype specific to the query example. The class probabilities are computed as the distance between a query and its prototype.

matching parameters when needed in a test-agnostic way. Unlike classifier-based methods, which need to train a classifier for every testing episode, matching-based approaches require no learning or adaptation at test time. They only need the pairwise matching between the query and each one of the support examples.

3.3   Chamfer++

We propose a new matching function, Chamfer++, which is non-temporal and achieves top performance while being parameter-free and intuitive. It is an extension of Chamfer similarity (Fig. 2a) with joint-matching over multiple shots (Fig. 2b) applied in conjunction with clip-feature-tuples instead of clip features (Fig. 2c). This section details its main components.

Chamfer. The Chamfer matching function f_Q is given by

$$f_Q(M) := \frac{1}{n} \sum_i \max_j m_{ij}. \tag{1}$$

Each clip of the query example is matched with its most similar one within the clips of a support example to produce the score between this specific query clip and this support example. The final video-to-video similarity is the average of all the query clip scores. The Chamfer matching function implies that each clip sampled from the query example contributes to the similarity score by matching its closest clip in the support example. One can transpose the temporal similarity matrix and derive the symmetric process where each clip from the support example needs to match a query clip. Then the matching function becomes

$$f_S(M) := \frac{1}{n} \sum_j \max_i m_{ij}. \tag{2}$$

Chamfer-S is equivalent to the Chamfer on the transposed matrix $M^\top$. Summing the two gives a symmetric Chamfer variant, where all clips from both the query and the support example are required to match:

$$f_{QS}(M) := f_Q(M) + f_S(M). \tag{3}$$

We refer to this symmetric variant as simply *Chamfer matching* in the context of few-shot action recognition. In the following, we also refer to Chamfer-Q as query-based and Chamfer-S as support-based.

Joint-Matching. As discussed in Sect. 3.2, the standard option to compute the query-to-class probabilities is by averaging the pairwise similarity score between the query example and all the support examples belonging to the class. Instead, we propose to match all support examples jointly. We concatenate the temporal similarity matrices between the query and all support examples. It creates the joint temporal similarity matrix M^+, with $M^+ \in \mathbb{R}^{n \times kn}$. Then, we compute the matching function on top of M^+ to obtain video-to-class similarity $S_{Q,c} := s(M^+)$. We evaluate the impact of this newly proposed joint-matching versus the standard single-matching in Sect. 4.3. We refer to this variant as Chamfer+ in the rest of the manuscript.

Clip Tuples. Besides the case where each clip feature is matched independently, Perret *et al.* [21] additionally propose matching feature *tuples*. Inspired by this, we extend Chamfer to enable matching of *clip* feature tuples formed by any clip subset of fixed length l. A clip feature tuple $\mathbf{t}^l$ contains l clip features, non-necessarily consecutive, but with the same relative order as in $Q = \{q_i\}$. For example, $\mathbf{t}^2 = \{(q_i, q_{j \neq i})\}$. Each clip feature tuple is concatenated and fed to the learnable projection head $\phi : \mathbb{R}^{ld} \rightarrow \mathbb{R}^D$. The resulting temporal similarity matrix is $M^{++} \in \mathbb{R}^{n' \times kn'}$, with $n' = \binom{n}{l}$. The clip tuples are sub-sequence representations on top of single-clip representations. By definition, clip tuples add additional temporal information to the representation. But the matching function remains non-temporal.

We also define non-temporally ordered clip tuples $\mathbf{t}_{all}^l$ as the permutations of l clip features. For example, $\mathbf{t}_{all}^2 = \{(q_i, q_{j \neq i})\}$. The resulting temporal similarity matrix is $M^{++} \in \mathbb{R}^{n' \times kn'}$ with $n' = n!$. The comparison between ordered and all tuples is presented in Table 3 and the appendix. In the following, the extension of Chamfer matching that jointly matches multiple shots and uses clip tuples is referred to as **Chamfer++**. Unless otherwise stated, we use ordered clip tuples.

4   Experiments

We report results on the three most commonly used benchmarks for few-shot action recognition, *i.e.* Kinetics-100 [35], Something-Something V2 (SS-v2) [12], and UCF-101 [24]. Kinetics-100 and UCF-101 contain videos collected from YouTube. Each video is trimmed to include only one coarse-grained human action such as "playing trumpet" or "reading book". SS-v2 contains egocentric videos where humans were instructed to perform predefined actions such as "pushing something from left to right" or "dropping something into something".

We use the train/val/test splits from [35] for all three datasets, containing 64/12/24 classes, respectively. We learn the parameters of the R(2+1)D backbone using the train split, similar to [33]. For matching-based approaches, we learn the projection ϕ together with any learnable parameters in the matching function f using episodic training also on the train split. We use the val split for hyper-parameter tuning and early stopping.

We evaluate on the common 1-shot and 5-shot setups. Unless otherwise stated, we use 5-way classification tasks. Training episodes are randomly sampled from the train set. We use the same fixed, predefined set of 10k test episodes sampled from the test set for all methods (prior work samples them randomly). We always evaluate *three* trained models and report mean and standard deviation.

4.1   Implementation Details

We use the publicly available code for TSL[3] for training the backbone as well as for reproducing the TSL method. We adapt the public TRX[4] codebase for learning parameters of the matching function and testing, as well as for reproducing the TRX and OTAM methods.

Table 1. Few-shot action recognition results. All methods are trained and evaluated by us and use the same R(2+1)D [28] backbone. TSL [33] learns a classifier at each episode during testing. All the rest are pairwise matching-based methods and are split into two categories: non-temporal, *i.e.* invariant to the temporal order of features, and temporal. [†] denotes Chamfer++ matching using *all* tuples. Best (second-best) results are presented in **bold** (underlined).

Method	SS-v2		Kinetics-100		UCF-101	
	1-shot	5-shot	1-shot	5-shot	1-shot	5-shot
Parametric classifier						
TSL [33]	60.6 ±0.1	79.9 ±0.0	93.6 ±0.0	98.0 ±0.00	97.1 ±0.0	<u>99.4</u> ±0.0
Non-temporal matching						
Mean	65.8 ±0.0	79.1 ±0.1	95.5 ±0.0	98.1 ±0.1	97.6 ±0.2	98.9 ±0.1
Max	65.0 ±0.2	79.0 ±0.0	95.3 ±0.1	<u>98.3</u> ±0.0	**97.9** ±0.1	98.9 ±0.0
Chamfer++[†]	67.0 ±0.3	80.8 ±0.1	**96.2** ±0.1	**98.4** ±0.1	<u>97.8</u> ±0.1	99.2 ±0.1
Chamfer++	**67.8** ±0.2	<u>81.6</u> ±0.1	<u>96.1</u> ±0.1	<u>98.3</u> ±0.0	97.7 ±0.0	99.3 ±0.0
Temporal matching						
Diagonal	66.7 ±0.1	80.1 ±0.0	95.3 ±0.1	98.1 ±0.1	97.6 ±0.2	99.0 ±0.0
Linear	66.6 ±0.1	80.1 ±0.2	95.5 ±0.1	98.1 ±0.0	97.6 ±0.1	98.9 ±0.0
OTAM [3]	67.1 ±0.0	80.2 ±0.2	95.9 ±0.0	**98.4** ±0.1	<u>97.8</u> ±0.1	99.0 ±0.0
TRX-{2,3} [21]	65.5 ±0.1	**81.8** ±0.2	93.4 ±0.2	97.5 ±0.0	96.6 ±0.0	**99.5** ±0.0
ViSiL [16]	67.7 ±0.0	81.3 ±0.0	95.9 ±0.0	98.2 ±0.0	<u>97.8</u> ±0.2	99.0 ±0.1

Learning the Backbone Parameters. We start from the publicly available 34-layer R(2+1)D backbone provided by the TSL codebase. This model is pre-trained on the large Sports-1M dataset [15]. We follow [33] and use a SGD optimizer with a constant learning rate of 0.001 for the backbone and 0.1 for the 64-class linear layer. We perform early-stopping using the 64-class validation dataset. Then, we use the backbone as a feature extractor to extract features from $n = 8$ uniformly sampled clips[5]. The input video clips are composed of 16 consecutive RGB frames with a spatial resolution of 112×112, and the dimensionality of the resulting feature vector is $d = 512$.

---

[3] https://github.com/xianyongqin/few-shot-video-classification.

[4] https://github.com/tobyperrett/trx.

[5] Note that although TSL uses randomly-sampled clips, we found that its performance is usually better when switching to uniformly-sampled clips.

Training Matching-Based Methods. We train the matching-based methods on the training episodes with an SGD optimizer and a constant learning rate of 0.001 for every method except for TRX where we use a learning rate of 0.01. Similar to prior work [3,21], we select the best model using early stopping by measuring performance on the validation set. We learn the projection ϕ jointly with any matching parameters. We set the projection dimension to $D = 1152$ if not stated otherwise. This is equivalent to the dimensionality that TRX [21] uses for its attention layer. We train all the matching methods with the cross-entropy loss that uses softmax with a learnable temperature τ.

Learning Classifiers for TSL. Instead of pairwise matching, TSL [33] learns classifiers at every test episode, using all available support examples. To reproduce TSL we follow [33] and use the Adam optimizer with a constant learning rate of 0.01 for 10 epochs. The original TSL approach uses $n = 10$ clips at test-time, but we set this number to $n = 8$ to keep it the same with all matching methods for a fair comparison. Preliminary experiments show that this choice doesn't affect the performance of TSL at all.

Data Augmentation. During training, videos are augmented with random cropping. We uniformly sample 8 clips from each video with temporal jittering, *i.e.* randomly perturb the starting point of each clip. Additionally, for the Kinetics-100 dataset, we also use random horizontal flipping as data augmentation. Since it is important for SS-v2 to distinguish between left-to-right and right-to-left, we do not use horizontal flipping for that dataset. We only apply a center crop for videos during validation and testing.

4.2   Results

In this section, we report and analyse our results. We first discuss the gains from using temporal representations and the comparison between matching-based and classifier approaches. We then discuss the use of temporal information during matching and present results when varying the number of classes per classification task (test episode). Finally, we compare Chamfer++ to other recently published methods and show that our approach achieves state-of-the-art performance.

Frame or Clip-Based Features? We start by evaluating a number of recent matching-based methods over temporal features. This is an important comparison that is missing from the current few-shot action recognition literature. In Fig. 3, we report one-shot performance for a number of matching methods under a common evaluation setup and using both frame-based (ResNet, blue points) and clip-based (R(2+1)D, orange points) features.

We clearly see that using a spatio-temporal backbone significantly boosts accuracy by more than 10% on all datasets. Interestingly, this is also true for the Kinetics-100 and the UCF-101 datasets that are known to be more biased towards spatial context [14]. Even for this case where context is important, we see that temporal dynamics remain a valuable cue for few-shot action recognition.

It is worth noting that the performance on UCF-101 and Kinetics-100 appears to be saturated when using spatio-temporal representations.

Pairwise Matching or Classifiers? Matching-based methods and classifiers are compared in Table 1 and Fig. 3. All results in the table are computed under a common framework, *i.e.* all methods share representations from an R(2+1)D backbone and are tested on the same set of episodes. We see that for all setups and datasets, several matching approaches outperform TSL. In the 1-shot regime, *most* matching-based methods outperform TSL.

How Useful is Temporal Matching? No significant difference in performance is observed between temporal and non-temporal matching approaches, as highlighted in Table 1 and Fig. 3. On the Kinetics-100 and UCF-101 datasets, where action classes are generally coarser and highly dependent on context [14], most methods we tested perform similarly well. Simply using $f(M) = \max_{ij} m_{ij}$ is enough to achieve top performance for UCF-101, while the proposed

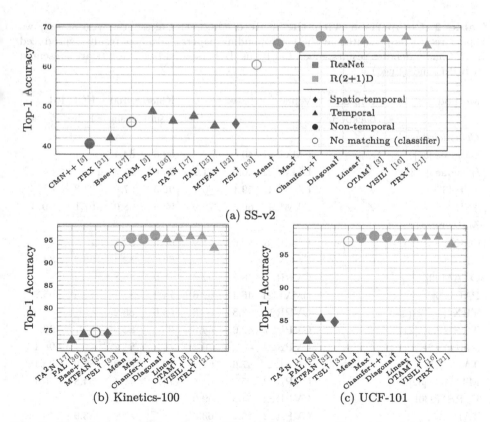

(a) SS-v2

(b) Kinetics-100   (c) UCF-101

Fig. 3. One-shot performance for different backbones, types of matching, or use of parametric classifiers. The different colors account for the different backbones. The different shapes account for the type of matching. † denotes methods reproduced in this study.

Chamfer++ outperforms all other methods on Kinetics-100. When it comes to the finer-grained SS-v2 dataset, we see that, although most non-temporal matching methods lag behind the temporal ones for the 1-shot case, the proposed Chamfer++ method achieves the highest performance without temporal matching. For 5-shot action recognition, TRX, ViSiL, and Chamfer++ perform similarly well, with the last being parameter-free, more intuitive, and faster.

Varying the Number of Classes in an Episode. Figure 4 shows the performance when extending the case of $C_f = 5$ classes to the maximum number of classes, $C_f = 24$, in the 1-shot and 5-shot regime. We see that for 1-shot the proposed non-temporal Chamfer++ method highly outperforms TRX and the classifier-based TSL method in all datasets.

Comparison to the State-of-the-Art. In Table 2, we compare the performance of the proposed Chamfer++ with the corresponding numbers reported in many recent few-shot action recognition papers. Although there is no direct

Table 2. Comparison with the state-of-the-art. Unless otherwise stated, we report results as presented in the corresponding papers. [†] denotes results reproduced in [37] and [‡] denotes results generated by us. Best (second-best) results are presented in **bold** (underlined).

Method	Clip-based backbone	Venue	SS-v2 1-shot	SS-v2 5-shot	Kinetics-100 1-shot	Kinetics-100 5-shot	UCF-101 1-shot	UCF-101 5-shot
Classifier-based								
Baseline [37]		BMVC21	40.8	59.2	69.5	84.4	-	-
Baseline + [37]		BMVC21	46.0	61.1	74.6	86.6	-	-
TSL [33]	✓	PAMI21	59.1	<u>80.1</u>	92.5	97.8	94.8	-
TSL [‡]	✓	PAMI21	<u>60.6</u>	79.9	<u>93.6</u>	<u>98.0</u>	97.1	**99.4**
Matching-based								
CMN++ [†] [35,37]		ECCV18	40.6	51.9	65.9	82.7	-	-
ARN [34]	✓	ECCV20	-	-	63.7	82.4	66.3	83.1
OTAM [3]		CVPR20	48.8	52.3	73.0	85.8	-	-
PAL [36]		BMVC21	46.4	62.6	74.2	87.1	85.3	95.2
TRX-{1} [21]		CVPR21	38.8	60.6	63.6	85.2	-	-
TRX-{2,3} [21]		CVPR21	42.0	64.6	63.6	85.9	-	96.1
STRM-{2} [27]		CVPR22	-	70.2	-	91.2	-	98.1
TA²N [17]		AAAI22	47.6	61.0	72.8	85.8	81.9	95.1
MTFAN [32]		CVPR22	45.7	60.4	74.6	87.4	84.8	95.1
HyRSM [30]		CVPR22	54.3	69.0	73.7	86.1	-	-
TAP [25]		CVPR22	45.2	63.0	-	-	83.9	95.4
CPM [19]		ECCV22	59.6	-	81.0	-	79.0	-
Chamfer++ [‡]	✓		**67.8**	**81.6**	**96.1**	**98.3**	**97.7**	99.3

comparison between all these methods, *i.e.*, no common setup or backbones, we present all results jointly to show the overall progress in the task.

4.3 Chamfer Matching Ablation and Interpretability

In Table 3, we present an ablative study with the 1-shot performance of all the different variants of Chamfer++ we discuss in Sect. 3.3 on the SS-v2 dataset. Combining support-based and query-based Chamfer matching helps learn a better projection ϕ, while both joint matching and the use of clip tuples improve performance compared to the vanilla variant. Although using ordered clip tuples improves for 1-shot on SS-v2, overall, we see in Table 1 that using all or ordered tuples results in more or less similar performance.

Table 3. Chamfer++ variants.

Method	l	Joint	Tupl.	1-shot	5-shot
Chamfer-Q	1			65.7 ±0.1	79.7 ±0.1
Chamfer-S	1			65.3 ±0.1	79.1 ±0.2
Chamfer	1			66.9 ±0.1	80.0 ±0.2
Chamfer+	1	✓		66.9 ±0.1	80.7 ±0.2
Chamfer++	2	✓	all	67.1 ±0.1	80.8 ±0.2
Chamfer++	2	✓	ord.	67.7 ±0.1	81.4 ±0.2
Chamfer++	3	✓	all	67.0 ±0.3	80.8 ±0.1
Chamfer++	3	✓	ord.	67.8 ±0.2	81.6 ±0.1

Which are the Most Informative Clips? As we presented in Sect. 3.1, we sample and encode a set of clips to represent each video. Not all clips are, however, equally valuable for the matching process. To provide some insight and illustrate the proposed method's interpretability, we study a qualitative example from SS-v2 in Fig. 5. When matching the query video to the two support videos belonging to the same class (top), we see many clip-to-clip correspondences with high magnitudes (shown as thicker lines). This is not the case when matching to the support videos belonging to a negative class (bottom) and we can only see a single strong correspondence. Upon better inspection, we saw that the motion of picking up the object in the fourth clip of the query video does exhibit a left-to-right motion locally that matches the negative class. Nevertheless, the other correspondences are weaker. Leveraging more than one clip helps Chamfer matching to disambiguate while, at the same time, its inherent selectivity makes it more robust against noisy correspondences. This example illustrates that not all the clip correspondences are equally valuable to compute a video-to-video similarity metric. The matching step is a dynamic way to select the most informative clip correspondences that highlight the similarity between two videos.

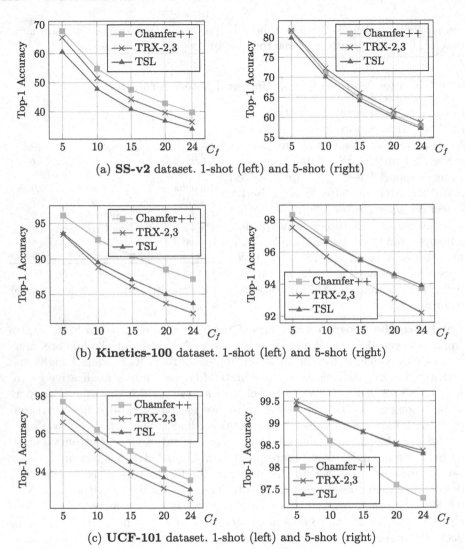

(a) **SS-v2** dataset. 1-shot (left) and 5-shot (right)

(b) **Kinetics-100** dataset. 1-shot (left) and 5-shot (right)

(c) **UCF-101** dataset. 1-shot (left) and 5-shot (right)

Fig. 4. Impact of the number of classes per episode (C_f) on three datasets.

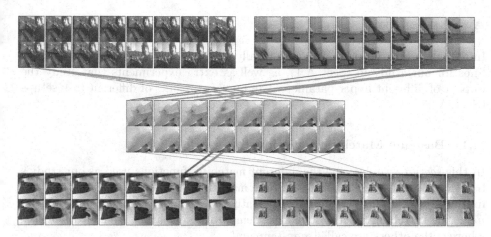

Fig. 5. Matching a query video (middle) from "Picking something up" with two support videos of "Picking something up" (top) and two support videos of "Pulling something from left to right" (bottom). Each video consists of 8 clips with 16 frames each. We only show the first and last frames on top of each other in the figure (see project page for animated versions). In grey, we draw the correspondences between the query and the support videos selected by Chamfer (query-based). In red, we draw the correspondence selected by max (single strongest correspondence). Line thickness corresponds to the pairwise similarity. (Color figure online)

5 Conclusion

A number of recent few-shot learning papers are abandoning the meta-learning protocol for representation learning [5,31,33,37] and show that adaptation from the best possible features leads to better performance. In the quest for rapid adaptation at test time, we also adopt this setup and show that, given strong visual representations, simple matching-based methods are really effective and able to beat both classifier-based and more complex matching-based approaches on many common benchmarks for few-shot action recognition. We further show that temporal information in the matching provides no particular benefit compared to the ability to learn or adapt from strong temporal features, and introduce an intuitive matching-based method that is not only parameter-free but also easy to visualize and interpret.

Acknowledgements. This work was supported by Naver Labs Europe, by Junior Star GACR GM 21-28830M, and by student grant SGS23/173/OHK3/3T/13. The authors would like to sincerely thank Toby Perrett and Dima Damen for sharing their early code and supporting us, Diane Larlus for insightful conversations, feedback, and support, and Zakaria Laskar, Monish Keswani, and Assia Benbihi for their feedback.

A   Appendix

In this appendix, we present more formally the matching functions used as baselines for our study in (Sect. A.1), as well as extra experiments that study the impact of different hyper-parameters and present results of different task setups (Sect. A.2).

A.1   Baseline Matching Functions

In this section, we describe the different matching functions used as our study's baseline, and depicted in Fig. 6. Some matching functions use temporal information by leveraging the absolute or relative position of the pairwise similarities m_{ij}. The matching functions that use temporal information are called *temporal* whereas the others are called *non-temporal*.

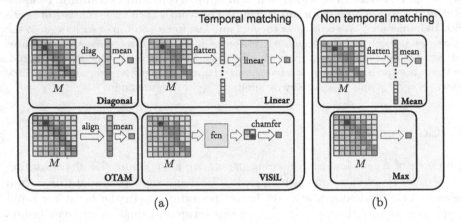

Fig. 6. Matching functions on the temporal similarity matrix M. We show how each method estimates a scalar video-to-video similarity given the input pairwise similarity matrix. The functions can be classified as a) temporal or b) non-temporal whether they use the temporal position of the features or not.

Temporal Matching Functions. We provide a list of the temporal matching functions implemented in this study as baselines. Some of them were already introduced in prior work.

Diagonal (Diag) is used as a baseline in prior work [3]. It is given by $s(M) = \sum_{ij} m_{ii}/n$. It assumes temporally aligned video pairs.

OTAM [3] uses and extends Dynamic Time Warping [20] to find an alignment path on M over which similarities are averaged to produce the video-to-video similarity. A differentiable variant is used for training.

Flatten+FC (Linear) is a simple baseline we use to learn temporal matching by flattening M and feeding it to a Fully Connected (FC) layer without bias and with a single scalar output. Video-to-video similarity is therefore given by $s(M) = \sum_{ij} w_{ij} m_{ij}$, where w_{ij} are learnable parameters which are n^2 in total.

Table 4. Impact of learning a feature projection on performance of matching-based methods. [†] denotes hand-crafted matching methods, *i.e.* no training is performed for the cases where a feature projection is not learned.

Method	Learned Proj.	SS-v2		Kinetics 100	
		1-shot	5-shot	1-shot	5-shot
Max[†]		63.40	75.80	94.90	97.50
Max	✓	64.97	78.97	95.27	98.30
		(↑ 1.57)	(↑ 3.17)	(↑ 0.37)	(↑ 0.80)
Chamfer++ (l=3)[†]		64.50	79.50	94.10	98.10
Chamfer++ (l=3)	✓	67.83	81.60	96.10	98.30
		(↑ 3.33)	(↑ 2.10)	(↑ 2.00)	(↑ 0.20)
Mean[†]		61.40	75.40	92.50	97.50
Mean	✓	65.77	79.13	95.53	98.10
		(↑ 4.37)	(↑ 0.73)	(↑ 0.03)	(↑ 0.00)
OTAM [3][†]		63.70	76.40	93.70	97.90
OTAM [3]	✓	67.10	80.23	95.93	98.37
		(↑ 3.40)	(↑ 3.83)	(↑ 2.23)	(↑ 0.47)

ViSiL [16] is an approach originally introduced for the task of video retrieval. We apply it to few-shot action recognition for the first time. A small Fully Convolutional Network (FCN) is applied on M. Its output is a filtered temporal similarity matrix, and the Chamfer similarity is applied on it. The filtering is performed according to the small temporal context captured by the small receptive field of this network.

Non-temporal Matching Functions. We provide a list of the non-temporal matching functions that were implemented in this study as baselines.

Mean is used as a baseline in prior work [3]. It is given by $s(M) = \sum_{ij} m_{ij}/n^2$. It supposes all the clip pairs should contribute equally to the similarity score.

Max is used as a baseline in prior work [3]. It is given by $s(M) = \max_{ij} m_{ij}$. It supposes that selecting the best matching clip pair is enough to recognize the action.

A.2   Additional Ablations and Impact of Hyper-parameters

In this section, we present additional ablations to evaluate the impact of the feature projection head, the ordering of the tuples, and the number of examples per class used in the support set. We also report the impact of using the different variants for the Kinetics-100 and UCF-101 datasets.

Table 5. Impact of the dimension size of the feature projection head for Chamfer++ using ordered-tuples and $l = 3$.

Method	D	SS-v2		Kinetics-100	
		1-shot	5-shot	1-shot	5-shot
Chamfer++ (l=3)		64.5 ±0.0	79.5 ±0.0	94.1 ±0.0	98.1 ±0.0
Chamfer++ (l=3)	512	67.3 ±0.2	81.2 ±0.0	95.9 ±0.1	98.2 ±0.1
Chamfer++ (l=3)	1024	67.8 ±0.3	81.3 ±0.1	96.1 ±0.2	98.3 ±0.0
Chamfer++ (l=3)	1152	67.8 ±0.2	81.6 ±0.1	96.1 ±0.1	98.3 ±0.0
Chamfer++ (l=3)	2048	67.9 ±0.0	81.8 ±0.2	96.1 ±0.1	98.3 ±0.0

Table 6. Chamfer++ variants on the three datasets: UCF-101, Kinetics-100 and SS-v2.

Method	l	Joint	Tupl.	SS-v2		Kinetics-100		UCF-101	
				1-shot	5-shot	1-shot	5-shot	1-shot	5-shot
Chamfer-Q	1			65.7 ±0.1	79.7 ±0.1	95.5 ±0.1	98.1 ±0.1	97.8 ±0.1	99.0 ±0.1
Chamfer-S	1			65.3 ±0.1	79.1 ±0.2	95.4 ±0.0	98.2 ±0.1	97.7 ±0.2	98.9 ±0.1
Chamfer	1			66.9 ±0.1	80.0 ±0.2	96.0 ±0.1	98.3 ±0.1	97.9 ±0.2	99.0 ±0.1
Chamfer+	1	✓		66.9 ±0.1	80.7 ±0.2	96.0 ±0.1	98.2 ±0.1	97.9 ±0.2	99.2 ±0.1
Chamfer++	2	✓	all	67.1 ±0.1	80.8 ±0.2	96.1 ±0.1	98.3 ±0.1	97.8 ±0.1	99.2 ±0.1
Chamfer++	2	✓	ord.	67.7 ±0.1	81.4 ±0.2	96.1 ±0.1	98.3 ±0.1	97.9 ±0.1	99.2 ±0.1
Chamfer++	3	✓	all	67.0 ±0.3	80.8 ±0.1	96.2 ±0.2	98.4 ±0.0	97.8 ±0.1	99.2 ±0.1
Chamfer++	3	✓	ord.	67.8 ±0.2	81.6 ±0.1	96.1 ±0.1	98.3 ±0.0	97.7 ±0.0	99.3 ±0.0

The Impact of the Projection Layer for matching methods is validated in Table 4. The performance is consistently improved on all setups and methods by including and learning a projection layer. Although the backbone is trained with TSL on the same meta-train set, the projection layer allows features and values in the temporal similarity matrix to better align with each matching process.

Projection Head Dimension. To match the work from [21], we set the projection dimension to $D = 1152$. This section evaluates the effect of using different values for D. The results are reported in Table 5. A minimum value of $D = 1024$ seems enough and could be used for future experiments.

Impact of Using Different Variants. We report the accuracy for the different variants of Chamfer++ for the Kinetics-100 and UCF-101 datasets in Table 6. As for SS-v2, both variants improve performances compared to the vanilla approach.

Impact of Ordering the Clip-Tuples. In this section, we evaluate the impact of using ordered clip feature tuples t^l versus using all the clip feature tuples t^l_{all}. The comparison between ordered tuples and all tuples is presented in Table 7. On the SS-v2 dataset, using ordered clip feature tuples boosts the accuracy. On the Kinetics-100 and the UCF-101 datasets, using ordered tuples doesn't provide a boost and can even slightly harm the performance. Since the number of tuples is significantly lower when they are in order, using the ordered clip feature tuples is preferable.

Table 7. Impact of using ordered clip feature tuples vs all the features for different values of l.

Method	l	Ordered Tuples	SS-v2		Kinetics-100		UCF-101	
			1-shot	5-shot	1-shot	5-shot	1-shot	5-shot
Chamfer++	2		67.10	80.80	96.10	98.33	97.80	99.23
Chamfer++	2	✓	67.73	81.40	96.10	98.33	97.87	99.23
			(↑ 0.63)	(↑ 0.80)	(↑ 0.00)	(↑ 0.00)	(↑ 0.07)	(↑ 0.00)
Chamfer++	3		67.03	80.83	96.17	98.37	97.77	99.23
Chamfer++	3	✓	67.83	81.60	96.10	98.30	97.73	99.27
			(↑ 0.80)	(↑ 0.77)	(↓ 0.07)	(↓ 0.07)	(↓ 0.04)	(↑ 0.04)

Impact of the Number of Examples per Class. The impact of k is shown in Fig. 7 by measuring performance for an increasing number of support examples per class while keeping the number of classes fixed, $C_t = C_f = 5$. We observe that TSL and TRX have inferior performances for the low-shot regime, while their performance increases faster with the number of shots. In the low-regime, Chamfer-QS++ outperforms the other methods and still keeps some benefits while the number of shots increases.

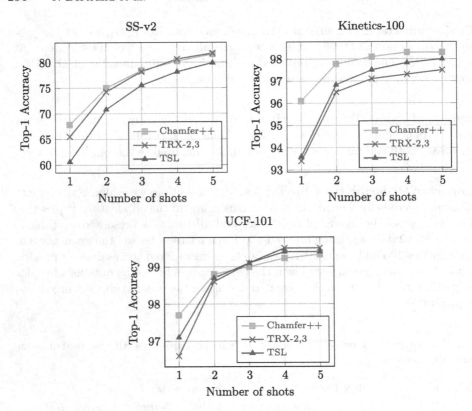

Fig. 7. Evolution of the accuracy with the number of examples per class in the support set for the three datasets.

References

1. Bishay, M., Zoumpourlis, G., Patras, I.: Tarn: temporal attentive relation network for few-shot and zero-shot action recognition. In: BMVC (2019)
2. Cao, K., Brbić, M., Leskovec, J.: Concept learners for few-shot learning. In: ICLR (2021)
3. Cao, K., Ji, J., Cao, Z., Chang, C.-Y., Niebles, J.C.: Few-shot video classification via temporal alignment. In: CVPR (2020)
4. Chang, C.-Y., Huang, D.-A., Sui, Y., Fei-Fei, L., Niebles, J.C.: D3tw: discriminative differentiable dynamic time warping for weakly supervised action alignment and segmentation. In: CVPR (2019)
5. Chen, W.-Y., Liu, Y.-C., Kira, Z., Wang, Y.-C.F., Huang, J.-B.: A closer look at few-shot classification. In: ICLR (2019)
6. Doersch, C., Gupta, A., Zisserman, A.: Crosstransformers: spatially-aware few-shot transfer. In: NeurIPS (2020)
7. Dvornik, M., Hadji, I., Derpanis, K.G., Garg, A., Jepson, A.: Aligning common signal between sequences while dropping outliers. In: NeurIPS, Drop-dtw (2021)
8. Dwivedi, S.K., Gupta, V., Mitra, R., Ahmed, S., Jain, A.: Towards few shot learning for action recognition. In: ICCVW, Protogan (2019)

9. Finn, C., Abbeel, P., Levine, S.: Model-agnostic meta-learning for fast adaptation of deep networks. In: ICML (2017)
10. Finn, C., Xu, K., Levine, S.: Probabilistic model-agnostic meta-learning. In: NeurIPS (2018)
11. Gidaris, S., Komodakis, N.: Dynamic few-shot visual learning without forgetting. In: CVPR (2018)
12. Goyal, R., et al.: The "something something" video database for learning and evaluating visual common sense. In: ICCV (2017)
13. Grant, E., Finn, C., Levine, S., Darrell, T., Griffiths, T.: Recasting gradient-based meta-learning as hierarchical bayes. In: ICLR (2018)
14. Huang, D.-A., et al.: What makes a video a video: Analyzing temporal information in video understanding models and datasets. In CVPR (2018)
15. Karpathy, A., Toderici, G., Shetty, S., Leung, T., Sukthankar, R., Fei-Fei, L.: Large-scale video classification with convolutional neural networks. In: CVPR (2014)
16. Kordopatis-Zilos, G., Papadopoulos, S., Patras, I., Kompatsiaris, I.: Fine-grained spatio-temporal video similarity learning. In: ICCV, Visil (2019)
17. Li, S., et al.: Ta2n: two-stage action alignment network for few-shot action recognition. In: AAAI (2022)
18. Lifchitz, Y., Avrithis, Y., Picard, S., Bursuc, A.: Dense classification and implanting for few-shot learning. In: CVPR (2019)
19. Huang, Y., Yang, L., Sato, Y.: Compound prototype matching for few-shot action recognition. In: ECCV (2022)
20. Müller, M.: Dynamic time warping. Information Retrieval for Music and Motion (2007)
21. Perrett, T., Masullo, A., Burghardt, T., Mirmehdi, M., Damen, D.: Temporal-relational cross transformers for few-shot action recognition. In: CVPR (2021)
22. Andrei, A., et al.: Meta-learning with latent embedding optimization. In: ICLR (2019)
23. Snell, J., Swersky, K., Zemel, R.: Prototypical networks for few-shot learning. In: NeurIPS (2017)
24. Soomro, K., Zamir, A.R., Shah, M.: Ucf101: a dataset of 101 human actions classes from videos in the wild. In: CRCV-TR-12-01 (2012)
25. Su, B., Wen, J.-R.: Temporal alignment prediction for supervised representation learning and few-shot sequence classification. In: ICLR (2022)
26. Sung, F., Yang, Y., Zhang, L., Xiang, T., Torr, P., Hospedales, T.: Learning to compare: Relation network for few-shot learning. In: CVPR (2018)
27. Thatipelli, A., Narayan, S., Khan, S., Anwer, R.M., Khan, F.S., Ghanem, B.: Spatio-temporal relation modeling for few-shot action recognition. In: CVPR (2022)
28. Tran, D., Wang, H., Torresani, L., Ray, J., LeCun, Y., Paluri, M.: A closer look at spatiotemporal convolutions for action recognition. In: CVPR (2018)
29. Vinyals, O., Blundell, C., Lillicrap, T., Kavukcuoglu, K., Wierstra, D.: Matching networks for one shot learning. In: NeurIPS (2016)
30. Wang, X., et al.: Hybrid relation guided set matching for few-shot action recognition. In: CVPR (2022)
31. Wang, Y., Chao, W.-L., Weinberger, K.Q., van der Maaten, L.: Simpleshot: Revisiting nearest-neighbor classification for few-shot learning. arXiv preprint arXiv:1911.04623 (2019)
32. Wu, J., Zhang, T., Zhang, Z., Wu, F., Zhang, Y.: Motion-modulated temporal fragment alignment network for few-shot action recognition. In: CVPR (2022)

33. Xian, Y., Korbar, B., Douze, M., Torresani, L., Schiele, B., Akata, Z.: Generalized few-shot video classification with video retrieval and feature generation. IEEE TPAMI (2021)
34. Zhang, H., Zhang, L., Qi, X., Li, H., Torr, P.H.S., Koniusz, P.: Few-shot action recognition with permutation-invariant attention. In: Vedaldi, A., Bischof, H., Brox, T., Frahm, J.-M. (eds.) ECCV 2020. LNCS, vol. 12350, pp. 525–542. Springer, Cham (2020). https://doi.org/10.1007/978-3-030-58558-7_31
35. Zhu, L., Yang, Y.: Compound memory networks for few-shot video classification. In: Ferrari, V., Hebert, M., Sminchisescu, C., Weiss, Y. (eds.) ECCV 2018. LNCS, vol. 11211, pp. 782–797. Springer, Cham (2018). https://doi.org/10.1007/978-3-030-01234-2_46
36. Zhu, X., Toisoul, A., Pérez-Rúa, J.-M., Zhang, L., Martinez, B., Xiang, T.: Few-shot action recognition with prototype-centered attentive learning. In: BMVC (2021)
37. Zhu, Z., Wang, L., Guo, S., Wu, G.: A new baseline and benchmark. In: MVC, A closer look at few-shot video classification (2021)

Accuracy of Parallel Distance Mapping Algorithms When Applied to Sub-Pixel Precision Transform

Daniel Ericsson[1], Åsa Detterfelt[1(✉)], and Ingemar Ragnemalm[2]

[1] Mindroad AB, Teknikringen 1F, Linköping 583 30, Sweden
{daniel.ericsson,asa.detterfelt}@mindroad.se
[2] Linköping University, Linköping 581 83, Sweden
ingemar.ragnemalm@liu.se

Abstract. The problem of distance mapping is a thoroughly studied topic in the field of image processing. Much focus has been spent on perfecting algorithms, calculating the Euclidean Distance Transform on binary images, in order to achieve the highest accuracy as efficiently as possible. Less focus has been spent on perfecting algorithms calculating the Euclidean Distance Transform on non-binary images where a much higher degree of precision is made possible by increasing the granularity of the input image. This study focuses on taking three different established methods for calculating distance mapping on binary images and applying them to non-binary images. We have found that the simpler algorithms, which simply propagate distance information according to a predetermined pattern work well with our new transform. The more analytical and in the case of the binary image, exact method, however, does not work as well with our new transform. Its way of propagating distance information is making too strict assumptions about the geometry of an image, which hold true for binary images, but not for non-binary images.

Keywords: Anti-Aliased Euclidean Distance Transform · AAEDT · Parallel Algorithms · SKW · JFA · PBA

1 Introduction

The binary Euclidean Distance Transform (EDT) is a widely studied operation in image processing, with several established efficient algorithms. The Anti-Aliased Euclidean Distance Transform (AAEDT) is an operation that solves the same problem as the binary EDT, but with sub-pixel precision. There are not as many established algorithms to implement AAEDT as there are for EDT.

This paper presents the resulting accuracy of AAEDT when implemented using three established parallel algorithms for EDT. These three algorithms are; SKW, Jump Flooding (JFA), and Parallel Banding (PBA).

Our results show that SKW and JFA produce comparable and good accuracy of the transform. The more complex PBA, however, relies on assumptions of the binary transform which do not hold for AAEDT, and produce much greater errors than the other algorithms.

© The Author(s), under exclusive license to Springer Nature Switzerland AG 2023
R. Gade et al. (Eds.): SCIA 2023, LNCS 13885, pp. 237–250, 2023.
https://doi.org/10.1007/978-3-031-31435-3_16

1.1 EDT

The Euclidean Distance Transform (EDT) is an operation that, given a black and white image, produces a map of Euclidean distances to the closest white pixel, also referred to as a foreground pixel. A predecessor to EDT was first presented by Rosenfeld and Pfalz [18]. Montanari [13] extended it to weighted metrics, which was further explored by Barrow [2] and Borgefors [3]. Danielsson [5] introduced the Euclidean Distance Transform. Danielsson's sequential algorithm did, however, produce some errors, which is why the problem of making the operation error-free has been explored thoroughly by Rutowitz [19], Ragnemalm [15], Paglieroni [14], Eggers [6], Meijster, Roerdink and Hesselink [12], and Maurer and Raghavan [11].

1.2 AAEDT

There are other similar transforms, devised to create distance maps. One such has been proposed by Gustavson and Strand [7], named the Anti-Aliased Euclidean Distance Transform (AAEDT), which uses grayscale information along the borders to provide sub-pixel precision. Gustavson and Strand showed that their transform produced results with greater precision than the binary transform. The algorithm was later extended to 3D by Ilic et al. [8]. Lindblad and Sladoje [9] made a related algorithm based on edge samples.

AAEDT is defined for grayscale images with anti-aliased edges. The intensity of a pixel corresponds to how much of a foreground object covers the space of the pixel. This information is used in AAEDT to approximate the location of, and distance to, an object's edge in the original source image, with sub-pixel precision.

1.3 This Study

The aim of this study is to adapt selected parallel algorithms, for computing the binary EDT, to the domain of AAEDT by Gustavson and Strand and evaluate the results. The implemented algorithms are made available online [1].

The performance of the adapted algorithms is not studied here. The only change to the original algorithms is a new kernel for calculating a new distance measure. This is assumed to give a linear offset in performance to the original algorithms as the propagation of information remains the same as in the original algorithms. The relative performance of these algorithms has been studied previously by Thanh-tung et al. [4].

2 Background

Several parallel algorithms exist for EDT. We will here focus on three, Jump Flooding (JFA) according to Danielsson [5] and Rong et al. [17], SKW by Ragnemalm [16] and Schneider, Kraus and Westermann [20], giving the algorithm its name from the three later author's initials, and Parallel Banding (PBA) by Thanh-tung et al. [4].

These three algorithms can all be used to first compute a pixel wise mapping to the Voronoi diagram of the input image's foreground pixels, as an intermediate step in computing the distance transform. This mapping is referred to as the discrete Voronoi diagram. A Voronoi diagram is a mapping from a discrete set of points to a set of regions where each region is the set of points with a given point, in the input set, as the closest point. When creating a Distance Transform, the Voronoi region of an image, which a pixel belongs to, can be used as an intermediate calculation to find the distance to the closest Voronoi site. This works for the EDT where we have a discrete set of foreground objects (the foreground pixels in the image). In the case of AAEDT we are not working with a discrete set of foreground objects in the same way, as we are trying to represent sloped edges of foreground objects, which consist of infinitely many points.

Each algorithm is preceded by an initialization where each object pixel is assigned itself as its closest object pixel in an incomplete discrete Voronoi diagram.

2.1 AAEDT

The anti-aliased Euclidean distance is defined as

$$d_{AAEDT}(p) = \min_{p_i \epsilon \Omega}(|d_E(p, p_i)| + d_f(p_i))$$

where Ω is the set of pixels with non-zero intensity, $d_E(p, p_i)$ is the Euclidean distance between the centers of pixels p and p_i, and d_f is the subpixel distance within p_i to the closest overlapping edge of a foreground object in a super-sampled image as seen in Fig. 1.

The version of the Anti-Aliased Euclidean Distance Transform, subject to the study in this paper, utilizes the gradient of the image in a pixel, and the intensity of the pixel to better approximate the location of an edge on a pixel. In Fig. 1 the gradient is denoted g.

The gradient g in Fig. 1 is defined by the 3×3 convolutional filters in Fig. 2. The gradient is assumed to be parallel to the line between two pixel centers when calculating the distance between the center of one pixel and the object edge on another pixel. A distance offset d_f from an edge pixel's center to an object's edge is added on to the distance to the edge pixel's center. The distance offset d_f is calculated based on the gradient, and intensity of the edge pixel.

The discrete Voronoi diagram used for the calculations when computing the EDT uses the centroid c in Fig. 1 as a foreground object for each foreground pixel. For our implementations we approximate a new point c' for each distance calculation. The location of this point is not consistent throughout the entire computation, due to the distance measure of AAEDT is dependent on the slope of the vector between a foreground and background pixel for which the distance between is measured.

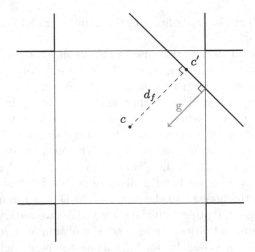

Fig. 1. Visualization of distance offset inside a pixel.

$$\begin{bmatrix} -1 & 0 & 1 \\ -\sqrt{2} & 0 & \sqrt{2} \\ -1 & 0 & 1 \end{bmatrix} \begin{bmatrix} -1 & -\sqrt{2} & -1 \\ 0 & 0 & 0 \\ 1 & \sqrt{2} & 1 \end{bmatrix}$$

Fig. 2. Convolutional filters for gradient, x-axis (left), and y-axis (right).

2.2 SKW

This algorithm performs two scans per dimension of the image. In two dimensions this results in four total scans.

The first scan starts with the second to left-most column and for each pixel in parallel, update their closest object pixel based on the closest object pixel in the three neighboring pixels in the previous column. The vector templates in Fig. 3 are applied to all pixels in the image. All columns are handled sequentially from left to right. This operation is then performed on each column, in the opposite direction, and then for each row, twice, as well.

SKW performs $\mathcal{O}(n^2)$ work with an n×n pixel image as input, which is optimal as every pixel needs to be processed at least once. Due to its operations sequential nature, it can, despite this optimal work, suffer from poor parallelization. This is less of an issue on larger images.

2.3 JFA

This algorithm performs a number of iterations equal to the base-2 logarithm of the images side length. In each iteration the template shown in Fig. 4 is applied to every pixel in the image, in parallel. The variable s starts as equal to half the image's side length and is halved each iteration. When the template is applied

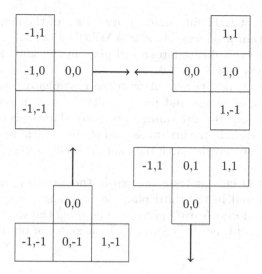

Fig. 3. Vector templates for SKW, with relative coordinates.

the best candidate for the closest object pixel is chosen from the pixels in the discrete Voronoi diagram corresponding to the pixels in the template.

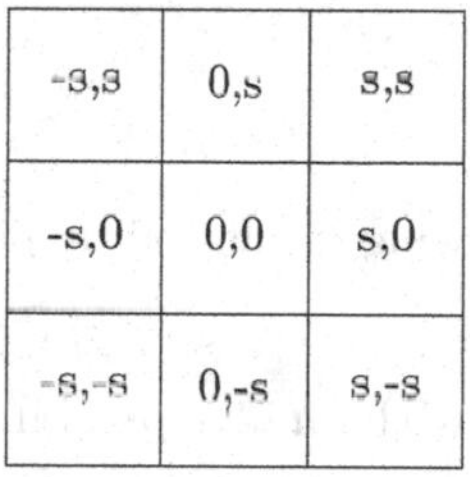

-s,s	0,s	s,s
-s,0	0,0	s,0
-s,-s	0,-s	s,-s

Fig. 4. Vector template for JFA, with relative coordinates.

Rong et al. [17] presented several versions of JFA, with various numbers of iterations and different orders for the various jump lengths. All variations of JFA perform $\mathcal{O}(n^2 \log(n))$ work with an n×n pixel image as input. It performs worse than optimal work, but parallelizes this work well, making it perform well on smaller images in particular.

2.4 PBA

The Parallel Banding Algorithm by Thanh-tung et al. [4] performs three distinct phases. In each phase a parallel divide and conquer approach is taken to maximize utilization of the parallel hardware. The algorithm performs optimal work and produces an exact EDT for binary images. Part of the aim of this study is to see if this algorithm produces an exact AAEDT as well.

In the first phase the closest foreground pixel in the same row is computed for each pixel in every row.

The second phase takes the result from the first phase, and constructs, for each column a stack of foreground pixels, which Voronoi region overlaps with the column. This is done by performing a geometrical analysis of the candidates generated for the column in the first phase and eliminating any candidates whose Voronoi region does not overlap with the column. An example of this can be seen in Fig. 5.

Given the stack of candidates generated in the second phase, the final distance map is generated in the third phase by stepping along the column and stack, assigning the closest feature pixels and popping the stack, until no better candidate is found and then proceeding to the next set of pixels in the column.

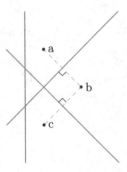

Fig. 5. Pixel b being eliminated by pixels a and c.

PBA performs $\mathcal{O}(n^2)$ work, with an n×n pixel image as input, and is designed to parallelize this work efficiently.

2.5   Problems with PBA for AAEDT

In order to efficiently parallelize the work in PBA, some assumptions of the geometry of the domain are made. Some of these assumptions break when moving to the sub-pixel domain of AAEDT.

From the first phase of PBA at most one pixel is selected from each row for each column as a candidate for the closest feature pixel to any pixel in that column. In the example illustrated in Fig. 6 we have a feature with an edge at 45°, located on the corner of a pixel a. In PBA information about pixels p_1 and p_2 are propagated to a, while information about p_0 is blocked in the first phase of the algorithm as information about p_1 takes precedence at p_1. This results in the distance assigned to a being $d_1 = d_2 = 1$ while the actual shortest distance is $d_0 = \dfrac{\sqrt{2}}{2}$ giving an error $e = 1 - \dfrac{\sqrt{2}}{2} \approx 0.29$.

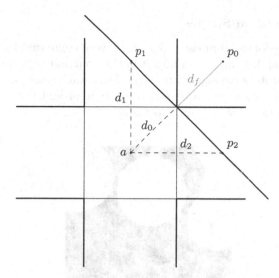

Fig. 6. Candidate distances for shortest distance to pixel a close to a sloped edge.

3   Method

3.1   Implementation

The three adaptations of the algorithms SKW, JFA and PBA, as well as a brute force algorithm were all implemented in C++ using CUDA. The implementations are not optimized for performance, but rather readability. The brute force implementation is used as a baseline for maximum accuracy. The brute-force implementation compares all possible distances for each pixel and assigns the shortest distance, as defined by the AAEDT distance measure.

SKW and JFA purely rely on the distance measure, defined by the transform by Gustavson and Strand [7], in all propagation of candidates for the closest object pixel. PBA, however, relies on a more complex analysis of homogeneous regions in the resulting discrete Voronoi diagram.

In the case of SKW and JFA, replacing the distance measure was a sufficient modification to compute the Anti-aliased EDT. This was not sufficient for PBA. In the second phase of PBA, candidates for the closest object pixels in each column are eliminated based on an exact geometric analysis of, in the case of the discrete transform, the center points of the candidate pixels, relative to the specific column. Our implementation projects a point for each candidate pixel on the imagined edge of an object, using the image's local gradient, for this analysis.

The version of JFA, adapted here, is the most basic version with halving jump length each iteration.

The adapted algorithms will be referred to as AAEDT-SKW, AAEDT-JFA and AAEDT-PBA. Implementations of these algorithms can be found online [1].

3.2   Test on Uniform Shapes

The resulting transforms from each algorithm were compared to two oracles, in this test; the actual distance to an object in the source image, and the resulting transform of the brute-force algorithm using the distance measure from the Anti-aliased EDT. Circles were used as feature objects in order to produce all possible orientations of edges, as well as an exact oracle.

Fig. 7. The input image with uniform shapes.

The experiments all used a single two-dimensional source image where objects were filled circles, as seen in Fig. 7. The source image was super-sampled uniformly using 16×16 samples per pixel for each input image. The input images were square with side lengths of 64, 128, 256, 512, 1024, and 2048 pixels.

3.3   Test on Non-Uniform Shapes

The three algorithms were applied to the Tattoo data set in this test. The data set was used by Linnér and Strand [10], and made available along with the algorithm implementations online [1]. The resulting transforms were compared to the transform generated by the brute-force implementation of the AAEDT. The data set consists of 28 images, each with a size of 512×512 pixels. The images were read with a gray-scale color-depth of 8-bits. One of the images and its corresponding AAEDT can be seen illustrated in Fig. 8.

The Tattoo data set was chosen due to its abundance of different types of edge shapes and features when compared to the image used in the test on uniform shapes, consisting solely of circle objects.

4   Results

4.1   Test on Uniform Shapes

The average absolute errors for the images produced by each algorithm, when compared to the brute-force implementation, are shown in Fig. 9. AAEDT-JFA

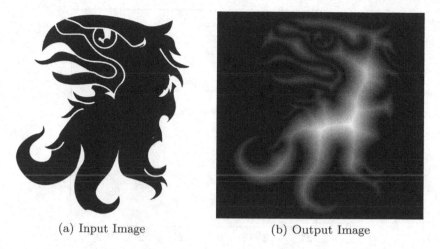

(a) Input Image   (b) Output Image

Fig. 8. One of the images in the Tattoo data set, 8(a) and the generated output image by AAEDT, 8(b) where lighter color in the output image corresponds to a longer distance to the white background in the input image.

and AAEDT-SKW show comparable accuracy, with AAEDT-JFA producing slightly closer to correct results. AAEDT-PBA, however, produces significant errors for all image sizes.

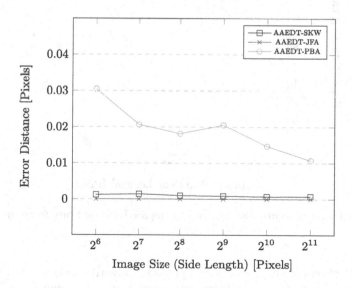

Fig. 9. Average absolute distance error compared to the brute-force implementation.

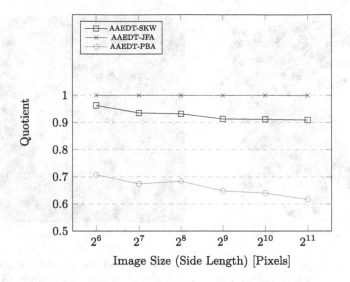

Fig. 10. The ratio of correct (according to the brute-force implementation) pixels to total pixels, for different image sizes.

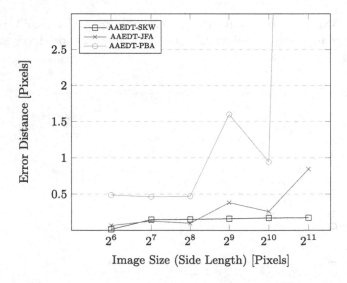

Fig. 11. Maximum absolute distance error compared to the brute-force implementation.

Figure 10 shows what proportion of pixels the algorithms generated the exact same distance for, as the brute-force implementation of the transform. The quotient takes into account all pixels in the image, including object pixels. The input image consists of approximately 40% object pixels, which are trivial to assign the correct value, meaning that AAEDT-PBA computes the correct distance for less than half of the non-trivial to-compute pixels. AAEDT-SKW and AAEDT-JFA perform better than AAEDT-PBA in this metric as well. AAEDT-SKW computes the correct distance for most pixels, and from Fig. 11 we know that those distances it does not compute correctly are still close to correct. AAEDT-JFA computes the correct distance for close to all pixels, but when errors occur, they can be larger than those AAEDT-SKW produces.

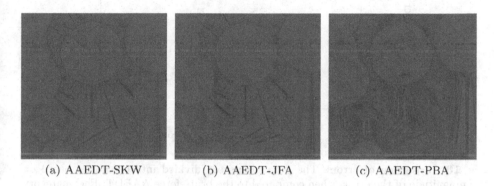

(a) AAEDT-SKW   (b) AAEDT-JFA   (c) AAEDT-PBA

Fig. 12. Error of AAEDT-SKW, AAEDT-JFA and AAEDT-PBA on a 512×512 image, compared to ideal transform. Medium gray is zero error, red is an overestimation of +0.1, and blue is an underestimation of -0.1. Errors are clamped to [-0.1, +0.1].

Error images for each algorithm can be seen in Fig. 12, where blue is an under-estimation, and red is an over-estimation of the distance for a specific pixel, according to the ideal transform of the source image. AAEDT-SKW and AAEDT-JFA show, once again, similar accuracy. AAEDT-PBA produces large areas of erroneous distances, radiating out from the top and bottom of the circles in the image. This phenomenon occurs because PBA assumes that the only candidate for the closest object pixel, for a column, on a given row is the object pixel closest to that column. This assumption is correct for the discrete case, but not for the Anti-aliased transform.

4.2 Test on Non-Uniform Shapes

The total number of errors of various sizes is shown in Fig. 13 for the three algorithms when applied to the Tattoo data set. The graph shows the total number of pixels the algorithms calculated the wrong distance for out of the 7.3 million in the complete data set.

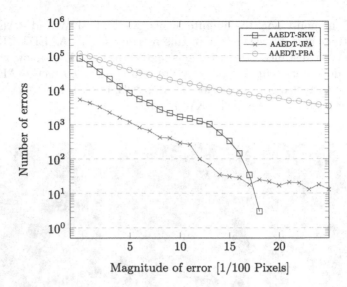

Fig. 13. Histogram of errors. The data points are divided into categories based on the magnitude of the error, when compared to the brute-force AAEDT. Each category covers a span of the length of one-hundredth of a pixel.

5   Discussion

5.1   Test on Uniform Shapes

The most prominent result as seen in the error images Fig. 12 is that AAEDT-PBA produces greater errors in streaks along the vertical axis, radiating out from foreground objects. One of the main differences between JFA & SKW and PBA, is that JFA & SKW perform statically determined distance calculations, while PBA calculates distance measures dynamically depending on how the stack of proximate foreground pixels is constructed in the second phase of the algorithm.

When calculating the distance measure in AAEDT-JFA & AAEDT-SKW, the background pixel from which the distance is measured and the target foreground pixel to which the distance is measured are always known. In AAEDT-PBA's second phase an approximation of the coordinates of a foreground object is performed based on the images gradient. This approximation causes issues with the construction of the stack of proximate sites. The issue is further affecting how the final distance transform is calculated in the third phase, as very strict assumptions of the distance measure are made in the third phase. In the third phase of AAEDT-PBA, it is assumed that for any pixel, as the stack of proximate pixels is traversed, as soon as a worse candidate than the previous is discovered, the best candidate is found. Due to the approximation of proximate site coordinates done in our implementation, some small errors are introduced to the distance measure, and this assumption breaks. The wrong candidate is then possibly assigned as the closest foreground pixel.

Another assumption of PBA which breaks for the AAEDT is that any foreground pixel is eliminated as a candidate for the proximate site by all pixels on the same row which are closer on the x-axis. This is true for EDT as all Voronoi sites are located on the centroids of pixels and thus on the same vertical height. This is not true for AAEDT since we approximate Voronoi sites with sub-pixel precision.

The fact that the input of AAEDT is an approximation of an image rather than a discrete grid as for EDT, is possibly the greatest limitation when it comes to what methods can be used to calculate the AAEDT.

5.2   Test on Non-Uniform Shapes

This test supports the results from the test on uniform shapes, namely that AAEDT-PBA has the highest error rate, and that AAEDT-JFA has the lowest. Figure 13 also shows that AAEDT-SKW produces errors with a lower maximum magnitude than the two other algorithms. Both AAEDT-JFA and AAEDT-SKW produce transforms with similar accuracy. AAEDT-JFA does produce a small number of errors with a higher magnitude than AAEDT-SKW, which is expected as studied by Rong and Tan [17]. The improvements to JFA presented by Rong and Tan to mitigate such errors are not explored in this study.

6   Conclusions

Our results show that the specified adaptation of PBA is in most terms of correctness and accuracy inferior to the two other algorithms, when applied to the new transform. This is because PBA is strictly optimized for the discrete EDT, and assumptions are made about the position of edge points of objects. These assumptions do not hold in the case of the Anti-aliased EDT and produce significant errors.

AAEDT-JFA produces images with slightly smaller average error than AAEDT-SKW but produces a greater maximum error in a few pixels, especially for larger image sizes.

Both AAEDT-JFA and AAEDT-SKW are valid algorithms with similar properties to their original counterparts JFA and SKW. Both algorithms have advantages, but which one should be used depends on demands on accuracy and performance as well as the input images being processed and the hardware the algorithm is executed on.

7   Future Work

We have shown that errors are produced with AAEDT-PBA. A further study should define what assumptions of PBA cause its reduced accuracy, and improvements to the adaptation of the algorithm, for calculating the AAEDT, presented here.

References

1. Aaedt git repo (2023). https://gitlab.com/mindroad-projects/gpgpu-aaedt.git
2. Barrow, H.: Parametric correspondence and chamfer matching: two new techniques for image matching. In: Proceedings IJCAI1977. pp. 659–663 (1977)
3. Borgefors, G.: Distance transformations in digital images. Comput. Vis. Graph. Image Process. **34**, 344–371 (1986)
4. Cao, T.-T., Tang, K., Mohamed, A., Tan, T.-S.: Parallel banding algorithm to compute exact distance transform with the GPU. In: Proceedings of 2010 Symposium on Interactive 3D Graphics and Games (I3D 2010), pp. 83–90. ACM (2010)
5. Danielsson, P.E.: Euclidean distance mapping. Comput. Graphics Image Process. **14**, 227–248 (1980)
6. Eggers, H.: Two fast Euclidean distance transformations in z2 based on sufficient propagation. Comput. Vis. Image Underst. **69**, 106–116 (1998)
7. Gustavson, S., Strand, R.: Anti-aliased Euclidean distance transform. Pattern Recogn. Lett. **32**(2), 252–257 (2011). https://doi.org/10.1016/j.patrec.2010.08.010
8. Ilic, V., Lindblad, J., Sladoje, N.: Precise Euclidean distance transform in 3d from voxel coverage representation. Pattern Recogn. Lett. **65**, 184–191 (2015)
9. Lindblad, J., Sladoje, N.: Exact linear time euclidean distance transforms of grid line sampled shapes. In: Benediktsson, J.A., Chanussot, J., Najman, L., Talbot, H. (eds.) ISMM 2015. LNCS, vol. 9082, pp. 645–656. Springer, Cham (2015). https://doi.org/10.1007/978-3-319-18720-4_54
10. Linnér, E., Strand, R.: A graph-based implementation of the anti-aliased Euclidean distance transform. In: 2014 22nd International Conference on Pattern Recognition, pp. 1025–1030 (2014). https://doi.org/10.1109/ICPR.2014.186
11. Maurer, C.R., Raghavan, V.: A linear time algorithm for computing exact Euclidean distance transforms of binary images in arbitrary dimensions. IEEE Trans. Pattern Anal. Mach. Intell. **25**, 265–270 (2003)
12. Meijster, A., Roerdink, J., Hesselink, W.: A general algorithm for computing distance transforms in linear time. Math. Morphol. Appl. Image Sig. Process. **18**, 331–340 (2000)
13. Montanari, U.: A method for obtaining skeletons using a quasi-Euclidean distance. J. ACM **15**, 600–624 (1968)
14. Paglieroni, D.: A unified distance transform algorithm and architecture. Mach. Vis. Appl. **5**, 47–55 (1992)
15. Ragnemalm, I.: Contour processing distance transforms. In: Progress in Image Analysis and Processing, pp. 204–212. World Scientific (1989)
16. Ragnemalm, I.: The Euclidean distance transform and its implementation on SIMD architectures. In: Proceedings 6th Scandinavian Conference on Image Analysis, Oulo, pp. 379–384 (1989)
17. Rong, G., seng Tan, T.: Jump flooding in GPU with applications to Voronoi diagram and distance transform. In: in Studies in Logical Theory, American Philosophical Quarterly monograph 2, pp. 109–116. ACM Press (2006)
18. Rosenfeld, A., Pfalz, J.L.: Sequential operations in digital picture processing. J. ACM **13**, 471–494 (1966)
19. Rutowitz, D.: Efficient processing of 2-d images. In: Progress in Image Analysis and Processing, pp. 229–253. World Scientific (1989)
20. Schneider, J., Kraus, M., Westermann, R.: GPU-based real-time discrete euclidean distance transforms with precise error bounds. In: In International Conference on Computer Vision Theory and Applications (VISAPP), pp. 435–442 (2009)

Distortion-Based Transparency Detection Using Deep Learning on a Novel Synthetic Image Dataset

Volker Knauthe[1]([✉])[iD], Thomas Pöllabauer[1,2][iD], Katharina Faller[1][iD],
Maurice Kraus[1][iD], Tristan Wirth[1][iD], Max von Buelow[1][iD], Arjan Kuijper[2][iD],
and Dieter W. Fellner[1,2,3][iD]

[1] Technical University of Darmstadt, Darmstadt, Germany
volker.knauthe@gris.informatik.tu-darmstadt.de
[2] Fraunhofer Institute for Computer Graphics Research IGD, Darmstadt, Germany
[3] CGV Institute, Graz University of Technology, Graz, Austria

Abstract. Transparency detection is a hard problem, as suggested by animals and humans flying or running into glass. However, humans seem to be able to learn and improve on the task with experience, begging the question, whether computers are able to do so too. Making a computer learn and understand transparency would be beneficial for moving agents, such as robots or autonomous vehicles. Our contributions are threefold: First, we conducted a perception study to obtain insights about human transparency detection methods, when borders of transparent objects are not visible. Second, based on our study insights we created a novel synthetic dataset called *DISTOPIA*, which focuses on the warping properties of transparent objects, placed in a variety of natural scenes and contains over 140 000 high resolution images. Third, we modified and trained a deep neural network classification model with an attention module to detect transparency through warping. Our results show that a neural network trained on synthetic data depicting only distortion effects can solve the transparency detection problem and surpasses human performance.

Keywords: Perception · Computer vision · Artificial intelligence · Scene understanding

1 Introduction

Transparency is a quite common phenomenon in modern day society. While it can be observed in abundance in urban scenes in the form of glass panes on buildings or cars, it is still not an easy task to recognize transparency for animals and humans. Notable examples are birds flying into windows, insects

Supplementary Information The online version contains supplementary material available at https://doi.org/10.1007/978-3-031-31435-3_17.

not being able to find their way back out through window gaps, dogs running into garden doors and even humans bumping into well cleaned shop fronts or doors. While this phenomenon is most probably due to the combination of different perceptual foci, the lack of transparency recognition seems to play a major role. A transparent material has distinguishable properties due to optical effects like distortion, light absorption or reflectance. While physical transparency can be measured with the appropriate tools and environments (objective measurements), perceived transparency is a vision task that requires information parsing from image input (perception). How this perception task works, is however still an open field in perceptual psychology. Understanding the human capability to detect transparency is not only interesting for psychological purposes, but also for bionic transfer applications. The recognition of transparency by machines and therefore the avoidance of e.g. closed glass doors, or the detection and mitigation of transparency related effects on general vision tasks become more important as moving agents, such robots and autonomous vehicles, become more prevalent. Although, x-junction (contour) detection is a viable cue for transparency detection, we deliberately exclude them from our data, arguing that contours are not a main effect of transparency. Most objects have perceivable contours, but there are use-cases without them, such as looking through a windows or safety glasses. This work provides new insights on the influence of two visual cues derived from transparent objects regarding perception, namely reflection and distortion. Furthermore, the impact of two different distortion objects, one resembling a glass pane, the other a lens, as well as the significance of urban and natural background scenes on human transparency detection were evaluated.

Furthermore, we introduce a large synthetic dataset using ray tracing. Due to our new insights about human perception we chose an image generation process, which enables us to extract varying degrees of the distortion properties of transparency and forgo reflectance. This is of special interest, as reflectance requires complex scene understanding and overlaps strongly with mirroring effects.

Finally we show that an ANN, trained on our dataset, is able to distinguish between transparency and transparency-free scenes. We evaluate our networks for the different scene and transparent object categories, as well as the necessary degree of distortion and elaborate our findings.

In summary our contributions are:

- A perception experiment showing that reflection and refraction play a crucial role in perceiving transparency setting a baseline for human transparency detection capabilities without x-junctions (contours).
- A novel dataset *DISTOPIA* for transparency detection via distortion, depicting a wide variety of scenes and levels of distortion, with more than 140 000 images with a high resolution of more than 5 Megapixels (2252 × 2252 pixels).
- We deploy our dataset to synthetically train an ANN and show that distortion is a sufficient cue to recognize transparancy in real world images. We achieve high classification accuracy of up to 83%, and investigate the performance difference of *Urban* versus *Nature* scenes, as well as difference between our two types of transparency and outperform the human baseline we established in our study.

2 Related Work

Physically, transparency is an optical material property that describes the ability to transmit electromagnetic waves depending upon its absorption and refraction [10]. Furthermore, the shape and background of a transparent object can yield additional degrees of visible distortions.

The question of how perceptual transparency works has been a topic of scientific research for over 150 years [1,2,19]. The most prominent findings in this area of research are X-junctions, luminance relations, and T-junctions, which occur when a line in the background passes behind a transparent object, undergoing a reduction in contrast [3,4,6]. [5] have further developed ordinal relations of transparency in regard to junctions. These junction phenomenons can be easily reproduced even in the absence of physical transparency and have been one of the most common techniques for depicting transparency in art from ancient Egyptian times to the present day [16]. However, those techniques often exploit human perception capabilities. Additional to the research on junctions and luminance relations, there has been research on the constraints of superimposed textures giving rise to perceptual transparency [8] and on color constraints occuring at an objects contours for transparency [9,11,12].

[14] showed a high dependency of the sensitivity to distortion on the local image structure and suggested that the detection of distortion is based on a higher-level representation of the image structure. [28] discuss that many shape-related properties of opaque objects cannot be simply transferred to transparent objects, because visual features become less perceivable. Furthermore, they showed that human subjects used distortion and specular reflection cues to compare similar transparent objects, with specular reflections being the dominantly used feature [22].

While stereopsis and motion can play a benefitial role for transparency detection, they are out of scope for our work [7,24]. This decision is based on the more constrained capturing setup for possible detection applications.

While computer vision-based transparency detection methods already exist, they focus on different aspects or data properties. Most algorithms use additional image information to complement rgb, including known object shapes [15,20,25], depth channels [18,31,36], structured light [23,26], light-fields [21,29] or polarization cues [13,38]. Furthermore, there are techniques that are exclusively based on rgb input, but regard general image depth estimation with included transparent objects [39] or transparency segmentation [30,32,34,41]. However, the networks of those methods are able to utilize transparency contours and reflection properties, which are excluded from our approach. One big challenge is, that only three public datasets which include transparency exist until now: Trans10k [32], ClearGrasp [31] and TRANSCG [37]. While they are all suitable for their intended purpose, no dataset we know of is applicable for our scenario, because they include objects with contours and do not isolate distortion.

3 Transparency Perception User Study

To detect transparency without border cues, we chose to investigate the impact of two transparency cues: specular reflection and distortion. To gather insights for machine vision solutions and to establish a baseline, we conducted a perceptual experiment. For this purpose, stimuli were created, evaluated with a small group of subjects and re-adjusted before the main experiment.

3.1 Experiment Design

We used a *yes-no task* setup, a standard method in psychological research to measure a subject's sensitivity to some particular sensory input. This design was selected because it is not prone to subjective scaling biases and provokes spontaneous reactions. The sensory input consisted of visual stimuli, showing a natural or urban scenes with a transparent object in front (signal) or without one. The intensity of each feature varied in steps from barely visible (index of refraction further written IOR of 1.1, reflectiveness 2%) to clearly visible (IOR 1.8, reflectiveness 16%). To eliminate the influence of X-junctions (borders), object contours were excluded, by showing the test subjects only the inner part of the image. Subsequently, the stimuli were presented to subjects, who were asked to decide whether they perceive transparency.

The experiment was designed using PsychoPy and conducted online with Pavlovia [42]. All four intensity values of the two cues were shown individually paired with both objects in front of all chosen background scenes. That leads to, $4 \cdot 2 \cdot 2 \cdot 5 = 80$ different stimuli in addition to images without transparency stimuli (*no object*). A proportional usage of images with and without stimuli would have lead to an unacceptable experiment duration, which would exceed the concentration capabilities of the test subjects. Therefore, only 20% of the stimuli did not contain a transparent object.

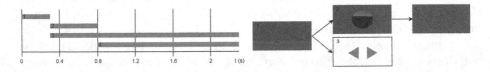

Fig. 1. Timeline of one trial: first, a white cross (1) was shown in the middle of the screen for 0.3 s, to prevent eye fixation and delayed responses. Then the stimulus (2) appeared for 0.5 s in total. After 0.5 s, the stimulus disappeared, and the gray window (4) remained empty for the last 1.5 s. Simultaneously with the stimulus, the subjects could react by pressing either the right arrow key for "Yes, there was an object." or the left arrow key for "No, there was no object." (3). A reaction immediately ended the trial, even if the stimulus has not been shown for the entire 0.5 s yet. If the subjects did not react, the trial was ended after 2.3 s, and the subsequent trial began.

Each of the 80 stimuli showing a transparent object were presented to subjects 21 times, while all five stimuli not containing an object were shown 84 times. This resulted in a total of 2100 trials, each lasting up to 2.3 s (see Fig. 1). Those 2100 trials per subject were split into seven blocks of equal size, giving the subjects the opportunity for a short break in between.

Each trial was conducted as described in Fig. 1. Stimuli were only shown for a short period of time to prevent subjects from consciously searching for abnormalities. This setup ensures that the reaction was as intuitive as possible.

Before conducting the actual experiment, an iterative preliminary experiment was performed to ensure a suitable process. The main study was conducted on 20 subjects consisted of eleven males and nine females with an average age of 25.6 (with a standard deviation of 7.4). Due to technical issues, the data showing the circular object in the *street* scene was excluded from the analysis.

3.2 Stimuli and Data Generation

In order to vary distortion and specular reflections of transparent objects separately, we generated stimuli with synthetic transparent objects. To make the experiment realistic, the objects were presented in different real-world scenes, represented as panorama images. Two of them were located in nature, and three in an urban environment. We used the Poly-Haven dataset, which contains panoramas as high dynamic range images [43].

When selecting the nature panoramas, the focus was to exhibit as few unnatural objects as possible, resulting in two different locations: *Versveldpas* (Fig. 2a) and *Desert* (Fig. 2c). Two different day times were deliberately selected, especially to get different lighting conditions and thus different specular reflections.

For the urban scenes, it was important that they were as diverse as possible and close to what we encounter in our everyday life. We chose the following three: *Street* (Fig. 2f), *Quattro Canti* (Fig. 2d) and *Hamburg* (Fig. 2e), as they represent different common municipal scenes. We used a ray tracer to generate the synthetic stimuli. The camera faces directly onto the transparent objects, while the panorama is placed as a texture around the scene. We decided to use two sophisticated, but common transparent objects: The first one being a quadratic slice, resembling a window (see Fig. 2a) that entails irregularities similar to the deformation of small waves in liquids with low viscosity, such as in water or older window glasses. The circular object (see Fig. 2b) is lens-shaped and shows the most deformation near the boundary regions of the object model.

Both objects have the same and constant material properties, except for their IOR and reflection ratio parameters. We decided to use a circular section of the stimuli, as seen in Fig. 1. This resulted in all outer points of the stimuli being equidistant from the center, thus providing similar times for saccadic eye movements. The background color on which the stimuli were presented was a neutral gray, to avoid unnecessary eye strain and avoid high contrasts. The outer stimuli boundary was blurred to de-emphasize borderlines.

(a) Versveld.   (b) Versveld.   (c) Desert   (d) 4 Canti   (e) Hamburg   (f) Street

Fig. 2. Scenes used in our experiments. (a) and (c) shows natural scenes, while (d), (e) and (f) show scenes from urban settings. (a) and (b) additionally show our two transparent object types.

3.3   Results

Data and Analysis. We calculated the mean correct answer ratios of all 20 participants, which is illustrated in Table 1. The data shows that images without transparent objects were recognized with a mean correct answer ratio of 92%. Furthermore, there is a difference in detectability between the stimuli in natural and urban scenes, as well as for our different transparent object types. This effect is especially visible in Fig. 3, which combines results from individual transparent object types and scenes. It shows that urban scenes with the squared object distortion cue leads to a significant increase in transparency perception.

For specular reflection, no apparent circumstantial differences can be perceived. There is one clear outlier, which is the quadratic object in *Quattro Canti*. This can be explained by the position of the light, resulting in few specular reflection stimuli. Figure 3 also reflects this behavior as the perceivability does differ marginally between transparent object types and scenes.

Transparency Perception Through Distortion. We evaluated the effect of different levels of distortion on the perceivability of transparent objects. A LEVENE tests shows that the variances from our experiment are inhomogeneous. A SHAPIRO-WILK test shows that our data is not normally distributed, which, however, can be neglected according to the central limit theorem as the number of samples is greater than 20 in our experiment. A robust WELCH test with a resulting p-value below .001 shows that there are significant differences between the mean values of the four groups. Given this observation, we further investigated how much these groups differ from each other using a GAMES-HOWELL test. This test reveals a significant difference between an IOR of 1.1 and the remaining distortion configurations. Additionally, it showed a significant difference between an IOR of 1.33 and 1.8, but no significant difference between 1.33 and 1.52 and IORs of 1.52 and 1.8, showing that increasing distortions are beneficial for detection, but the distance between distortion values might be to close.

Similar to this analysis, we examined whether our different transparent object types influence the perceivability of transparency. Therefore, we also performed a WELCH test ($p < 0.001$), indicating a strong influence on perception. Analogous, the scene type influences the perceivability of transparency ($p < 0.001$).

Table 1. Mean correct answer ratios of all 20 subjects for each stimulus. Values close to 50% indicate mere guessing. 0% represents no correct prediction and 100% means only correct prediction. × denotes images without transparent objects. The circles and Squares denote the type of transparent object.

	×	Distortion (IOR)				Specularity			
		1.10	1.33	1.52	1.80	2%	4%	8%	16%
Hamb. □	92	16	78	85	89	10	43	90	94
Hamb. ○	92	10	17	18	23	9	7	28	78
4 Canti □	93	9	53	76	83	7	7	6	9
4 Canti ○	93	8	11	17	21	7	12	90	96
Street □	86	18	67	84	90	26	22	79	95
Vers. □	94	8	10	18	34	6	7	51	89
Vers. ○	94	7	5	6	6	7	8	25	74
Desert □	94	6	11	18	36	8	16	77	92
Desert ○	94	10	7	9	10	9	22	77	93
μ	92	10	29	37	44	10	16	58	80

We can show that the presence of straight line segments in the stimuli, that differs strongly between urban and nature scenes, explains part of the difference between transparency perceivability. Therefor, we calculate the ratio of pixels, that lie on straight line segments according to the deep learning detector *M-LSD* [17,33], for all distortion levels in each scene. The resulting ratios are depicted in Fig. 2, which correlate with our perception results.

While the perception of transparency is weaker in natural scenes, it might be overcome by machine vision, as humans supposedly do not have an adequate training level for transparent objects in natural environments. Furthermore this cue seems viable, as distortion effects are characteristic and seldom false flags.

Table 2. Normalized differences between the lines in original and the distorted images. The circles and Squares denote the type of transparent object. Natural scenes are at 0, as there are no lines to be lost due to distortion.

Distortion (IOR)	Hamburg		4 Canti		Street	Versveldpas		Desert	
	□	○	□	○	□	□	○	□	○
1.10	0.18	0.04	0.22	0.11	0.09	0	0	0.01	0.01
1.33	0.3	0.14	0.63	0.32	0.48	0	0	0.01	0.01
1.52	0.35	0.07	0.85	0.25	0.51	0	0	0.01	0.01
1.80	0.42	0.13	1	0.2	0.55	0	0	0.01	0.01

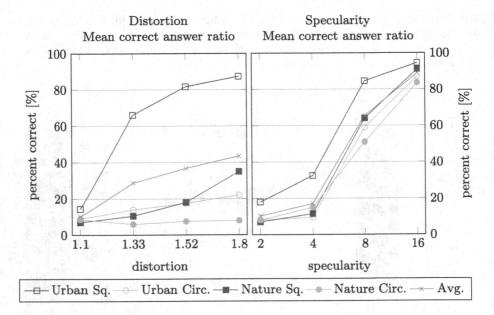

Fig. 3. Mean correct answer ratio of distortion and reflection values grouped by scene and the transparent object type.

Transparency Perception Through Specular Reflections. Similar to the distortion, Fig. 3 shows the mean ratios of correct answers for the specular reflection stimuli. A significant LEVENE test indicates data inhomogeneity. Furthermore, a WELCH test with p-value below 0.001 indicates that the results differ significantly between the different groups. In contrast to our previous analysis on different distortion levels, the succeeding GAMES-HOWELL test revealed significant differences between all selected reflection levels.

It can be concluded that the levels of specular reflection significantly affect the perceiveability of transparent objects. However, while the results indicated that humans are able work with this cue, specular reflection can also occur due to metallic properties and might not be a strong indicator for transparency. Furthermore, specular reflections require specific light setups to be perceivable.

Object and Scene Influence. A comparison of the influence of the chosen transparent object on the participants' ability to perceive the transparent objects showed no significance. Due to an insignificant LEVENE test, we conducted a one-way analysis of variance, which revealed an insignificant p-value of 0.595. For a comparison of the perceptual influence of the different scenes, we performed a WELCH test, which indicated insignificant differences. The marginal effects of specularity changes can be taken from Fig. 3.

4   Artificial Transparency Detection Method

The insights from our perception study lead us to the conclusion, that distortion is a viable cue for transparency detection, due to its characteristic and rare property. We discard specular reflection and reflection as a whole, due to its dependency on correct lighting and notion to be exhibited by various non transparent objects. To evaluate our idea, we generated a large synthetic dataset, shown in Sect. 4.1 and we evaluate how well a neural network can detect transparency when trained only on distortion in Sect. 4.2.

4.1   *DISTOPIA* Dataset

Following the design choices from our perception study, we synthetically create a large dataset, which only depicts distortion cues. To obtain a broader background base, we decided on a total of 315 panorama scenes (170 urban and 145 nature). To eliminate unintended background distortions we disregarded any scene containing substantial amounts of inherent transparent structures. Furthermore we filtered out natural scenes with large urban influences and vice versa. We kept the window structure and lens object, to have a direct comparison to our study.

For rendering we again adopt a ray tracer (Blender's Cycles). We place the camera within a HDRI background sphere and orient it towards our transparent objects. Next we rotate the camera in $5°$ steps on a horizontal axis and generate one image without transparent object and up to three images per object for each step. Different to our study, the IOR was chosen randomly for each generated image, and sampled from an uniform distribution. This results in a fairer learning set, as humans are also trained on a variety of IOR values. We repeated this procedure for all 315 scenes and two objects, resulting in over 140 000 high resolution images. To test for generalization to real-world photographs, we collect a small set of photos and report results on it alongside our synthetic data.

Table 3. Number of images within our training, validation, and test datasets. Training, Validation, and Testing splits depict disjunct sets of scenes. We collect a small set of photographs to evaluate the generalization to real world images.

	Both (A)	Circle (B)	Rect (C)	RectC (D)	Real
Training	250.426	125.192	125.218	876.514	-
Validation	5.539	3.075	3.079	34.302	363
Testing	5.850	3.250	3.250	96.601	-

4.2   Classification

To show that distortion is a sufficient cue for transparency detection we use the common CNN-based architecture ResNet, adopting the parameterization found in state-of-the-art GAN literature [35]. We provide each network block with a (bilinearly) down-sampled version of the current tensor, and a residual connection. Our network predicts transparency presence, transparency type (if any), distortion levels. Also, for regularization reasons, we predict whether an image shows a *Nature* or *Urban* scene. We use (binary) cross entropy as loss function for the first three and L1-loss for the distortion strength prediction.

In addition we argue that features relevant for distortion detection without visible corners and edges are located in the earlier layers of the feature extractor. Therefore we add bottleneck attention modules [27] after the first ResNet block. Combining attention with a bottleneck should allow the network to focus on relevant cues and discard the rest. Multiple blocks allow each to attend to different features. Also we use different non-linearities, that is average pooling for half of the blocks and max pooling for the other. We provide a detailed description of the architecture in the appendix for easy reproducibility.

We train ResNet networks for four different splits of our dataset. Next we evaluate our new network architecture (with attention) to the two best performing experiments and compare their performance on real data. The experiments A, B, C, and D are as follows: First, we train a classifier to differentiate between scenes containing rectangular transparent objects, circular transparent objects and scenes containing no transparencies (experiment A, *Both*), second, we train another network to distinguish circular transparencies from images with no transparency (B, *Circle*), a third to classify rectangular with only central crops (C, *Rect*), and finally a fourth for rectangular with additional random crops (D, *RectC*). We central crop the image, to remove the transparency's edges. Not removing the edges would provide exploitable features for the classifier and most likely reduce its generalization capabilities. For experiment D (rectangular with random cropping, before scaling to 512×512), after applying the central crop, we create an additional 7 random crops (after first removing the edge via central crop) to get different image details. For each crop in all experiments we also include the same image portion without transparencies, giving us a total of 876 513 images. For task A we include circular transparency, for which the circular distortion is an important cue. Therefore we restrict our data processing to only central cropping whenever circles are included. We end up with a total of 250 425 images for experiment A (*Both*). Finally, we want to see the individual performance when training a classifier each to just recognize one kind of transparency (B: circles only, and C: rectangles only). They have 125 192 and 125 218 images for training. The exact number of training, validation, and test images are to be found in Fig. 3. In addition we validate on a set of photographs to evaluate the generalization of our distortion-only training to real world images.

We use conventional image augmentation during training, such as flipping along y, rotating, re-scaling, adjusting brightness, contrast, hue, and saturation. Our experiments showed best performance with small batch sizes and we ended up using 16 images per batch, together with small learning rates between 0.0001 to 0.00025. As for our attention modules we varied their number and tried as few as 1 and up to 8, and report our results with 8. We adopt an early stopping strategy, showing our classifiers 50 million images, testing every 0.5 million views against our validation set, and using the best checkpoint on our test data. Our real data is a first portion of our ongoing effort to record a real-world dataset we plan to publish together with our synthetic data.

4.3   Evaluation

Ambiguous Scenes. The quantitative results of our experiments can be seen in Fig. 4. We provide a prediction accuracy for each possible individual subgroup derived from our data splits, concerning the background and object. Furthermore, we chose to show different buckets depending on the IOR each object had and additional buckets for all images without transparency (*None*), as well as all images with transparency (*All*). The reasoning is, that different IORs result in differently hard to detect warping effects. This can be observed as a general trend in our data. While there are maxima in each row (bold), that are not in the $\lfloor 1.7 \rfloor$ category, the prediction differences between those two values are only up to 0.05. Furthermore, the $\lfloor 1.1 \rfloor$ bucket always contains the lowest predictions with a large gap to the $\lfloor 1.2 \rfloor$ and following buckets, suggesting that this IOR is comparatively hard to predict. The emphasized values indicate the highest prediction value over all buckets, including the *None* category. These overlap with the best bucket predictions for the Circle object (B). In all other cases, the None prediction outperform the warping predictions. Especially the Both object (A) category shows large differences of up to 0.54 between the best bucket and None prediction. These findings suggest, that it might be easier to learn the detection of no transparency over learning two different object warping effects for (A). Our findings from the perception experiment, that different background scenery types might influence predictions, did not directly translate to our data.

While the background scene type shows no particular effect on the overall prediction accuracy, the object shape seems to be of high relevance for *All* predictions. The Rect object (C) allows for an accuracy of up to 83%, with the cropped version dropping to 75%. The Circle object (B) however, reaches only up to 59%. This seems to be reflected in the Both (A) set, where only about 62% of images are classified correctly. Furthermore, all sets containing circles exhibit either strong prediction accuracies in the bucket categories, or the *None* category. This behaviour is not observed in the Rect sets, where evenly high accuracies are reached over all categories. In total, this leads to the conclusion, that object shapes and therefore different warping characteristics play a major role in detecting transparency for our setup.

Real Data. To check for generalization to real data we validate our checkpoints against our set of photographs during training. We visualize our results in Fig. 4. Looking at the results, first, we note strong generalization early on. Second, we see clear outperformance of our model compared to the ResNet version especially for longer training times, as well as a smoothing out of the curve. Finally, when comparing both of our models between experiments, we see that experiment C (without crops) comes close to experiment D (with crops) after around 17 to 18 million views, stays in a similar range till around 27 to 29 million views and than drops to a similar level as the ResNet version. We argue that this is because of overfitting, since experiment D has around seven times the amount of images and much more variation. All together the drop in performance between our synthetic and our real data is around 15%.

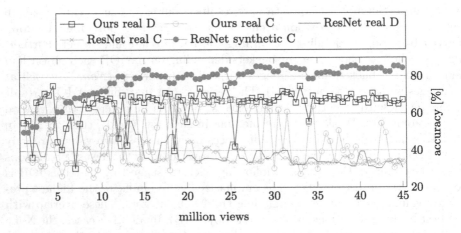

Fig. 4. Accuracy on photographs with ResNet architecture against ours with increasing number of training images (1 to 45 million) shown on experiments C and D. For comparison we include the results of experiment C on synthetic data.

Failure Cases. We present failure cases of our classifiers, focusing on images with strong distortions, classified incorrectly as depicting no transparency, as well as images with no transparency, classified as containing transparency. Also, we show some interesting effects, for instance, although filtering the scenes beforehand, we missed some nature scenes containing man-made structures, which led to crops containing predominantly urban scenery being classified in the training data as *Nature*. Interestingly, the classifiers predicted some of those images correctly on the test set (that is as *Urban*). In addition, we have some scenes containing transparencies, such as windows. Since we only labeled images as containing transparencies when we ourselves placed some in the rendering, these were also inconsistently labeled. Again, the classifiers predicted some of those confusing samples correctly (containing transparency).

Figure 5 shows incorrect classifications for circles, rectangles with and rectangles without random cropping. We notice particular problems with skies and landscapes without much detail (such as snowscapes, fields, and wood), as well as with views showing nothing but forest. We also note that strong backlighting (sun, human-made lights) leads to errors. Finally, we include a sample of circle-classifier failures on an image depicting vignetting-like characteristics (top right), incorrectly classifying the image as containing a circular transparency.

Table 4. For each of our experiments A (rectangular and circular transparency), B (circle only), C (rectangular only), and D (rectangular with random cropping) we report quantified results for both scenes as well as their combination. We subdivide the range of possible refraction parameter choices into buckets $\lfloor \delta \rfloor = [\delta, \delta + 0.1)$. The table also includes performance on images containing no transparency (×) and over all buckets (∀). The emphasized value in each row denotes the highest prediction value in a row and bold values exclude ×.

| Data | | Index of Refraction (IOR) | | | | | | | | |
Object	Scene	×	⌊1.1⌋	⌊1.2⌋	⌊1.3⌋	⌊1.4⌋	⌊1.5⌋	⌊1.6⌋	⌊1.7⌋	∀
Both (A)	Both	*0.83*	0.26	0.37	0.36	0.39	0.44	**0.45**	0.44	0.61
	Urban	*0.92*	0.23	0.26	0.27	0.3	0.36	**0.42**	0.38	0.62
	Nature	*0.73*	0.3	0.48	0.48	0.48	**0.53**	0.48	0.51	0.6
Circle (B)	Both	0.19	0.87	0.93	0.94	0.93	0.93	0.96	*0.96*	0.56
	Urban	0.15	0.86	0.91	0.92	0.91	0.9	0.93	*0.94*	0.53
	Nature	0.23	0.88	0.96	0.96	0.96	0.96	0.98	*0.98*	0.59
Rect (C)	Both	*0.97*	0.45	0.61	0.65	0.68	0.76	**0.79**	0.76	0.82
	Urban	*0.97*	0.47	0.64	0.66	0.65	0.76	**0.85**	0.8	0.83
	Nature	*0.97*	0.43	0.59	0.64	0.72	**0.76**	0.72	0.72	0.81
RectC (D)	Both	*0.98*	0.43	0.49	0.53	0.57	0.62	0.63	**0.63**	0.75
	Urban	*0.99*	0.41	0.49	0.54	0.57	0.62	**0.64**	0.62	0.73
	Nature	*0.97*	0.44	0.49	0.52	0.57	0.62	0.62	**0.65**	0.76

Although we handpicked our scenes to have a clean split between scenes depicting *Urban* structures and those, showing *Nature* such as forests and plains, some samples slipped through our filtering. This becomes especially apparent when looking at our use case D (rectangles with cropping) in which a small man-made structure (such as a house) in the background, can become the central element of the image. An interesting effect we witnessed concerns transparencies within the background image, as can be seen in Fig. 5: Our classifier (correctly) predicts the presence of the transparancy, even though our ground truth label says otherwise. To combat this problem, one needs to filter even more rigorously for use case D.

(a) Building (*Nature* scene)   (b) Tree (*Nature* scene)   (c) Brick wall (*Urban* scene)

Fig. 5. Although we checked by hand, some scenes of our dataset contain some ambiguity, which makes it interesting to see, what our classifiers do with such samples. For instance, the first image (a) shows a building within a *Nature* scene, blurring the line between our two splits of *Nature* versus *Urban* scenes. This becomes a problem when, during our random cropping, the house becomes the main subject of the resulting image. Similarly, the second image (b) shows a tree (*Nature*) in front of a brick wall (*Urban*). The last image (c) shows strong transparencies in the background (glass).

5   Conclusion and Future Work

We motivated the problem of transparency detection for e.g. intelligent moving agents. In order to investigate whether specular reflections and distortions are possible cues for transparency detection and to establish a baseline, we designed and carried out a perception experiment. Our subsequent analysis shows that the two cues have a significant positive effect on the perceivability of transparency. While specular reflection is the more robust cue for humans, it is also strongly dependent on scene understanding, lighting conditions and overlaps with reflection properties. Therefore, in our second experiment we investigate how well computers can detect transparency when trained on distortion only. We created *DISTOPIA*, our novel dataset for distortion-only-based transparency detection containing more than 140 000 high-resolution images. This dataset will be released, to allow the reproducibility of our findings and further research by the community. We then evaluated the performance of a convolutional neural network on our dataset and introduce a modification to better suit our applications and improve results on real data. We achieve high performance, both on our synthetic test set, as well as on our real photographs. This leads us to conclude, that distortion is also a viable cue to achieve high classification accuracy for computer-based vision models, that surpasses human transparency perception.

In the future, we want to expand our dataset by including more kinds of transparent objects, especially regarding their overall shape and material properties (e.g. acrylic glass and toned glass) and create a bigger and more diverse real world dataset for testing. Also, after demonstrating the validity of using distortion-only for transparency detection via a CNN, we want to adopt additional network architecture such as the vision transformer and new developments such as FocalNets [40] to improve upon our results.

Acknowledgements. Part of the research in this paper was funded by the Deutsche Forschungsgemeinschaft (DFG, German Research Foundation) project number 407 714 161. We thank Frank Jäkel for his generous support and the anonymous reviewers whose comments helped improve this manuscript.

References

1. von Hermann, H.: Allgemeine Encyklopädie der Physik / 9 Handbuch der physiologischen Optik, volume 9. Leopold Voss. https://doi.org/10.3931/E-RARA-21259
2. Beatrix Tudor-Hart. Beiträge zur psychologie der gestalt. (1). https://doi.org/10.1007/bf00492012
3. Metelli, F.: An algebraic development of the theory of perceptual transparency. (1). https://doi.org/10.1080/00140137008931118
4. Metelli, F.: The perception of transparency. (4). https://doi.org/10.1038/scientificamerican0474-90
5. Adelson, E.H., Anandan, P.: Ordinal characteristics of transparency
6. Watanabe, T., Cavanagh, P.: Transparent surfaces defined by implicit x junctions. (16). https://doi.org/10.1016/0042-6989(93)90111-9
7. Anderson, B.L., Julesz, B.: A theoretical analysis of illusory contour formation in stereopsis (4) https://doi.org/10.1037/0033-295x.102.4.705
8. Watanabe, T., Cavanagh, P.: Texture laciness: the texture equivalent of transparency? (3). https://doi.org/10.1068/p250293
9. D'Zmura, M., Colantoni, P., Knoblauch, K., Laget, B.: Color transparency. (4). https://doi.org/10.1068/p260471
10. Tavel, M.: What determines whether a substance is transparent? for instance, why is silicon transparent when it is glass but not when it is sand or a computer chip. http://www.scientificamerican.com/article/what-determines-whether-a/
11. Faul, F., Ekroll, V.: Psychophysical model of chromatic perceptual transparency based on substractive color mixture. (6). https://doi.org/10.1364/josaa.19.001084
12. Fulvio, J.M., Singh, M., Maloney, L.T.: Combining achromatic and chromatic cues to transparency. (8). https://doi.org/10.1167/6.8.1
13. Thilak, V., Voelz, D.G., Creusere, C.D.: Polarization-based index of refraction and reflection angle estimation for remote sensing applications. (30). https://doi.org/10.1364/ao.46.007527
14. Bex, P.J.: (in) sensitivity to spatial distortion in natural scenes. (2). https://doi.org/10.1167/10.2.23
15. Klank, U., Carton, D., Beetz, M.: Transparent object detection and reconstruction on a mobile platform. In: 2011 IEEE International Conference on Robotics and Automation, pp. 5971–5978. IEEE. https://doi.org/10.1109/icra.2011.5979793
16. Sayim, B., Cavanagh, P.: The art of transparency. (7). https://doi.org/10.1068/i0459aap
17. von Gioi, R.G., Jakubowicz, J., Morel, J.-M., Randall, G.: Lsd: a line segment detector. https://doi.org/10.5201/ipol.2012.gjmr-lsd
18. Alt, N., Rives, P., Steinbach, E.: Reconstruction of transparent objects in unstructured scenes with a depth camera. In: 2013 IEEE International Conference on Image Processing, pp. 4131–4135. IEEE. https://doi.org/10.1109/icip.2013.6738851

19. Koffka, K.: Principles Of Gestalt Psychology. Routledge. https://doi.org/10.4324/9781315009292
20. Lysenkov, I., Rabaud, V.: Pose estimation of rigid transparent objects in transparent clutter. In: 2013 IEEE International Conference on Robotics and Automation, pp. 162–169. IEEE. https://doi.org/10.1109/icra.2013.6630571
21. Maeno, K., Nagahara, H., Shimada, A., Taniguchi, R.-I.: Light field distortion feature for transparent object recognition. In: 2013 IEEE Conference on Computer Vision and Pattern Recognition, pp. 2786–2793. IEEE. https://doi.org/10.1109/cvpr.2013.359
22. Schlüter, N., Faul, F.: Are optical distortions used as a cue for material properties of thick transparent objects? 14(14), 2. https://doi.org/10.1167/14.14.2
23. Han, K., Wong, K.-Y.K., Liu, M.: A fixed viewpoint approach for dense reconstruction of transparent objects. In: 2015 IEEE Conference on Computer Vision and Pattern Recognition (CVPR), pp. 4001–4008. IEEE. https://doi.org/10.1109/cvpr.2015.7299026
24. Kawabe, T., Maruya, K., Nishida, S.: Perceptual transparency from image deformation. (33). https://doi.org/10.1073/pnas.1500913112
25. Phillips, C.J., Lecce, M., Daniilidis, K.: Seeing glassware: from edge detection to pose estimation and shape recovery. In Robotics: Science and Systems XII, vol. 3, p. 3. Michigan, USA, Robotics: Science and Systems Foundation. https://doi.org/10.15607/rss.2016.xii.021
26. Qian, Y., Gong, M., Yang, Y.-H.: 3d reconstruction of transparent objects with position-normal consistency. In: 2016 IEEE Conference on Computer Vision and Pattern Recognition (CVPR), pp. 4369–4377. IEEE. https://doi.org/10.1109/cvpr.2016.473
27. Park, J., Woo, S., Lee, J.-Y., Kweon, I.S.: Bam: Bottleneck attention module. In: British Machine Vision Conference. arXiv. 1048550/ARXIV.1807.06514
28. Schlüter, N., Faul, F.: Visual shape perception in the case of transparent objects. (4). https://doi.org/10.1167/19.4.24
29. Tsai, D., Dansereau, D.G., Peynot, T., Corke, P.: Distinguishing refracted features using light field cameras with application to structure from motion. (2). https://doi.org/10.1109/lra.2018.2884765
30. Mei, H., et al.: Don't hit me! glass detection in real-world scenes. In: 2020 IEEE/CVF Conference on Computer Vision and Pattern Recognition (CVPR), pp. 3687–3696. IEEE. https://doi.org/10.1109/cvpr42600.2020.00374
31. Sajjan, S., et al.: Clear grasp: 3d shape estimation of transparent objects for manipulation. In: 2020 IEEE International Conference on Robotics and Automation (ICRA), pp. 3634–3642. IEEE. DOI: https://doi.org/10.1109/icra40945.2020.9197518
32. Xie, E., Wang, W., Wang, W., Ding, M., Shen, C., Luo, P.: Segmenting transparent objects in the wild. In: Vedaldi, A., Bischof, H., Brox, T., Frahm, J.-M. (eds.) ECCV 2020. LNCS, vol. 12358, pp. 696–711. Springer, Cham (2020). https://doi.org/10.1007/978-3-030-58601-0_41
33. Gu, G., Ko, B., Go, S., Lee, S.-H., Lee, J., Shin, M.: Towards light-weight and real-time line segment detection. https://doi.org/10.48550/ARXIV.2106.00186
34. He, H., et al.: Enhanced boundary learning for glass-like object segmentation. In: 2021 IEEE/CVF International Conference on Computer Vision (ICCV), pp. 15859–15868. IEEE. https://doi.org/10.1109/iccv48922.2021.01556
35. Karras, T., et al.: Alias-free generative adversarial networks. In: Proceedings NeurIPS. arXiv. 1048550/ARXIV.2106.12423

36. Zhu, L., et al.: Rgb-d local implicit function for depth completion of transparent objects. In: 2021 IEEE/CVF Conference on Computer Vision and Pattern Recognition (CVPR), pp. 4649–4658. IEEE. https://doi.org/10.1109/cvpr46437.2021.00462

37. Fang, H., Fang, H.-S., Xu, S., Lu, C.: Transcg: a large-scale real-world dataset for transparent object depth completion and a grasping baseline. (3). https://doi.org/10.1109/lra.2022.3183256

38. Mei, H., et al.: Glass segmentation using intensity and spectral polarization cues. In: 2022 IEEE/CVF Conference on Computer Vision and Pattern Recognition (CVPR), pp. 12622–12631. IEEE. https://doi.org/10.1109/cvpr52688.2022.01229

39. Wirth, T., Jamili, A., von Buelow, M., Knauthe, V., Guthe, S.: Fitness of general-purpose monocular depth estimation architectures for transparent structures. In: Pelechano, N., Vanderhaeghe, D., (eds.) Eurographics 2022 - Short Papers. The Eurographics Association. https://doi.org/10.2312/EGS.20221020

40. Yang, J., Li, C., Dai, X., Yuan, L., Gao, J.: Focal modulation networks. https://doi.org/10.48550/ARXIV.2203.11926

41. Yu, L., et al.: Progressive glass segmentation. https://doi.org/10.1109/tip.2022.3162709

42. Pavlovia. https://pavlovia.org/

43. Poly haven. https://polyhaven.com/

Regenerated Image Texture Features for COVID-19 Detection in Lung Images

Ankita Sharma$^{(\boxtimes)}$ and Preety Singh🆔

The LNM Institute of Information Technology, Jaipur 302031, India
{ankita.sharma.y19pg,preety}@lnmiit.ac.in
https://www.lnmiit.ac.in/

Abstract. Covid-19 pandemic led to a worldwide pandemic and brought tremendous strain on patient testing facilities. In this paper, we aim to provide an automated rapid detection of Covid-19 infection using lung images. This will reduce the manual efforts required for speedy diagnosis. To achieve this we extract texture features from CT-Scans or X-ray images of suspected patients. These features are regenerated using graph-based techniques and employed for ascertaining the infection. Results show that our proposed approach achieves high accuracy along with high sensitivity for both types of radiology images.

Keywords: Covid-19 · Texture features · Artificial neural network

1 Introduction

Coronavirus-2 (SARS-CoV-2) is a virus strain leading to the Covid-19 infection. It is airborne and fatal if not tested and treated early at its onset [17,18]. Extensive research is being done to incorporate artificial intelligence in its rapid diagnosis from processing of radiological images [14,15,19]. Radiography examinations such as chest X-rays or computed tomography (CT) scans are employed for examining the extent of the infection in the lungs. These radiological images play a vital role in the diagnosis of the disease [20,21]. Covid-19 spreads rapidly leading to a substantial volume of patients in a short duration of time. Amid the critical crisis of a pandemic, manual identification of the infection in the images becomes a colossal and tedious task, prone to delays and human errors. Hence, the necessity of an automated system for detection of the disease in lung images becomes an exigency. The main requirement of developing an intelligent system for detection of Covid-19 is to have a high accuracy along with low false negatives and false positives. False negatives results in infected people going undiagnosed which leads to spread of the virus in the community. A false positive may cause unnecessary stress to the person diagnosed with the infection erroneously as they need to be isolated.

In this paper, we propose an automated system to detect Covid-19 from radiological images. The main contribution of our paper is the use of regenerated image texture features. Texture features can be instrumental in detecting irregularities in images. These features are regenerated using graph techniques.

R. Gade et al. (Eds.): SCIA 2023, LNCS 13885, pp. 268–278, 2023.
https://doi.org/10.1007/978-3-031-31435-3_18

The modified features are employed for classification using artificial neural networks. The paper is organized as follows: Sect. 2 outlines few research works in related domain. Section 3 discusses our proposed methodology. Section 4 presents the experiments and results. Section 5 concludes the chapter.

2 Literature Survey

These are the existing papers to understand the problem statement and the research which has been done in the same domain.

In [1] Gunraj et al. introduced COVIDNet-CT, a deep convolutional neural network architecture for detection of COVID-19 cases from chest CT images via a machine-driven design exploration. The approach COVIDNet-CT achieves a high COVID-19 sensitivity of 97.3%.

Apostolopoulos and Mpesiana have used a dataset of X-ray images from patients with bacterial pneumonia, confirmed Covid-19 infection and normal cases to detect whether a patient is Covid-19 positive or not by utilizing transfer learning with convolutional neural networks [2]. They have reported a best accuracy, sensitivity, and specificity of 96.78%, 98.66%, and 96.46% respectively.

In [3] Wehbe et al. developed an ensemble of convolutional neural networks, DeepCOVID-XR, to detect COVID-19 on frontal chest radio-graphs. The goal of using an ensemble of convolutional neural networks for this task is to improve the accuracy and reliability of the detection process. They obtain an accuracy of 83% with an AUC of 0.90.

In [4] Sharma and Singh have designed a model which uses graph neural network along with Artificial Neural Network on the pixel intensities of the images of dataset. The dataset consists of CT-scan images of Covid-19 positive and negative patients. The authors have reconstructed the original pixel intensities of the images in the dataset and used the reconstructed features for classification. The authors have achieved an accuracy of 99% for classification of CT-scan images of the Covid-19 dataset.

3 Proposed Methodology

We propose a model for automated identification of Covid-19 infection in lung images of a suspected patient. The methodology of our research is shown in Fig. 1. Our dataset consists of CT-scans and X-rays having Covid-19 positive and negative labelled images. We extract texture features, representing the spatial arrangement of intensities in the image. These features are regenerated and classified using an artificial neural network. This is discussed in detail in the subsequent subsections.

3.1 Extraction of Texture Features

We have used the Gray Level Co-occurrence Matrix (GLCM) for representing the texture features. GLCM is a histogram of co-occurring gray-scale values at

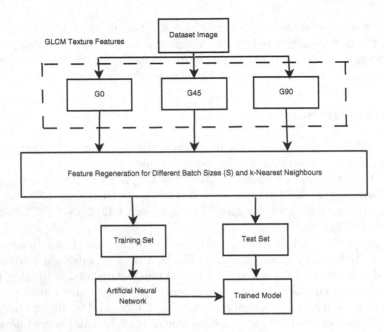

Fig. 1. Proposed methodology.

a given offset and this technique is popularly used to extract the second-order statistical texture features of images [16]. The computed matrix G is of size $l \times l$, where l is the number of intensity levels in the image. An entry G_{ij} shows the frequency of pixels with an intensity i having a horizontal, vertical, or diagonal neighbouring pixel with intensity j in the input image. We constructed GLCM matrices using an offset of 5 and in three different directions, viz. 0°, 45° and 90°. Offset refers to the distance between the considered pixels. Each matrix is of size 256×256 as dataset images are gray level images. The computed matrices are G_0 (considering neighbours at 0° direction), G_{45} (for neighbours at 45° direction) and G_{90} (with neighbours at 90° direction). These texture features are used for further processing.

3.2 Feature Regeneration

The extracted GLCM features are regenerated using an algorithm proposed by Yu et al. [8] and outlined below:

1. Extract features from each image. Feature set of each image j is a column vector and denoted by f_j. Consider each feature set of an image as one node.
2. Divide image feature sets into batches. The ith batch of features is denoted by F_i. Let N be the total number of images and S denote the batch size (number of images in each batch). Thus, the number of batches, n, will be:

$$n = \lceil S/N \rceil$$

Each batch of images can be represented as:

$$F_i = [f_1, \cdots, f_j, \cdots, f_s]^T \qquad (1 \leq j \leq S)$$

The set of all batches, F, is then given by:

$$F = [F_1, \cdots, F_i, \cdots, F_n] \qquad (1 \leq i \leq n)$$

3. Compute Euclidean distance between each node to the remaining nodes within each batch. This distance between two points p and q in an m-dimensional space is defined as:

$$d(p, q) = \sqrt{\sum_{i=1}^{m} (q_i - p_i)^2}$$

4. Find k nearest neighbours in a batch, for each node, on basis of computed distance.
5. Compute adjacency matrix, A_i, for each batch i. For a simple graph, this matrix contains values 1 or 0, where a value 1 at position (m, n) indicates that nodes m and n are neighbouring nodes.
   - **Initialization:** The initialization of variables is done as follows:
     - A_i of size S, initialized to zero:

$$A_i = Zeros(S, S)$$

   - Distance matrix $Distance$ of size S, initialized to infinity:

$$Distance = Infinity(S, S)$$

   - Sorted distance matrix $SortDist$ of size S, initialized to zero. This represents the nearest neighbours in ascending order:

$$SortDist = Zeros(S, S)$$

   - Index matrix $Index$ of size S, initialized to zero. It shows the index of other nodes in order of the smallest distance from the current node:

$$Index = Zeros(S, S)$$

   - **Distance calculation:** For each image (node) p in a given batch, its distance with every other image (node) q in the same batch is computed.
   - **Computing values in A_i:** Computation of values in the adjacency matrix is done based on k nearest neighbours of the node.
     for $p = 1 : S$
     $SortDist(p, :), Index(p, :) = sort(Distance(p, :))$
     for $c = 1 : k$
     $A_i[p, Index(p, c)] = 1$
     end
     end
     Assign a value 1 in an adjacency matrix A at positions that represent the first k neighbours. Remaining values are marked 0.

6. Normalize A_i by first creating a degree matrix $D \in R^{S \times S}$:

$$D(j, h) = \sum_{h=1}^{S} A_i(j, h)$$

where:

$$D(j, h) = \begin{cases} k, & if \quad j = h \\ 0, & if \quad j \neq h \end{cases}$$

We then normalize A_i as:

$$\tilde{A}_i = D^{-\frac{1}{2}}(A_i + I)D^{-\frac{1}{2}}$$

where, I is the identity matrix.

7. Regenerate each feature f_j in batch F_i as:

$$f_j = \tilde{A}_i^j F_i$$

where, $\tilde{A}_i^j$ is the jth row of $\tilde{A}_i$, the normalized adjacency matrix. Multiplying the normalized adjacency matrix $\tilde{A}_i$ with F_i gives us the regenerated feature batch F_i':

$$F_i' = \tilde{A}_i F_i$$

The regenerated features are used for classification using an Artificial Neural Network (ANN). Classification was performed using the original GLCM texture features and the regenerated GLCM features.

4   Experiments and Results

As discussed in Sect. 3, we have extracted the GLCM features from the lung CT-scan and X-ray images. Each feature set is reconstructed and the modified features are used to classify the presence of Covid-19 infection using an Artificial Neural Network (ANN). The experiments comprise of two sets:

- Classification of texture features using ANN
- Classification of reconstructed texture features using ANN

The metrics used for classification are Precision, Sensitivity, Accuracy, F1-score, Specificity and Area under the ROC curve (AUC). These metrics were used in the classification results for different approaches. Details are presented in the following subsections.

4.1   Parameters of ANN Architecture

Our Artificial Neural Network (ANN) comprises of fully connected (FC) and drop out layers. The dropout layers were added to reduce overfitting. The architecture of the ANN used in our experiments is:

- First layer - FC layer, activation function Relu, 128 neurons.
- Second layer - Drop out layer, 0.2 dropout rate.
- Third layer - FC layer, activation function Relu, 64 neurons.
- Fourth layer - Drop out layer, 0.2 dropout rate.
- Fifth layer - FC layer, activation function Relu, 32 neurons.
- Sixth layer - Drop out layer, 0.2 dropout rate.

The final classification layer uses the sigmoid activation function. The RMSprop optimizer and 2e-5 learning rate have been employed. A batch size of 64 has been used during training with 200 epochs.

4.2   CT-scan Images

For the CT-scan images, we have used the SARS-CoV-2 CT-scan dataset [5]. This contains 1252 Covid-19 positive and 1230 negative labelled gray images. Sample images from the dataset are shown in Fig. 2. Few blurry images were removed. The dataset was augmented using horizontal flipping resulting in total of 4800 images with equal number of images from both classes. Stratified splitting was used to split the dataset into train, validation and test sets in the ratio 70:15:15. Distribution of Covid-19 positive and negative images is shown in Table 1. GLCM features were extracted from the images yielding three matrices G_0, G_{45}, and G_{90} for each image. These features were classified using the ANN. Individual feature sets were then regenerated for different values of S of the regeneration algorithm. For each S, value of k is taken as the approximate square root of S. The regenerated feature sets were also used for classification using an ANN with architecture as specified in Sect. 4.1.

Table 1. Distribution of Covid-19 positive and negative images in training, validation and test sets.

Image Label	Train	Validation	Test
Covid-19 Positive	1680	360	360
Covid-19 Negative	1680	360	360

GLCM Texture Feature Classification with ANN. For images in the dataset, GLCM features G_0, G_{45} and G_{90} features were classified using the ANN. The results are shown in Table 2.

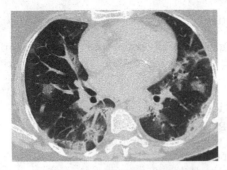 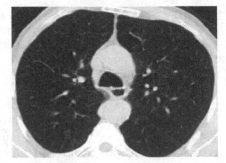

(a) Covid-19 Positive CT-scan   (b) Covid-19 Negative CT-scan

Fig. 2. Sample images from CT-scan dataset

Table 2. Evaluation metrics for classification of GLCM texture features using ANN.

Feature	Precision (%)	Sensitivity (%)	Accuracy (%)	F1-score	Specificity (%)	AUC
G_0	85	65.55	81	0.81	81	0.814
G_{45}	85	60.27	80	0.79	80	0.797
G_{90}	82	76.38	82	0.82	82	0.819

Texture Feature Regeneration and Classification with ANN. The GLCM features G_0, G_{45} and G_{90} were regenerated as per the regeneration algorithm in Sect. 3.2 and classified using ANN. The ROC-AUC curves observed for few combinations of S and k are shown in Fig. 3. The other evaluation metrics are shown in Tables 3, 4 and 5.

Table 3. Results for classification of regenerated G_0 features with ANN

S	k	Precision (%)	Sensitivity (%)	Accuracy (%)	F1-score	Specificity (%)	AUC
2400	49	86	86	86	0.86	85	0.864
1200	35	99	99	99	0.99	98.3	0.99
600	25	**100**	**100**	**100**	**1**	**99.44**	**0.996**
300	18	99	99	99	0.99	98.33	0.99
150	13	98	99	99	0.99	98.61	0.988
75	9	97	97	97	0.97	95.83	0.969

4.3   X-ray Images

For the X-ray images, we have taken the COVID-19 Radiography Database [6,7]. The dataset contains a total of 13,808 images with 3616 positive and 10,192 negative images. Sample images are shown in Fig. 4. The dataset was then split

Table 4. Results for classification of regenerated G_{45} features with ANN

S	k	Precision (%)	Sensitivity (%)	Accuracy (%)	F1-score	Specificity (%)	AUC
2400	49	86	86	86	0.86	83.8	0.857
1200	35	96	96	96	0.96	95.56	0.964
600	25	**98**	**98**	**98**	**0.98**	**96.94**	**0.976**
300	18	98	97	97	0.97	96.11	0.975
150	13	97	97	97	0.97	95.83	0.968
75	9	97	97	97	0.97	95.83	0.968

Table 5. Results for classification of regenerated G_{90} features with ANN

S	k	Precision (%)	Sensitivity (%)	Accuracy (%)	F1-score	Specificity (%)	AUC
2400	49	86	86	86	0.86	84.1	0.86
1200	35	96	96	96	0.96	93.33	0.958
600	25	**96**	**96**	**96**	**0.96**	**95**	**0.958**
300	18	97	97	97	0.97	95.83	0.974
150	13	96	96	96	0.96	96.67	0.963
75	9	94	94	94	0.94	90.83	0.936

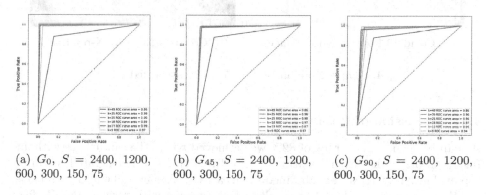

(a) G_0, S = 2400, 1200, 600, 300, 150, 75

(b) G_{45}, S = 2400, 1200, 600, 300, 150, 75

(c) G_{90}, S = 2400, 1200, 600, 300, 150, 75

Fig. 3. ROC-AUC curves for G_0, G_{45} and G_{90}

into into train, validation and test sets in the approximate ratio 70:15:15 using stratified splitting. The distribution of Covid-19 positive and negative images of the X-ray dataset in training, validation and test is shown in Table 6.

Table 6. Distribution of Covid-19 positive and negative images of X-ray dataset in training, validation and test set.

Image Label	Train	Validation	Test
Covid-19 Positive	2532	542	542
Covid-19 Negative	7134	1529	1529

While testing the CT-scan images, we observed that GLCM features G_0 resulted in good results (refer Table 3). Thus, these features only were used for performing experiments over the X-ray dataset. The G_0 features from X-ray images were regenerated using a batch size S as 3452 (25% of the dataset) and value of k was set to 59 (square root of S). The regenerated features were classified using ANN. The classification results are shown in Table 7.

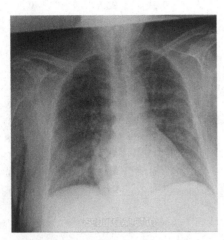 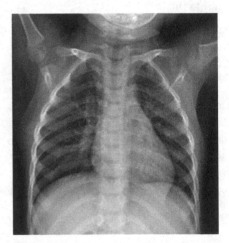

(a) Covid-19 Positive X-ray Image   (b) Covid-19 Negative X-ray Image

Fig. 4. Sample X-ray images of Covid-19 suspected patients

4.4   Discussion of Results

As can be seen, an accuracy of 82% was achieved with CT-scan images without regeneration of the texture features (refer Table 2). With regenerated features, top evaluation metrics were obtained in all cases of regenerated G_0, G_{45} and G_{90} for $S = 600$ and $k = 25$. Best results were obtained with regenerated G_0 features. These gave an accuracy of 100% along with a sensitivity value of 100%. The results of the proposed methodology will be tested on a larger dataset in future. For X-ray images, the accuracy obtained is 99% with sensitivity of 99.54%. High accuracy along with high sensitivity will ensure higher true positive rate for Covid-19 so that no case goes undetected. Also, high specificity ensures that healthy patients will not be incorrectly diagnosed with Covid-19. Our proposed

Table 7. Results for classification of regenerated G_0 features with ANN for X-ray dataset

S	k	Precision (%)	Sensitivity (%)	Accuracy (%)	F1-score	Specificity (%)	AUC
3452	59	99	99.54	99	0.99	99	0.984

model shows improved results than Covid-19 detection results in [1–4]. Our proposed model shows good performance compared to state-of-art Convolutional Neural Networks (CNN) also, which are popularly used in medical image processing [4]. As number of dataset images are less in number, use of regenerated features along with neural networks has advantages over Convolutional Neural Networks which require large datasets. However, these results need to be verified on a larger dataset before it can be used for critical medical domains.

5   Conclusions

In our paper, we propose an automated system for detection of Covid-19 from lung images of suspected patients. We extract texture features from the images which are then regenerated using the concept of nearest neighbours and graph techniques. The regenerated features are classified using an Artificial Neural Network. For the CT-scan images, an accuracy of 100% with a specificity of 99.44% is obtained whereas for X-ray images, an accuracy and specificity of 99% is achieved. Our approach works well for both types of radiology images and can be explored for detection of other lung-related diseases as well.

References

1. Gunraj, H., Wang, L., Wong, A.: COVIDNet-CT: a tailored deep convolutional neural network design for detection of COVID-19 cases from chest CT images. Front. Medicine 7 (2020)
2. Apostolopoulos, I.D., Mpesiana, T.A.: Covid-19: automatic detection from X-ray images utilizing transfer learning with convolutional neural networks. Phys. Eng. Sci. Med. **43**(2), 635–640 (2020). https://doi.org/10.1007/s13246-020-00865-4
3. Wehbe, R.M., et al.: DeepCOVID-XR: an artificial intelligence algorithm to detect COVID-19 on chest radiographs trained and tested on a large U.S. Clinical Data Set. Radiology (2021)
4. Sharma, A., Singh, P.: Detection of COVID-19 in Lung CT-Scans using Reconstructed Image Features. In: Convolutional Neural Networks for Medical Image Processing Applications, Ozturk, S. (Ed.), CRC Press 154–169 (2022)
5. Soares, E., Angelov, P., Biaso, S., Froes, M.H., Abe, D.K.: SARS-CoV-2 CT-scan dataset: a large dataset of real patients CT scans for SARS-CoV-2 identification. medRxiv (2020)
6. Chowdhury, M.E.H., et al.: Can AI help in screening Viral and COVID-19 pneumonia? IEEE Access **2**, 132665–132676 (2020)
7. Rahman, T., et al.: Exploring the effect of image enhancement techniques on COVID-19 detection using chest X-ray images. Comput. Biol. Med. 132 (2021)
8. Yu, X., Wang, S., Zhang, Y.-D.: CGNet: a graph-knowledge embedded convolutional neural network for detection of pneumonia. Info. Process. Manage. 58(1) (2021)
9. Müller, D., Rey, I.S., Kramer, F.: Automated chest CT image segmentation of COVID-19 lung infection based on 3D U-Net. arXiv:2007.04774 (2020)
10. Saha, P., Mukherjee, D., Singh, P.K., Ahmadian, A., Ferrara, M., Sarka, R.: Graph-CovidNet: a graph neural network based model for detecting COVID-19 from CT scans and X-rays of chest. Scientific Reports 11(8304) (2021)

11. Ni, Q., et al.: A deep learning approach to characterize 2019 coronavirus disease (COVID-19) pneumonia in chest CT images. Euro. Radiol. **30**(12), 6517–6527 (2020)
12. Liu, C., Wang, X., Liu, C., Sun, Q., Peng, W.: Diferentiating novel coronavirus pneumonia from general pneumonia based on machine learning. BioMedical Eng. OnLine **19** (2020)
13. Wang, S.-H., Zhang, Y.-D.: DenseNet-201-based deep neural network with composite learning factor and precomputation for multiple sclerosis classification. ACM Trans. Multimedia Comput. Commun. Appl. **16**(25), 1–19 (2020)
14. Öztürk, Ş, Özkaya, U., Barstuğan, M.: Classification of Coronavirus (COVID-19) from X-ray and CT images using shrunken features. Int. J. Imaging Syst. Technol. **31**, 5–15 (2021)
15. Naudé", W.: Artificial Intelligence against COVID-19: an early review. Tech. Report, IZA Inst. Labor Econ., Maastricht, The Netherlands, 13110 (2020)
16. Haralick, R.M., Shanmugam, K., Dinstein, I.: Textural features for image classification. IEEE Trans. Syst. Man Cybernet. SMC **3**(6), 610–621 (1973)
17. Li, K., et al.: The clinical and Chest CT features associated with severe and critical COVID-19 pneumonia. Invest. Radiol. **55**(6), 327–331 (2020)
18. Chen, N., et al.: Epidemiological and clinical characteristics of 99 cases of 2019 novel coronavirus pneumonia in Wuhan, China: a descriptive study. Lancet **395**(10223), 507–513 (2020)
19. Gjesteby, L., Yang, Q., Xi, Y., Zhou, Y., Zhang, J., Wang, G.: Deep learning methods to guide CT image reconstruction and reduce metal artifacts. Med. Imaging 2017: Phys. Med. Imaging **10132**, 752–758 (2017)
20. Sun, W., Pang, Y., Zhang, G.: CCT: lightweight compact convolutional transformer for lung disease CT image classification. Front. Physiol., 13 (2022)
21. Uddin, K.M.M., Dey, S.K., Babu, H.M.H., Mostafiz, R., Uddin, S., Shoombuatong, W., Moni, M.A.: Feature fusion based VGGFusionNet model to detect COVID-19 patients utilizing computed tomography scan images. Sci. Rep., 12, 21796 (2022)

Depth-Aware Image Compositing Model for Parallax Camera Motion Blur

German F. Torres[✉] and Joni Kämäräinen

Tampere University, Tampere, Finland
{german.torresvanegas,joni.kamarainen}@tuni.fi
https://github.com/germanftv/ParallaxICB

Abstract. Camera motion introduces spatially varying blur due to the depth changes in the 3D world. This work investigates scene configurations where such blur is produced under parallax camera motion. We present a simple, yet accurate, Image Compositing Blur (ICB) model for depth-dependent spatially varying blur. The (forward) model produces realistic motion blur from a single image, depth map, and camera trajectory. Furthermore, we utilize the ICB model, combined with a coordinate-based MLP, to learn a sharp neural representation from the blurred input. Experimental results are reported for synthetic and real examples. The results verify that the ICB forward model is computationally efficient and produces realistic blur, despite the lack of occlusion information. Additionally, our method for restoring a sharp representation proves to be a competitive approach for the deblurring task.

Keywords: blur formation · image compositing blur · neural representations · deblurring

1 Introduction

Motion blur is a common problem in photography and in certain computer vision tasks such as feature matching [22] or object detection [16]. In essence, motion blur occurs when either the camera or the scene objects, or both, are in motion during exposure. Recovering the edges and textures of the latent sharp image, *i.e.* deblurring, remains as an open problem since there are infinite latent sharp sequences consistent with the generated blur.

In conventional deblurring approaches, there is a model that describes the formation of the blur, coupled with an image prior that regularizes the solution space of the optimization problem. A major part of the research has been conducted upon suitable image priors that characterize natural images [3,14,15,20,24,41]. However, the applicability in real scenarios also depends on the accuracy of the assumed blur formation model. Pioneering works assume that the blur results from the shift-invariant convolution of the sharp image with an unknown Point Spread Function (PSF) [7,18,19,39]. For this to be precise, either of the two possible scenarios must hold, apart from the scene being

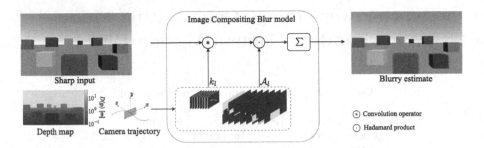

Fig. 1. Our blur formation model accurately describes parallax motion blur. By providing the depth and camera trajectory, we can fit a set of motion kernels k_l and alpha-matte terms $\mathcal{A}_l$, which are used to blend the blur from different layers.

static: 1) the camera shake only involves in-plane translation while the scene is either planar or sufficiently far from the camera, 2) the focal length of the camera is large and there is no in-plane rotation [38]. Otherwise, camera shake generally induces non-uniform (spatially varying) blur.

Several deblurring algorithms have been proposed to deal with spatially-varying blur that is produced by more realistic 6D camera motion [8,10,30,38, 41]. Nevertheless, these works fail at modeling the induced blur in 3D scenes, especially at depth discontinuities. With the advances in deep learning, several network architectures have been proposed to handle multiple types of blur by learning from data [4,5,17,23,32,34,35,42–44]. They benefit from not requiring an explicit description of the blur formation process. In such works, a neural network is trained over large-scale datasets to restore the sharp image. Deep deblurring represents state-of-the-art on multiple benchmarks, but their performance depends on the type of blur that is present in the training set.

Due to the parallax effect, objects positioned at different depths from the camera produce spatially varying blur, when the camera moves during capture. Following this line, a number of works have incorporated the depth in their deblurring methods as an extra auxiliary input [21,25,27] or by a joint estimation process [11,40,47], but they do not provide a concrete blur model.

In this work, we study the impact of the depth variation on the motion blur focusing on *parallax camera motion*, *i.e.* when the camera moves parallel to the image plane. By analyzing the geometry of this type of camera motion, we identify two realistic scene types where depth plays a significant role in the produced blur: 1) *Macro* Photography and 2) *Trucking* Photography. For such configurations, we propose a tractable Image Compositing Blur (ICB) model for parallax motion assuming that the depth and camera trajectory are available. This model accurately approximates the camera blur under parallax motion (Fig. 1). In addition, we provide evidence that our ICB model, in conjunction with coordinate-based Multi-Layer Perceptron (MLP) models, can be used to extract a sharp neural representation from a single blurry image.

In summary, the main contributions are: **1)** insight analysis about the scene configurations and capture settings for which depth becomes meaningful in the blur formation; **2)** a simplified, yet accurate enough, Image Compositing Blur (ICB) model for parallax depth-induced camera motion blur; **3)** an alternative approach to restore sharp images without the need for training over large datasets; and **4)** one synthetic and one real dataset of realistic scenes that include pairs of blurry and sharp images with depth maps and camera trajectories.

2 Related Work

Blur Formation Models. Blur formation models have been studied in the context of image deblurring. Arguably, the simplest model assumes uniform behavior over the whole image. In this case, the blurred image is presumed to be the result of shift-invariant convolution with a Point Spread Function (PSF) [7]. However, this model only holds for very limited practical scenarios.

For the more general case of spatially-varying blur, some works are based on the projective motion path of the camera shake [30]. Gupta *et al.* [8] assume that the blur can be accurately modeled by in-plane camera translation and rotation. White *et al.* [38] focus on the blur produced by 3D rotations. Furthermore, Hirsch *et al.* [10] model the blur as the linear combination of patch-based blur kernel basis. Nevertheless, none of these models precisely determine the blur generation in 3D scenes, especially around abrupt changes in depth

Image Deblurring. Conventional methods for image deblurring are optimization frameworks that tackle the blur produced by the camera motion. To handle the well-known ill-posed nature, previous works enforced different image priors in their solutions, such as Total-Variation [3], normalized sparsity prior [15], L_0-norm regularization [41], dark channel prior [24], or discriminative prior [20].

With the advances in deep learning, several Convolutional Neural Network (CNN) architectures have been proposed. These architectures only take the blurred image as input and produce the estimated sharp image. Su et al. [29] used an encoder-decoder architecture for video deblurring. Nah et al. [23] incorporated the multi-scale processing approach in their deep network. Following the multi-scale principle, numerous CNN-based methods have been introduced including components such as Generative Adversarial Networks (GAN) [16,17], Long-Short Term Memory (LSTM) [32], scale-iterative upscaling scheme [42], half instance normalization [4], multi-scale inputs and outputs [5], blur-aware attention [34], and multi-stage progressive restoration [44]. More recently, progress on Transformer [37] and MLP models demonstrate the ability to handle global-local representations for image restoration tasks [35, 43].

The problem with conventional methods is that they do not use depth in deblurring and they are computationally expensive. On the contrary, deep deblurring performance strongly depends on the training data which, in the case of the above works, do not contain spatial blur induced by depth variations.

Depth-Aware Deblurring. The involvement of the depth cue in the motion blur, although not widely studied, it is not new. Xu and Jia [40] proposed the first work on this track. They used a stereopsis setup to estimate the depth information and subsequently perform layer-wise deblurring. Optimization-based solutions have been introduced for the joint estimation of the scene depth and sharp image, employing either expectation-maximization [11] or energy-minimization [25] methods. Sheng et al. [27] proposed an algorithm that iteratively refines the depth and estimates the latent sharp image from an initial depth map and a blurry image, using belief propagation and Richardson-Lucy algorithm, respectively. Park and Lee [26] proposed and alternating energy-minimization algorithm for the joint dense-depth reconstruction, camera pose estimation, super-resolution, and deblurring. However, their method requires an image sequence instead of a single image.

On the deep learning side, Zhou et al. [47] proposed a stereo deblurring network that internally estimates bi-directional disparity maps to convey information about the spatially-varying blur that is caused by the depth variation. Moreover, Li et al. [21] introduced a depth-guided network architecture for single-image deblurring, which both refines an initial depth map and restores the sharp image.

The above depth-aware deblurring methods properly acknowledge that depth changes produce spatially-varying blur, but it is not clear in which cases this holds. The depth is used as an additional cue for deblurring but, on the other hand, they do not address the spatial blur due to scene depth variation. In contrast, we first identify practical scenarios where the depth variations certainly yield to non-uniform blur. We then characterize how depth and camera motion result in regions with different blur behavior.

3   Geometry of Camera Motion Blur

3.1   Fundamentals

Projective Motion Path Blur Model. For static scenes, image blur comes from the motion of the camera during the exposure time. More precisely, the captured blurry image $\mathbf{y}$ is the summation of the transient sharp images $\{\mathbf{x}_m\}_{m=1}^{M}$ seen by the camera in the poses $\{\vartheta_m\}_{m=1}^{M}$ that follow its trajectory. Assuming there is a linear transformation $\mathcal{T}_{\vartheta_m}$ that warps the latent sharp image $\mathbf{x}$ to any transient image $\mathbf{x}_m$, the blurred image $\mathbf{y}$ can be expressed as:

$$\mathbf{y} = \sum_{m=1}^{M} w_m \mathcal{T}_{\vartheta_m}(\mathbf{x}) + \eta \ , \tag{1}$$

where the weight w_m indicates the time the camera stays at pose ϑ_m and η error in the model. The transformation $\mathcal{T}_{\vartheta_m}$ is induced by a homography H_m such that a pixel $\mathbf{p}$ from the latent image $\mathbf{x}$ is mapped to the pixel $\mathbf{p}'_m$ in the transient image $\mathbf{x}_m$. In homogeneous coordinates, $[\mathbf{p}'_m]_h = H_m[\mathbf{p}]_h$, where $[\cdot]_h$ denotes the conversion from Cartesian to homogeneous coordinates.

For a camera following a 6D motion trajectory, the homography H_m that relates pixels from the latent image $\mathbf{x}$ to the transient image $\mathbf{x}_m$, which are captured from a planar scene at depth D, has the form:

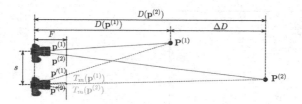

Fig. 2. Blur induced by camera translation of length s for two 3D points $\mathbf{P}^{(1)}$ and $\mathbf{P}^{(2)}$ with their depth difference of ΔD.

$$H_m = C(R_m + \frac{1}{D}T_m[0,0,1])C^{-1} \, , \tag{2}$$

where R_m and T_m stand for the rotation and translation components, and C is the intrinsic camera matrix. Equation (2) reveals that there is non-uniform blur caused by the depth-dependence of the translation component, as well as when rotations are introduced. Notwithstanding, the homography model only holds for fronto-parallel scenes since the warping operator would require an estimation of the occluded areas that become visible, particularly at the depth discontinuities.

Pixel-Wise Blur (PWB) Model. In general, image blur has a spatially-varying nature. To take this into account, the blurred image $\mathbf{y}$ can be modeled via convolutions with pixel-wise kernels $\mathbf{k}(\mathbf{p}, \mathbf{u})$:

$$\mathbf{y}(\mathbf{p}) - \mathbf{x}(\mathbf{p}) * \mathbf{k}(\mathbf{p}, \mathbf{u}) + \eta \, , \tag{3}$$

where $*$ denotes to the convolution operator, $\mathbf{p} = (i, j)$ are pixel coordinates and $\mathbf{u} = (u, v)$ the kernel coordinates. One can blur an image by computing the Empirical Probability Density Function (EPDF) of pixel displacements $\Delta\mathbf{p}'_m = \mathbf{p}'_m - \mathbf{p}$. This model is used as a baseline in our experiments, and its limitations against the proposed blur formation model are demonstrated.

In the remainder of this section, we take a closer look at the influence of depth in the blur generation for in-plane camera translations. Here, we provide insights of what are the scenarios, and to what extent, the depth should be considered in the deblurring problem.

3.2   In-plane Camera Motion

Let us first consider a pin-hole camera with a uniform in-plane motion in the horizontal axis of length s during the exposure time, and two trivial 3D points $\mathbf{P}^{(1)}$ and $\mathbf{P}^{(2)}$ such that the former represents the closest point to the camera in the depth direction and the latter is the farthest as depicted in Fig. 2. On the one hand, $\mathbf{P}^{(1)}$ and $\mathbf{P}^{(2)}$ are respectively mapped to the points $\mathbf{p}^{(1)}$ and $\mathbf{p}^{(2)}$ in the latent image $\mathbf{x}$. On the other hand, they are seen, by the camera at pose ϑ_m, on $\mathbf{p}'^{(1)}$ and $\mathbf{p}'^{(2)}$. In this case, the induced homography $H_m(\mathbf{p})$ is given by

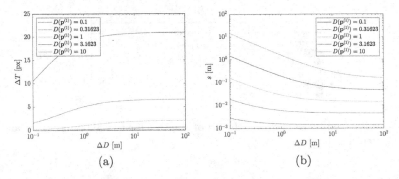

Fig. 3. *Blur variation* determined by Eq. 6, at different depths of the closest point $D(\mathbf{p}^{(1)})$ (in meters): (a) *blur variation* in pixels as function of the depth difference (fixed camera displacement baseline of $s = 3[mm]$); (b) the camera displacement as a function of the depth difference (fixed blur variation $\Delta T = 10$ pixels [px]). Camera focal length is $F{=}2.8[mm]$ and pixel size $4[\mu m]$ which correspond to settings that can be found in mobile phone cameras (ultrawide lenses).

$$H_m(\mathbf{p}) = \begin{bmatrix} 1 & 0 & T_m(\mathbf{p}) \\ 0 & 1 & 0 \\ 0 & 0 & 1 \end{bmatrix} \quad , \tag{4}$$

where $T_m(\mathbf{p})$ is the image plane translation component that is dependent on the pixel depth $D(\mathbf{p})$ as

$$T_m(\mathbf{p}) = \frac{sF}{D(\mathbf{p})} \quad , \tag{5}$$

where F denotes the focal length of the camera. Due to the simplicity of the motion, the *blur extent* of an arbitrary 3D point $\mathbf{P}$ in the blurry image $\mathbf{y}$ is given by $T_m(\mathbf{p}) = \mathbf{p}'_x - \mathbf{p}_x$, where x denotes the horizontal component. Noteworthy, this is equivalent to the disparity in stereo vision.

Blur Variation. Since there is a difference in depth $\Delta D = D(\mathbf{p}^{(2)}) - D(\mathbf{p}^{(1)})$, there must be difference in the blur extent for $\mathbf{P}^{(1)}$ and $\mathbf{P}^{(2)}$, as illustrated in Fig. 2. Thus, we define the *blur variation* ΔT as *the difference in blur extent between two points at different depths*. Expressively, $\Delta T = T_m(\mathbf{p}^{(1)}) - T_m(\mathbf{p}^{(2)})$. ΔT measures the non-uniform behavior of the blur caused by the depth and under in-plane camera movements. By replacing terms, we get

$$\Delta T = \frac{sF}{D(\mathbf{p}^{(1)}) \left[\frac{D(\mathbf{p}^{(1)})}{\Delta D} + 1 \right]} \quad . \tag{6}$$

To gain intuition of the *blur variation* in practical scenarios, we describe its behavior in Fig. 3 by assuming $F{=}2.8[mm]$ and pixel size of $4[\mu m]$.

Macro Photography Scenes. Figure 3(a) illustrates the *blur variation* ΔT as a function of the depth difference ΔD, at different depths of the closest point $D(\mathbf{p}^{(1)})$, while keeping a fixed camera displacement $s=3$[mm] (a reasonable choice for natural hand shake). It can be seen that whereas the *blur variation* is negligible for far-field scenes no matter what is the depth variation, non-uniform blur becomes significant for near-field macro scenes (the closest target ≤ 0.1 m from the camera) even with rather low depth variation (≥ 0.1 m). Although the *blur variation* increases as the depth difference gets higher, there is an upper bound that is determined by $\Delta T < \frac{sF}{D(\mathbf{p}^{(1)})}$. In conclusion, spatially-variant blur is particularly affected by the proximity of the scene whenever there is any variation in depth. Consequently, depth plays a significant role for *Macro Photography* scenes. In this setting, images suffer from defocus blur due to the limited depth-of-field of optics, but defocus blur is a separate issue addressed in other works [1,2,46].

Trucking Photography Scenes. From another perspective, Fig. 3(b) shows the camera displacement s as a function of the depth difference ΔD, by assuming a constant blur variation $\Delta T = 10$ pixels, for different depths of the closest point $D(\mathbf{p}^{(1)})$. In other words, this plot tells us how much the camera should be moved to produce a *blur variation* of 10 pixels. In this case, it is observed that a few millimeters are sufficient to produce such *blur variation* for near-field scenes, regardless of the depth difference. In contrast, in the case of far-field scenes, such a level of blur variation can only be achieved through a camera displacement that ranges from tens of centimeters to a few meters, depending on the depth difference. Such intense movement is unlikely to happen in natural hand shake, but appears in cases where the camera is placed on a fast-moving object. For example, when capturing pictures from inside a moving car. We dub this as *Trucking Photography* scenes.

4   Image Compositing Blur (ICB) Model

From Fig. 3(a), we see that there are depth ranges that yield to nearly the same amount of blur. Hence, pixels in a particular depth range share a common 2D convolutional kernel that characterizes the blur. Inspired by the defocus blur formation models of Hassinoff *et al.* [9] and Ikoma *et al.* [12], we present a new parallax motion Image Compositing Blur (ICB) model that takes the depth into account:

$$\mathbf{y} = \sum_{l=0}^{L-1} (\mathbf{x} * k_l) \cdot \mathcal{A}_l + \eta \ , \tag{7}$$

where $\{\mathcal{A}_l\}_{l=0}^{L-1}$ and $\{k_l\}_{l=0}^{L-1}$ are the set of alpha-matting terms and blur kernels, respectively; and "$\cdot$" is pixel-wise multiplication. We define each alpha matte as:

$$\mathcal{A}_l = \frac{\hat{\mathcal{R}}_l \cdot \mathcal{M}_l}{C} \ , \tag{8}$$

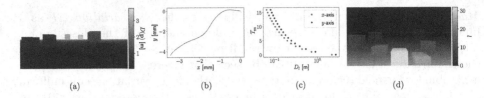

Fig. 4. Spatially-varying blur from depth: (a) full depth map of the latent image $\mathbf{x}$, (b) in-plane camera trajectory, (c) depth sequences in the x and y axis that delimit the regions with the same amount of blur (see Eq. 13) and (d) the region indicators from 0 to 32 that denote the amount of blur from less than 1 to 16 pixels in both dimensions.

where C is a normalization constant over the L depth layers (*i.e.* $C := \sum_{l=0}^{L-1} \hat{\mathcal{R}}_l \cdot \mathcal{M}_l$). $\mathcal{M}_l$ are the z-buffers from far to near layers:

$$\mathcal{M}_l = \prod_{l'=l+1}^{L-1} (1 - \hat{\mathcal{R}}_l') \ . \tag{9}$$

$\hat{\mathcal{R}}_l$ is the smooth spatially-extended version of the depth region $\mathcal{R}_l$. $\hat{\mathcal{R}}_l$ is defined as $\hat{\mathcal{R}}_l := (\mathcal{R}_l \oplus \text{supp}\, k_l) * G_{\sigma,\text{supp}\, k_l}$, with $\oplus$ denoting the dilation operator, and $G_{\sigma,\text{supp}\, k_l}$ is a Gaussian smoothing window with the standard deviation σ and a window size of $\text{supp}\, k_l$. $\{\mathcal{R}_l\}_{l=0}^{L-1}$ comes from the discretization of the depth map, but dilation and smoothing of $\hat{\mathcal{R}}_l$ are used to approximate the mixed blur around the depth discontinuities, and therefore allows to omit explicit estimation of the occluded pixels. Specifically, $\mathcal{R}_l$ is determined by the scene depth as

$$\mathcal{R}_l = \begin{cases} \mathbf{p} \in \Omega | D(\mathbf{p}) \geq D_0 & , l = 0 \\ \mathbf{p} \in \Omega | D_{l-1} < D(\mathbf{p}) \leq D_l & , l = 1, \ldots, L-1 \end{cases} , \tag{10}$$

where Ω refers to the pixel domain in the latent image $\mathbf{x}$, and $\{D_l\}_{l=0}^{L-1}$ is the sequence of depth values that define the regions with "uniform" blur. In particular, D_0 represents the depth limit value, the depth values from D_0 to ∞, for which pixels seem not to move at all.

Next, we derive how to compute the depth sequence $\{D_l\}_{l=0}^{L-1}$ and the respective kernels k_l, for the known camera trajectory s and depth map $D(\mathbf{p})$.

4.1   Depth-Dependent Regions

The image regions for which the blur behaves in the same way are completely defined by the depth sequence $\{D_l\}_{l=0}^{L-1}$. Without loss of generality, let us consider a one-dimensional camera movement whose maximum absolute displacement is denoted by $s_{\max}$. As introduced above, we consider D_0 the depth limit where pixels do not move, namely those pixels whose blur extent is less than one pixel (half a pixel for rounding issues, in practice). The pixels must satisfy

$$\frac{\delta}{2} = \frac{s_{\max} F}{D_0} \ , \tag{11}$$

where δ denotes the pixel size. This means that $D_0 = 2\kappa$ with $\kappa = \frac{s_{\max}F}{\delta}$.

For the rest elements of the sequence, we take into account our definition of *blur variation* presented in Sect. 3.2. The next element in the sequence is characterized as *the depth that produce a blur variation of n pixels* [1]. In other words, the blur extent varies n pixels from $\mathcal{R}_{l-1}$ to $\mathcal{R}_l$. This is expressed as:

$$\Delta T = \frac{s_{\max}F}{D_l} - \frac{s_{\max}F}{D_{l-1}} = n\delta \ . \tag{12}$$

Similarly, by reorganizing the terms, we find an equation for the l-th element of the sequence. It can be proven by induction that

$$D_l = \frac{2\kappa}{2ln+1} \ , \ \text{where} \ \kappa = \frac{s_{\max}F}{\delta} \ . \tag{13}$$

To extend this methodology to 2D motion, we simply compute the component-wise sequences $D_{l(x)}$ and $D_{l(y)}$, which are obtained by replacing different values of $s_{\max}$ and δ for the x and y components of the movement. Then, the complete sequence D_l is the sorted vector of the set union $\{D_{l(x)} \cup D_{l(y)}\}$. Figure 4 exemplifies the discrete regions automatically obtained using the above procedure for a synthetically generated image, with $n = 1$. Figure 4(a) and (b) show the full depth and the 2D camera trajectory, respectively. Figure 4(c) illustrates the depth sequences $D_{l(x)}$ and $D_{l(y)}$ computed by using the aforementioned procedure. Lastly, the set of regions $\{\mathcal{R}_l\}_{l=0}^{L-1}$ whose blur behaves similarly within each layer is shown in Fig. 4(d). It is worth mentioning that the total number L of regions is completely adaptive to the scene configuration. One only needs to compute the sequences until $D_{L-1(x)}$ and $D_{L-1(y)}$ cover the minimum depth.

4.2   Blur Kernels Synthesis

Having the time-dependent in-plane camera trajectory $s(t)$ and the depth map $D(\mathbf{p})$, the pixel-wise motion blur kernel is given by

$$\mathbf{k}(\mathbf{p}) = \hat{f}\left(\left\lfloor \frac{-s(t)F}{\delta D(\mathbf{p})} \right\rceil\right) , \tag{14}$$

where $\lfloor \cdot \rceil$ denotes the rounding operation and $\hat{f}$ is the operator that computes the EPDF for the discretized values in the argument. Instead of computing kernels $\mathbf{k}(\mathbf{p})$ at pixel level, we compute a smaller set $\{k_l\}_{l=0}^{L-1}$ where every kernel is paired with a region $\hat{\mathcal{R}}_l$. The pixels in the region $\hat{\mathcal{R}}_l$ share the same motion blur kernel k_l:

$$k_l = \hat{f}\left(\left\lfloor \frac{-s(t)F}{\delta D_l^*} \right\rceil\right) , \tag{15}$$

---

[1] σ and n correspond to hyper-parameters in our blur formation model. Ablation studies on those can be found in the supplementary material.

where D_l^* is the optimal depth value in the range $[D_l, D_{l-1}]$ that minimizes the mean-square error in $D(\mathbf{p})$ for $\mathbf{p} \in \mathcal{R}_l$. The depths $D(\mathbf{p})$ in $[D_l, D_{l-1}]$ follow a random variable ζ with PDF $f(\zeta)$ and the mean-square error is determined by

$$\int_{D_l}^{D_{l-1}} (\zeta - D_l^*)^2 f(\zeta) d\zeta . \tag{16}$$

It can be proven that the mean depth $\bar{D}(\mathbf{p})$ in the range minimizes (16).

5   Neural Representations from Blur

Advances in implicit neural representations demonstrate that MLPs can learn the high-frequency details in 2D images [28,31]. In those works, a coordinate-based MLP Φ_θ optimizes its parameters θ to fit a sharp image, $i.e.$ $\Phi_\theta : \mathbf{p} \mapsto \mathbf{x}$. We propose a different approach where Φ_θ fits the sharp image $\mathbf{x}$ from its corresponding blurred one $\mathbf{y}$, by embedding a blur function $b : \mathbf{x} \mapsto \mathbf{y}$ defined by either the PWB model (3) or our ICB model (7). This provides an alternative solution for deblurring from a single blurred image. Since b is differentiable, we can use gradient-descent methods to optimize θ with the following loss:

$$\mathcal{L} = \sum_{\mathbf{p}} \|b(\Phi_\theta(\mathbf{p})) - \mathbf{y}(\mathbf{p})\|_2^2 + \lambda \|\nabla_{\mathbf{p}} \Phi_\theta(\mathbf{p})\|_1^1 , \tag{17}$$

where λ is a hyper-parameter that controls the smoothness of the gradients. This method is similar to the approach presented by Ulyanov et $al.$ [36], with the exception that we utilize a coordinate-based MLP rather than a CNN for fitting $\mathbf{x}$. In practice, we use the SIREN architecture [28] for its ability to fit derivatives robustly.

6   Experiments

6.1   Evaluation Datasets

Synthetic Dataset. We constructed the Virtual Camera Motion Blur (Virtual-CMB) dataset, where the ground-truth latent images and depth maps are rendered from the 3D scene models. We utilized the Unity engine [33] for rendering 3D scenes in HD resolution. The dataset was built using five high-quality scenes available in the unity asset store. The viewpoints were manually selected to represent a virtual snapshot camera and motion blur for the three studied cases: **1)** *Macro Photography*, **2)** *Trucking Photography* and **3)** *Standard Photography*. Table 1 summarizes the number of images captured for each case. Macro and Trucking represent practical settings where depth contributes to blurring (see Sect. 3.2). Standard Photography is the typical setting where all scene objects are far from the camera and thus depth-agnostic models work well. In all cases, the camera was moved through pre-defined trajectories. For the Macro and Standard cases, we randomly selected six trajectories from the Kohler dataset [13].

Table 1. Summary of captured images: i) parallax motion, ii) with *pan-tilt* rotations (w/ *xy* rotations), and iii) 6-DoF.

Scene	Macro			Trucking			Standard		
	i)	ii)	iii)	i)	ii)	iii)	i)	ii)	iii)
VikingVillage	26	28	26	23	22	23	–	–	30
IndustrialSet	–	–	–	60	60	60	–	–	30
ModularCity	–	–	–	60	60	60	–	–	30
ModernStudio	58	55	55	–	–	–	–	–	30
LoftOffice	56	50	51	–	–	–	–	–	30
Total: 983	**140**	**133**	**132**	**143**	**142**	**143**	**–**	**–**	**150**

For the Trucking Photography cases, six linear trajectories with a constant speed in the xy plane were generated. The purpose is to mimic photography from a moving object (e.g., inside a car). To test our method beyond motion parallax, also camera motions of *pan-tilt* rotations and full 6-DoF camera motion were recorded. Overall, 983 blurred images with corresponding latent sharp images and depth maps were rendered.

Real Dataset. For evaluation with real images, we used the iOS app introduced by Chugunov *et al.* [6] to capture synchronized RGB, LiDAR depth maps, and camera poses. These videos match with the *Macro photography* case, where a static object is recorded by a hand-held smartphone camera. As preprocessing, RGB frames are down-scaled to the depth map resolution (256×192), blurry frames are obtained by temporal average, while the sharp and depth correspondences are taken from the middle point in the camera trajectory. Accordingly, we built the Real Camera Motion Blur (RealCMB) dataset, comprised of 58 pairs of blurry and sharp images, as well as depth and camera motion; from which 48 come from our own recordings and 10 are available in [6].

6.2   Model Validation

Parallax Motion Blur. Our ICB model in Sect. 4 was particularly designed for the parallax motion of a camera. Thus, we first evaluate the model under in-plane camera motion in the VirtualCMB dataset. For comparison, we considered the PWB model (3). The standard image quality metrics: PSNR and SSIM, and the perceptual quality metric LPIPS [45], are used for performance evaluation.

Parallax results are in the "Parallax" column of Table 2. In the terms of PSNR, SSIM, and LPIPS, the proposed ICB model outperforms the baseline PWB model in both of the main cases: Macro and Trucking, except for the LPIPS in the Trucking case. Although the difference between SSIM and LPIPS is marginal in practice. By nature, PWB cannot properly trace the generated blur over the depth discontinuities where occluded areas become visible during the motion. Conversely, our ICB model merges blur from different depth layers more effectively, resulting in a more realistic blur. Figure 5(a) and (b) illustrates this finding for the two types of blur, Macro, and Trucking. The error images reveal that the proposed model is more precise at the object edges.

Table 2. Blur formation results in VirtualCMB.

| | Parallax | | | | w/ xy rotation | | | | 6 DoF | | | | | |
| | Macro | | Trucking | | Macro | | Trucking | | Macro | | Trucking | | Standard | |
	Ours	PWB	Ours	PWB	Ours	PWB	Ours	PWB	Ours	PWB	Ours	PWB	Ours	PWB
↑PSNR	**42.48**	41.54	**37.42**	36.59	**42.16**	41.11	**36.99**	36.21	**38.84**	38.37	**37.08**	36.29	**37.38**	36.97
↑SSIM	**0.993**	0.992	**0.985**	0.984	**0.990**	0.989	**0.984**	0.983	**0.984**	0.982	**0.984**	0.983	**0.978**	0.978
↓LPIPS (×10⁻⁵)	**4.639**	5.049	5.870	**5.238**	**5.399**	6.316	6.158	**5.676**	**8.883**	9.744	6.088	**5.613**	**11.17**	13.01

Table 3. Blur formation results for RealCMB (avg over 58 test images).

	↑PSNR	↑SSIM	↓LPIPS (×10⁻⁵)
PWB	36.31	0.984	6.826
Ours	**38.21**	**0.990**	**4.484**

Table 4. Results in terms of computational resources.

| | RealCMB | | VirtualCMB | |
	Memory [MB]	Run time [s]	Memory [MB]	Run time [s]
PWB	57.39	**1.95**	2799	13.41
Ours	**1.74**	2.78	**48.97**	**7.27**

Out-of-Plane Rotations and 6-DoF. Non-uniform blur does not only come from motion parallax but also rotations. Assuming a large focal length as in [38], *pan-tilt* rotations can be approximated by xy translations that are non-depth dependent. Thus, we can compute a global uniform kernel which is added on top of the kernels in Sect. 4.2. In particular, this approximation works well in narrow-lens devices. This approach was adopted to handle motion camera blur beyond motion parallax, neglecting the effect of z translation and *roll* rotation. The results beyond parallax motion (xy-rotation and 6-DoF) in Table 2 are similar to the parallax motion experiment. Consequently, the used approximation works well for the captured images in the VirtualCMB dataset.

Real Images. Surprisingly, the proposed ICB model performs clearly better on real 6-DoF motion in the RealCMB dataset (Table 3) than in the previous experiment with synthetic data. These results indicate that 1) parallax motion can be more dominating in real data than in our simulated cases and 2) our model is robust to depth and trajectory noise that appears in real data. Moreover, as the depth measurements are non-linearly quantized in ICB (see Sect. 4.1), there is no need for high-resolution depth maps. Figure 5(d) shows a visual example of the blur generation in the RealCMB dataset.

Computational Resources. Table 4 reports the averaged run time and memory size of our Pytorch implementations of PWB and ICB. It turns out that the proposed ICB model is slightly slower in the RealCMB dataset but significantly faster in VirtualCMB. Most importantly, ICB demonstrates considerably greater efficiency in terms of memory consumption, with reductions of ×32 and ×50 in the RealCMB and VirtualCMB datasets, respectively.

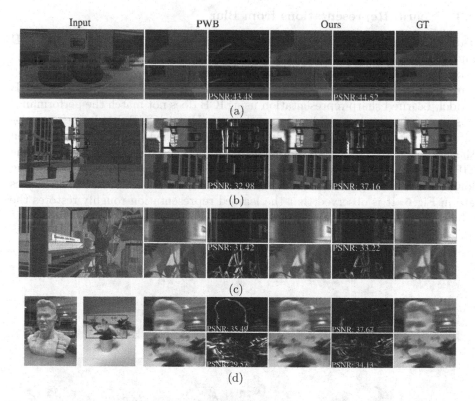

Fig. 5. Examples of the blur formation results. (a) Macro, (b) Trucking (c) Macro with 6-DoF motion, and (d) Real images.

Table 5. Comparison of sharp restoration results.

	VirtualCMB			RealCMB		
	↑PSNR	↑SSIM	↓LPIPS ($\times 10^{-4}$)	↑PSNR	↑SSIM	↓LPIPS ($\times 10^{-4}$)
SRN [32]	29.92	0.9135	7.423	28.26	0.9146	8.492
SIUN [42]	29.75	0.9114	**7.235**	28.33	0.9139	7.231
HINet [4]	29.86	0.9133	8.651	28.32	0.9133	7.627
BANet [34]	29.77	0.9099	9.007	28.34	0.9140	8.241
MIMO-UNet++ [5]	28.79	0.8964	13.12	27.92	0.9106	10.33
MPRNet [44]	30.01	0.9146	8.062	28.79	0.9178	7.769
MAXIM [35]	30.34	**0.9186**	7.418	28.89	0.9207	5.695
Restormer [43]	**30.41**	0.9174	7.318	29.56	0.9243	6.887
PWB + SIREN	27.08	0.7975	37.25	30.61	0.9429	5.398
Ours + SIREN	27.14	0.8001	36.56	**31.92**	**0.9546**	**4.032**

6.3   Neural Representations from Blur

Table 5 summarizes the results of the sharp implicit representations with ICB and PWB models. In addition, we evaluated SOTA deep-deblurring methods. For the task of learning implicit representations from a single blurry image, our ICB model produces superior reconstruction results compared to the PWB model. Learned sharp representation using ICB does not match the performance of state-of-the-art deep deblurring methods on VirtualCMB, but it performs significantly better than others on RealCMB. The variation in performance between the two datasets can be attributed to the difference in image resolution. The SIREN architecture utilized in the experiments may be better suited to handling low-resolution images, such as those in RealCMB. Visual restoration examples are in Fig. 6. It is observed that the learned representation roughly restores the edges, but global noise remains in the VirtualCMB example. On the contrary, an accurate sharp representation is obtained in the RealCMB case.

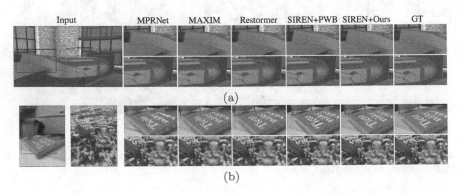

Input   MPRNet   MAXIM   Restormer   SIREN+PWB   SIREN+Ours   GT

(a)

(b)

Fig. 6. Examples of the deblurring results in (a) VirtualCMB, (b) RealCMB datasets.

7   Conclusion

This work provides analytical and experimental results about the scene configurations in which the scene depth affects the camera motion blur. In particular, we identified two types of scenes that appear in consumer photography: "Macro" and "Trucking". Primarily, we presented an Image-Compositing Blur (ICB) model that efficiently and accurately describes the induced blur in those cases. Experimental validation was performed in our introduced synthetic and real datasets. Interestingly, we demonstrated the effectiveness of the ICB model to learn sharp neural representations from a single blurry image. Our findings and the new datasets help to develop better deblurring approaches.

Limitations. Although our ICB model is derived for parallax motion, the model was found accurate enough under certain scene configurations, *e.g.*, Macro and Trucking photography. Besides, the model is computationally efficient and robust against occlusions due to abrupt depth changes. Regarding the deblurring task, our results are still far from being practical. In real scenarios, the depth maps and camera trajectories need to be estimated and that would need a careful study of the suitability of IMU-based odometry and depth sensors in the current hand-held devices.

Acknowledgements. This project was supported by a Huawei Technologies Oy (Finland) project. We also thank Jussi Kalliola for building the iOS app [6] for data collection.

References

1. Akpinar, U., Sahin, E., Meem, M., Menon, R., Gotchev, A.: Learning wavefront coding for extended depth of field imaging. IEEE Trans. Image Process. **30**, 3307–3320 (2021)
2. Anwar, S., Hayder, Z., Porikli, F.: Depth estimation and blur removal from a single out-of-focus image. In: BMVC, vol. 1, p. 2 (2017)
3. Chan, T.F., Wong, C.K.: Total variation blind deconvolution. IEEE Trans. Image Process. **7**(3), 370–375 (1998)
4. Chen, L., Lu, X., Zhang, J., Chu, X., Chen, C.: HINet: half instance normalization network for image restoration. In: Proceedings of the IEEE/CVF Conference on Computer Vision and Pattern Recognition, pp. 182–192 (2021)
5. Cho, S.J., Ji, S.W., Hong, J.P., Jung, S.W., Ko, S.J.: Rethinking coarse-to-fine approach in single image deblurring. In: Proceedings of the IEEE/CVF International Conference on Computer Vision, pp. 4641–4650 (2021)
6. Chugunov, I., Zhang, Y., Xia, Z., Zhang, X., Chen, J., Heide, F.: The implicit values of a good hand shake: Handheld multi-frame neural depth refinement. In: Proceedings of the IEEE/CVF Conference on Computer Vision and Pattern Recognition, pp. 2852–2862 (2022)

7. Fergus, R., Singh, B., Hertzmann, A., Roweis, S.T., Freeman, W.T.: Removing camera shake from a single photograph. In: ACM SIGGRAPH 2006 Papers, pp. 787–794 (2006)
8. Gupta, A., Joshi, N., Lawrence Zitnick, C., Cohen, M., Curless, B.: Single image deblurring using motion density functions. In: Daniilidis, K., Maragos, P., Paragios, N. (eds.) ECCV 2010. LNCS, vol. 6311, pp. 171–184. Springer, Heidelberg (2010). https://doi.org/10.1007/978-3-642-15549-9_13
9. Hasinoff, S.W., Kutulakos, K.N.: A layer-based restoration framework for variable-aperture photography. In: 2007 IEEE 11th International Conference on Computer Vision, pp. 1–8. IEEE (2007)
10. Hirsch, M., Schuler, C.J., Harmeling, S., Schölkopf, B.: Fast removal of non-uniform camera shake. In: 2011 International Conference on Computer Vision, pp. 463–470. IEEE (2011)
11. Hu, Z., Xu, L., Yang, M.H.: Joint depth estimation and camera shake removal from single blurry image. In: Proceedings of the IEEE Conference on Computer Vision and Pattern Recognition, pp. 2893–2900 (2014)
12. Ikoma, H., Nguyen, C.M., Metzler, C.A., Peng, Y., Wetzstein, G.: Depth from defocus with learned optics for imaging and occlusion-aware depth estimation. In: 2021 IEEE International Conference on Computational Photography (ICCP), pp. 1–12. IEEE (2021)
13. Köhler, R., Hirsch, M., Mohler, B., Schölkopf, B., Harmeling, S.: Recording and playback of camera shake: benchmarking blind deconvolution with a real-world database. In: Fitzgibbon, A., Lazebnik, S., Perona, P., Sato, Y., Schmid, C. (eds.) ECCV 2012. LNCS, vol. 7578, pp. 27–40. Springer, Heidelberg (2012). https://doi.org/10.1007/978-3-642-33786-4_3
14. Krishnan, D., Fergus, R.: Fast image deconvolution using hyper-Laplacian priors. Adv. Neural. Inf. Process. Syst. 22, 1033–1041 (2009)
15. Krishnan, D., Tay, T., Fergus, R.: Blind deconvolution using a normalized sparsity measure. In: CVPR 2011, pp. 233–240. IEEE (2011)
16. Kupyn, O., Budzan, V., Mykhailych, M., Mishkin, D., Matas, J.: DeblurGAN: blind motion deblurring using conditional adversarial networks. In: Proceedings of the IEEE Conference on Computer Vision and Pattern Recognition, pp. 8183–8192 (2018)
17. Kupyn, O., Martyniuk, T., Wu, J., Wang, Z.: DeblurGAN-v2: deblurring (orders-of-magnitude) faster and better. In: Proceedings of the IEEE/CVF International Conference on Computer Vision, pp. 8878–8887 (2019)
18. Levin, A., Weiss, Y., Durand, F., Freeman, W.T.: Understanding and evaluating blind deconvolution algorithms. In: 2009 IEEE Conference on Computer Vision and Pattern Recognition, pp. 1964–1971. IEEE (2009)
19. Levin, A., Weiss, Y., Durand, F., Freeman, W.T.: Efficient marginal likelihood optimization in blind deconvolution. In: CVPR 2011, pp. 2657–2664. IEEE (2011)
20. Li, L., Pan, J., Lai, W.S., Gao, C., Sang, N., Yang, M.H.: Blind image deblurring via deep discriminative priors. Int. J. Comput. Vision 127(8), 1025–1043 (2019)
21. Li, L., Pan, J., Lai, W.S., Gao, C., Sang, N., Yang, M.H.: Dynamic scene deblurring by depth guided model. IEEE Trans. Image Process. 29, 5273–5288 (2020)
22. Mustaniemi, J., Kannala, J., Särkkä, S., Matas, J., Heikkilä, J.: Gyroscope-aided motion deblurring with deep networks. In: IEEE Winter Conference on Applications of Computer Vision (WACV) (2019)
23. Nah, S., Kim, T.H., Lee, K.M.: Deep multi-scale convolutional neural network for dynamic scene deblurring. In: Proceedings of the IEEE Conference on Computer Vision and Pattern Recognition, pp. 3883–3891 (2017)

24. Pan, J., Sun, D., Pfister, H., Yang, M.H.: Blind image deblurring using dark channel prior. In: Proceedings of the IEEE Conference on Computer Vision and Pattern Recognition, pp. 1628–1636 (2016)
25. Pan, L., Dai, Y., Liu, M.: Single image deblurring and camera motion estimation with depth map. In: 2019 IEEE Winter Conference on Applications of Computer Vision (WACV), pp. 2116–2125. IEEE (2019)
26. Park, H., Lee, K.M.: Joint estimation of camera pose, depth, deblurring, and super-resolution from a blurred image sequence. In: Proceedings of the IEEE International Conference on Computer Vision, pp. 4613–4621 (2017)
27. Sheng, B., Li, P., Fang, X., Tan, P., Wu, E.: Depth-aware motion deblurring using loopy belief propagation. IEEE Trans. Circuits Syst. Video Technol. **30**(4), 955–969 (2019)
28. Sitzmann, V., Martel, J., Bergman, A., Lindell, D., Wetzstein, G.: Implicit neural representations with periodic activation functions. Adv. Neural. Inf. Process. Syst. **33**, 7462–7473 (2020)
29. Su, S., Delbracio, M., Wang, J., Sapiro, G., Heidrich, W., Wang, O.: Deep video deblurring for hand-held cameras. In: Proceedings of the IEEE Conference on Computer Vision and Pattern Recognition, pp. 1279–1288 (2017)
30. Tai, Y.W., Tan, P., Brown, M.S.: Richardson-Lucy deblurring for scenes under a projective motion path. IEEE Trans. Pattern Anal. Mach. Intell. **33**(8), 1603–1618 (2010)
31. Tancik, M., et al.: Fourier features let networks learn high frequency functions in low dimensional domains. Adv. Neural. Inf. Process. Syst. **33**, 7537–7547 (2020)
32. Tao, X., Gao, H., Shen, X., Wang, J., Jia, J.: Scale-recurrent network for deep image deblurring. In: Proceedings of the IEEE Conference on Computer Vision and Pattern Recognition, pp. 8174–8182 (2018)
33. Technologies, U.: Unity real-time development platform. https://unity.com/
34. Tsai, F.J., Peng, Y.T., Tsai, C.C., Lin, Y.Y., Lin, C.W.: Banet: a blur-aware attention network for dynamic scene deblurring. IEEE Trans. Image Process. **31**, 6789–6799 (2022)
35. Tu, Z., et al.: Maxim: Multi-axis MLP for image processing. In: Proceedings of the IEEE/CVF Conference on Computer Vision and Pattern Recognition, pp. 5769–5780 (2022)
36. Ulyanov, D., Vedaldi, A., Lempitsky, V.: Deep image prior. In: Proceedings of the IEEE Conference on Computer Vision and Pattern Recognition, pp. 9446–9454 (2018)
37. Vaswani, A., et al.: Attention is all you need. Advances in Neural Information Processing Systems 30 (2017)
38. Whyte, O., Sivic, J., Zisserman, A., Ponce, J.: Non-uniform deblurring for shaken images. Int. J. Comput. Vision **98**(2), 168–186 (2012)
39. Xu, L., Jia, J.: Two-phase kernel estimation for robust motion deblurring. In: Daniilidis, K., Maragos, P., Paragios, N. (eds.) ECCV 2010. LNCS, vol. 6311, pp. 157–170. Springer, Heidelberg (2010). https://doi.org/10.1007/978-3-642-15549-9_12
40. Xu, L., Jia, J.: Depth-aware motion deblurring. In: 2012 IEEE International Conference on Computational Photography (ICCP), pp. 1–8. IEEE (2012)
41. Xu, L., Zheng, S., Jia, J.: Unnatural l0 sparse representation for natural image deblurring. In: Proceedings of the IEEE Conference on Computer Vision and Pattern Recognition, pp. 1107–1114 (2013)
42. Ye, M., Lyu, D., Chen, G.: Scale-iterative upscaling network for image deblurring. IEEE Access **8**, 18316–18325 (2020)

43. Zamir, S.W., Arora, A., Khan, S., Hayat, M., Khan, F.S., Yang, M.H.: Restormer: efficient transformer for high-resolution image restoration. In: Proceedings of the IEEE/CVF Conference on Computer Vision and Pattern Recognition, pp. 5728–5739 (2022)
44. Zamir, S.W., et al.: Multi-stage progressive image restoration. In: Proceedings of the IEEE/CVF Conference on Computer Vision and Pattern Recognition, pp. 14821–14831 (2021)
45. Zhang, R., Isola, P., Efros, A.A., Shechtman, E., Wang, O.: The unreasonable effectiveness of deep features as a perceptual metric. In: Proceedings of the IEEE Conference on Computer Vision and Pattern Recognition, pp. 586–595 (2018)
46. Zhang, X., Wang, R., Jiang, X., Wang, W., Gao, W.: Spatially variant defocus blur map estimation and deblurring from a single image. J. Vis. Commun. Image Represent. **35**, 257–264 (2016)
47. Zhou, S., Zhang, J., Zuo, W., Xie, H., Pan, J., Ren, J.S.: DAVANet: stereo deblurring with view aggregation. In: Proceedings of the IEEE/CVF Conference on Computer Vision and Pattern Recognition, pp. 10996–11005 (2019)

Detection, Recognition, Classification, and Localization in 2D and/or 3D

Affine Moment Invariants of Tensor Fields

Jan Flusser[1], Tomáš Suk[1(✉)], Matěj Lébl[1], Roxana Bujack[2],
and Ibrahim Ibrahim[3]

[1] Czech Academy of Sciences, Institute of Information Theory and Automation,
Pod Vodárenskou věží 4, 182 08 Praha 8, Czech Republic
`{flusser,suk,lebl}@utia.cas.cz`
[2] Data Science at Scale Team, Los Alamos National Laboratory, P.O. Box 1663,
Los Alamos, NM 87545, USA
`bujack@informatik.uni-leipzig.de`
[3] MR Unit, Department of Diagnostic and Interventional Radiology,
Institute for Clinical and Experimental Medicine IKEM, Vídeňská 1958/9,
140 21 Praha 4, Czech Republic
`ibib@ikem.cz`

Abstract. Tensor fields (TF) are a special kind of multidimensional
data, in which a tensor is given for each point in space. Often, it is a
3×3 array in each voxel. To detect the patterns of interest in the field,
special matching methods must be developed. We propose a method for
the description and matching of TF patterns under an unknown affine
transformation of the field. Transformations of TFs act not only in the
spatial coordinates but also on the field values, which makes the detection more challenging. To measure the similarity between the template
and the field patch, we propose original invariants with respect to affine
transformations designed from moments. Their performance is demonstrated by experiments on real data from diffusion tensor imaging.

Keywords: Tensor field · affine invariants · template matching

1 Introduction

A gray-level image can be described by a scalar function. Sometimes, we need a
vector defined in each point of the space, then we talk about a vector field. An
example can be wind or water flow in a river. There are even more complicated
cases, where we need a tensor of rank 2 or higher in each point. An example is the
Cauchy stress tensor, where in each point of space, we need information not only

This work was supported by the Czech Academy of Sciences through *Praemium
Academiae*, by the Czech Science Foundation under the grant No. GA21-03921S, and
by the MH CZ - DRO "Institute for Clinical and Experimental Medicine - IKEM",
IN 00023001 and by the National Nuclear Security Administration (NNSA) Advanced
Simulation and Computing (ASC) Program. The data were acquired and provided
by the Institute for Clinical and Experimental Medicine (IKEM) in Prague, Czech
Republic.

R. Gade et al. (Eds.): SCIA 2023, LNCS 13885, pp. 299–313, 2023.
https://doi.org/10.1007/978-3-031-31435-3_20

about magnitude of the inner force and its direction, but also about transverse components of the force that try to turn the inner part of the material.

Another example comes from diffusion tensor imaging (DTI). DTI is a modern technique based on magnetic resonance imaging (MRI) for an examination of tissues with internal anisotropic structure, such as neural axons of white matter in the brain and peripheral nerve fibres. It reconstructs the diffusion of water molecules in each voxel by measuring their movement in several distinct directions. This measurement is accomplished via several diffusion-weighted acquisitions, each obtained with a different orientation of the diffusion sensitizing gradients. After obtaining a complete set of such measurements (six diffusion-encoding gradient directions are the minimum needed to calculate the diffusion tensor; usually 30, 64, or more gradient directions are used), a symmetric second-rank 3×3 tensor is calculated in each voxel. This tensor image is an extremely useful modality, because it offers a possibility to detect a subtle pathology in the brain, to track neural tracts through the brain (this process is called tractography), to examine the integrity of peripheral nerves, and to diagnose of many neurological diseases [1, 9, 30].

In this paper, we deal with the template matching problem. A template, extracted from a reference image, shall be localized in the sensed image. However, there might be a deformation between the template and the corresponding patch. Template matching is an important part of registration of data taken at different times and in localization of regions of interest. Due to the tensorial nature of DTI data, common algorithms known from scalar image template matching cannot be used directly.

To measure the similarity between the template and the field patch, we need special kinds of descriptors, that are invariant to particular deformation of the template, to the template size and orientation. In this paper, we model the template variations by a *total affine transform*, which means that the transformation acts in the coordinate domain as well as in the value (vector or tensor) domain. This model is sufficiently general to capture most of the situations that appear in practice and, at the same time, it is still sufficiently simple to be handled mathematically.

2   Literature Survey

Our research is a follow-up of the previous work on the rotation and affine invariants of images and vector fields. The rotation invariants of vector fields were first studied by Schlemmer et al. [23]. Liu and Ribeiro [19] used them to detect singularities on meteorological satellite images showing wind velocity and Liu and Yap [18] applied them to the indexing and recognition of fingerprint images.

A generalization to more than two dimensions using tensor contraction was proposed by Langbein and Hagen [17]. Bujack et al. [5] showed that the invariants can be derived also by means of the field normalization approach. Yang et al. improved the numerical stability of the invariants by using orthogonal Gaussian-Hermite [33] and Zernike [32] moments. Recently, Bujack [3] introduced a *flexible basis* of the invariants to avoid moments that vanish on the given templates. In [6], Bujack et al. propose the systematic approach to the generation of the tensor field invariants.

In contrast to the above group of papers on vector field rotation invariants, *affine moment invariants* (AMI) of graylevel images have been studied in hundreds of papers and books [13,21,22,24,25,27]. Special AMIs were proposed for color images [7,20,26]. The most recent paper on this field is [16], where affine invariants of 2D vector fields are proposed.

In this paper, we focus on the case of 3D tensor fields. We restrict ourselves to the case of the second-rank symmetric tensors and we assume the inner (tensor values) and outer (coordinate) affine transformations are the same. Both assumptions are implied by physics of DTI. However, the presented theory of invariants could be developed in a more general way even without these limitations.

3   Affine Tensor Field Moment Invariants

Intuitively speaking, a tensor is an array of numbers, where the number of indices is called its *rank* and their range of the indices its *dimension*. If a tensor has a rank r and a dimension d, it has d^r components[1]. Unlike usual arrays, the tensors have two types of indices, contravariant and covariant. They differ in behavior under affine transformations of the space. The tensors are multiplied with the matrix of the direct transformation on behalf of each covariant index and by the matrix of the inverse transformation on behalf of each contravariant index. Formally, the tensor σ in affine transformation behaves

$$\sigma'^{i_1 i_2 \cdots i_n}_{j_1 j_2 \cdots j_m} = a^{\ell_1}_{j_1} a^{\ell_2}_{j_2} \cdots a^{\ell_m}_{j_m} \bar{a}^{i_1}_{k_1} \bar{a}^{i_2}_{k_2} \cdots \bar{a}^{i_n}_{k_n} \sigma^{k_1 k_2 \cdots k_n}_{\ell_1 \ell_2 \cdots \ell_m} ,$$

$$i_1, i_2, \ldots, i_n, j_1, j_2, \ldots, j_m, k_1, k_2, \ldots, k_n, \ell_1, \ell_2, \ldots, \ell_m = 1, 2, \ldots, d,$$

(1)

where a^k_j are elements of the matrix of the direct affine transformation $\mathbf{A}$ and $\bar{a}^i_\ell$ are elements of the matrix of the inverse affine transformation $\mathbf{A}^{-1}$. Here $i_1, i_2, \ldots, i_n$, $k_1, k_2, \ldots, k_n$ are contravariant indices, n is *contravariant rank*, $j_1, j_2, \ldots, j_m$, $\ell_1, \ell_2, \ldots, \ell_m$ are covariant indices, m is *covariant rank* and $r = n + m$ is *total rank* or just *rank*.

Special cases include scalars, which are tensors of rank zero, vectors, which are tensors of rank one, and matrices, which are tensors of rank two. The dimension and rank of tensor fields used in practice is limited. The most common tensor fields in physics are Cauchy stress tensor, viscous stress tensor, diffusion tensor and Maxwell stress tensor. All of them have dimension three and contravariant rank two, i.e. they form 3×3 arrays in each point of the 3D space

$$\sigma \in \mathbb{R}^{3 \times 3}.$$

(2)

---

[1] This rank differs from the rank of matrix in linear algebra. Alternatively, it is called "order", but it can be confused with moment order. In this paper, we work with the moment order, but not with matrix rank, therefore we use rank here for the number of tensor indices.

The Cauchy stress tensor describes internal stress at a point inside a solid material. It is symmetric, i.e. $\sigma^{ij} = \sigma^{ji}$, so, it contains only six degrees of freedom. The diagonal components express magnitude and direction, while the other components of the tensor express the transverse components of the inner stress. A description of the tensors and operations with them can be found in [2] or in [8]. A good explanation can also be found in [10] or in its English translation [11].

Unlike matrices, tensors are multiplied in following fashion:

$$\sigma_p{}^{ijk\ell} = \sigma_1{}^{ij}\sigma_2{}^{k\ell}, \quad i,j,k,\ell = 1,\ldots,3, \tag{3}$$

where $\sigma_p \in \mathbb{R}^{3\times3\times3\times3}$ i.e. each component of the first tensor is multiplied with each component of the second tensor. The tensor product is noted as

$$\sigma_p = \sigma_1 \otimes \sigma_2. \tag{4}$$

The product has four indices, thus it is not a second rank tensor, but it satisfies the general definition of a tensor.

Two Cauchy stress tensors can be added

$$\sigma_s{}^{ij} = \sigma_1{}^{ij} + \sigma_2{}^{ij} \quad i,j = 1,\ldots,3, \tag{5}$$

i.e. only corresponding components are added. The result $\sigma_s = \sigma_1 + \sigma_2$ is again a second rank tensor. Please note that this product and sum extend to tensors of all ranks. For a formal introduction to general tensors, we recommend [4].

The viscous stress tensor is analogous to the Cauchy stress tensor in fluids. Unlike the Cauchy stress tensor, it can have an antisymmetric component and is generally not symmetric. The Maxwell stress tensor is the analogon of the Cauchy stress tensor for electromagnetic forces. It is also symmetric.

An example from biology comes from diffusion tensor imaging. It is a way of using magnetic resonance imaging (MRI), which measures the restricted diffusion of water molecules in the tissue. The diffusion tensor $\mathbf{D} \in \mathbb{R}^{3\times3}$ is a second rank three-dimensional symmetric tensor, like the Cauchy stress tensor.

3.1 Covariant and Contravariant Indices

Generally, tensors have two types of indices - covariant and contravariant. The covariant indices are notated as subscripts, e.g. ν_{ij}, the contravariant indices are notated as superscripts, e.g. ν^{ij}.

The range of the indices equals the dimension d of the space, i.e. $i = 1, 2$ in 2D and $i = 1, 2, 3$ in 3D. Let $A \in \mathbb{R}^{d\times d}$ be a matrix representing an affine transformation. A tensor of covariant rank two behaves under the transformation A as

$$\nu'_{ij} = \sum_{k=1}^{d}\sum_{\ell=1}^{d} \mathbf{A}_i^k \mathbf{A}_j^\ell \nu_{k\ell}. \tag{6}$$

Similarly for a tensor of contravariant rank two, we have

$$\nu'^{ij} = \sum_{k=1}^{d}\sum_{\ell=1}^{d} (\mathbf{A}^{-1})_k^i (\mathbf{A}^{-1})_\ell^j \nu^{k\ell}. \tag{7}$$

The most popular second rank tensor, probably known from linear algebra, is a matrix denoting a linear transform. It has both covariant and contravariant indices and transforms via

$$\nu'^j_i = \sum_{k=1}^{d}\sum_{\ell=1}^{d} A^k_i (A^{-1})^j_\ell \nu^\ell_k, \tag{8}$$

which is equivalent to the common matrix transformation $A\nu A^{-1}$.

3.2 Contraction

There is another important operation with tensors - the contraction. It is the sum over two indices, one covariant and one contravariant. Let us take such a tensor ν^j_i. Its contraction equals

$$c = \sum_{i=1}^{d} \nu^i_i, \tag{9}$$

which is equivalent to the trace of the matrix. In so-called Einstein notation [31], the symbol of sum is omitted and we write just $c = \nu^i_i$. The contraction is sometimes noted

$$c = \sum_{(i,j)} \nu^j_i, \tag{10}$$

It means the sum is performed over the summands satisfying $i = j$.

They key property used in this paper is the invariance of the total contraction to affine transformations. If we observe the contraction of a tensor ν^j_i subject to an affine transformation, we obtain

$$\sum_{i=1}^{d}\nu'^i_i = \sum_{i=1}^{d}\sum_{k=1}^{d}\sum_{\ell=1}^{d} A^k_i (A^{-1})^i_\ell \nu^\ell_k = \sum_{i=1}^{d}\nu^i_i = c. \tag{11}$$

Thanks to the common index i, the matrices A and A^{-1} are multiplied as matrices, the result is an identity matrix and the contraction remains unchanged regardless the transformation. The contraction is the way to affine invariants. When we can compute a total contraction (i.e. contractions over all indices) of a tensor, it is an affine invariant.

3.3 Transformations of Tensor Fields

Now, let us look at an actual tensor field

$$\sigma(x,y,z) \in \mathbb{R}^{3\times 3} \tag{12}$$

assigning a tensor to each point in space. Sometimes, it is noted as

$$\sigma(\mathbf{x}), \tag{13}$$

where $\mathbf{x} = (x,y,z)^T = (x^1, x^2, x^3)^T$ is the vector of coordinates.

As mentioned above, for transforming tensor fields (as for vector field) we need to define two transformations. One for transforming the coordinate system, the second one transforms the tensor (vector) values. Usually these transformations are identical resulting in intuitive transformation, e.g. rotating the tensor (vector) field also rotates the directions of tensors (vectors). Formally let $\mathbf{A}$, $\mathbf{B}$ be these two affine transformations acting on a tensor field

$$\boldsymbol{\sigma}'(\mathbf{x}') = \mathbf{B}(\boldsymbol{\sigma}(\mathbf{A}^{-1}(\mathbf{x}))). \tag{14}$$

We can write the transformations using their 3×3 matrix representations, which we also denote by $\mathbf{A}$, $\mathbf{B}$. Here, the inner transformation of the coordinates takes the form

$$\mathbf{x}'^i = \sum_{(j,k)} (\mathbf{A}^{-1} \otimes \mathbf{x})_k^{ij} = (\mathbf{A}^{-1})_j^i \mathbf{x}^j, \tag{15}$$

which coincides with the standard matrix vector product in matrix notation $\mathbf{A}^{-1}\mathbf{x}$. The outer transformation of the tensor values is written as

$$\sigma'^{kl} = \sum_{\substack{(i,m) \\ (j,n)}} (\mathbf{B} \otimes \mathbf{B} \otimes \boldsymbol{\sigma})_{ij}^{klmn} = \{\mathbf{B}_i^k \mathbf{B}_j^\ell \sigma^{ij}\} \tag{16}$$

using contraction and Einstein notation.

The case $\mathbf{A} = \mathbf{B}$ is called the total transformation, cases where transformations differ are rare. The same way the tensor multiplication together with contraction in the inner transformation (15) equals the matrix multiplication, the outer transformation (16) can be rewritten to the matrix multiplication as

$$\boldsymbol{\sigma}' = \mathbf{B}\boldsymbol{\sigma}\mathbf{B}^T. \tag{17}$$

Note that affine transformation does not change the rank of a tensor.

3.4   Moment Tensors

Geometric moments of a real valued function $f(x,y)$ have been introduced to pattern recognition in [14]

$$m_{pq} = \int\limits_{-\infty}^{\infty} \int\limits_{-\infty}^{\infty} x^p y^q f(x,y) \, \mathrm{d}x \, \mathrm{d}y. \tag{18}$$

The sum $o = p+q$ is called *order* of the moment. For a 3D tensor field, we simply extend (18) by the third spatial coordinate and replace the scalar function f by our tensor valued function σ^{ij}

$$m_{pqr}^{(ij)} = \int\limits_{-\infty}^{\infty} \int\limits_{-\infty}^{\infty} \int\limits_{-\infty}^{\infty} x^p y^q z^r \sigma^{ij}(x,y,z) \, \mathrm{d}x \, \mathrm{d}y \, \mathrm{d}z. \tag{19}$$

The moments of order o can be arranged to the *moment tensor* $^o\mathbf{M}$. For general tensors we have

$$^o\mathbf{M}_{j_1\ldots j_m}^{k_1\ldots k_o i_1\ldots i_n} = \int_{\mathbb{R}^d} x^{k_1}\cdots x^{k_o}\sigma_{j_1\ldots j_m}^{i_1\ldots i_n}\left(x^1\cdots x^d\right)\mathrm{d}^d x, \tag{20}$$

where o is the order of the moment tensor, m is the covariant rank of the tensor field and n is its contravariant rank[2]. For example, the moment tensor of a Cauchy stress tensor is

$$^o\mathbf{M}^{k_1\ldots k_o ij} = \int_{-\infty}^{\infty}\int_{-\infty}^{\infty}\int_{-\infty}^{\infty} x^{k_1}\cdots x^{k_o}\sigma^{ij}(x^1, x^2, x^3)\,\mathrm{d}x^1\mathrm{d}x^2\mathrm{d}x^3. \tag{21}$$

Also note that the components of the moment tensor equal the geometric moments

$$^o\mathbf{M}_{j_1\ldots j_m}^{k_1\ldots k_o i_1\ldots i_n} = m_{p_1\ldots p_d(j_1\ldots j_m)}^{(i_1\ldots i_n)} \tag{22}$$

iff p_ℓ many of the indices $k_1,\ldots,k_o$ equals ℓ for all $\ell = 1,\ldots,d$.

3.5 Construction of the Invariants

The affine invariants can be constructed as total contractions of tensor products of moment tensors and permutation tensors [11]. The *permutation tensor* ε in 2D takes the form

$$\varepsilon_{ij} = \begin{pmatrix} 0 & 1 \\ -1 & 0 \end{pmatrix}. \tag{23}$$

In 3D, it is the $3\times 3\times 3$ cube with slices

$$\varepsilon_{ij1} = \begin{pmatrix} 0 & 0 & 0 \\ 0 & 0 & 1 \\ 0 & -1 & 0 \end{pmatrix}, \ \varepsilon_{ij2} - \begin{pmatrix} 0 & 0 & -1 \\ 0 & 0 & 0 \\ 1 & 0 & 0 \end{pmatrix}, \ \varepsilon_{ij3} - \begin{pmatrix} 0 & 1 & 0 \\ -1 & 0 & 0 \\ 0 & 0 & 0 \end{pmatrix}. \tag{24}$$

If the index values create a cyclic shift of 123, the value is 1, if it is a cyclic shift of 321, the value is -1. In the remaining 21 positions, the value is 0.

An example of such an invariant is

$$I = \sum_{\substack{(i_1,i_2)(j_1,j_2)(j_1,j_2) \\ (k_1,k_2)(\ell_1,\ell_2)(m_1,m_2) \\ (n_1,n_2)(o_1,o_2)(p_1,p_2)}} {}^2\mathbf{M}^{i_1 j_1 k_1 \ell_1} \otimes {}^1\mathbf{M}^{m_1 n_1 o_1} \otimes {}^0\mathbf{M}^{p_1 q_1} \otimes \varepsilon_{i_2 k_2 n_2} \otimes \varepsilon_{j_2 m_2 p_2} \otimes \varepsilon_{\ell_2 o_2 q_2} =$$

$$= {}^2\mathbf{M}^{ijk\ell}\, {}^1\mathbf{M}^{mno}\, {}^0\mathbf{M}^{pq}\varepsilon_{ikn}\varepsilon_{jmp}\varepsilon_{\ell oq}. \tag{25}$$

If we need to generate all the affine invariants of a tensor field, we need to generate all total contractions of the type of Eq. (25), i.e. all tensor products of all moment tensors and permutation tensors, where each index is used exactly twice, once in the moment tensor and once in the permutation tensor.

---

[2] Here x^{k_1} is not power, but upper index (superscript), i.e. if $k_1 = 3$, then $x^{k_1} = z$. We multiply o coordinates in the integral.

3.6   Tensor Field Affine Moment Invariants and Quadri-Layer Hypergraphs

Not all combinations are needed to unambiguously describe a template. We are looking for a subset that is *complete*, i.e., it has enough invairiants to discern two templates that differ something other than an affine transform, but has as little elements as possible to maximize efficiency. When we want to generate a complete set of affine invariants, we must generate all possible combinations of moment tensors and permutation tensors and also all possible total contractions on the given combinations.

We can help us with the idea of graphs, where each node corresponds to a moment tensor and each edge to a permutation tensor. When we generate all the graphs with the given parameters and compute the corresponding invariants, we obtain the complete set. In the case of tensor field affine moment invariants (TFAMIs), we need so-called *quadri-layer hypergraphs*. Let $G = (\mathcal{V}; E))$ be a graph consisting of a set of vertices (nodes) $\mathcal{V}$ and a set of edges E.

In standard graphs, each edge connects two nodes. In the hypergraph, each edge can connect multiple nodes. Similarly, in the standard graph, all edges are qualitatively equal but sometimes we need more types of edges. Such graphs are called multilayer graphs.

In the case of symmetric 3D tensor fields, we need quadri-layer hypergraphs, where each edge connects three nodes. We note it

$$G = (\mathcal{V}; E_1, E_2, E_3, E_4).$$

Further we denote

$$G_k = (\mathcal{V}; E_k)$$

the *k-th layer* of the graph G.

An arbitrary invariant can be represented by a quadri-layer graph as follows. Each moment tensor in the product (25) corresponds to a graph node. Each permutation tensor ε_{ijk} corresponds to an edge connecting three nodes. It connects the moment tensor with the index i, the moment tensor with the index j, and the moment tensor with the index k.

The edges from E_1 use only coordinate indices, the edges from E_2 connect two coordinate indices and one value index, the edges from E_3 use one coordinate index and two value indices, and the edges from E_4 connect only value indices. The plotting of the triple edges of four types is not easy; we decided to denote them as tripods with color arms. The black arm means the coordinate index, while the contravariant value index is noted by magenta color. The examples of the color combinations are in Fig. 1.

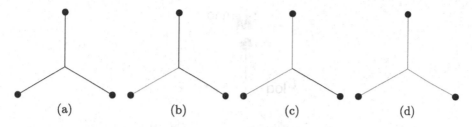

Fig. 1. Examples of triple hyperedges of individual types: (a) E_1 connecting only coordinate indices, (b) E_2 connecting two coordinate indices and one value index (c) E_3 connecting one coordinate index and two value indices, and (d) E_4 connecting only value indices.

We can use different types of graphs, but this type proved its efficiency in the invariant generation. In Fig. 2, we can see the graph representing the invariant (25). We can observe a node with two black edges, a node with one black edge and a node without black edges. They correspond to the moments of orders 2, 1, and 0 respectively. All the nodes has two magenta edges corresponding to the value indices, because we work with the tensors of the second rank.

An algorithm for a systematic generation of all such graphs can be found on the webpage [28]. In Table 1, there are the numbers of the generated invariants. Here, # edges is the number of edges permitted, # graphs denotes the overall possible number of configurations that produce affine invariants, # invariants represents the number to which we could reduce the set while maintaining its completeness, and # independent is the theoretically possible lower limit of independent affine invariants.

Table 1. The numbers of graph edges, graphs, all invariants, and independent invariants.

# edges	# graphs	# invariants	# independent
3	635	1	1
4	14941	12	12
5	404448	41	40
6	11862154	2123	152

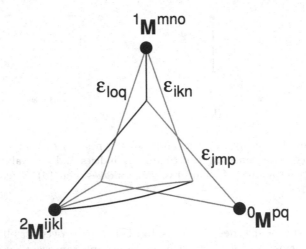

Fig. 2. The graph representing invariant from Eq. (25).

4   Numerical Experiment

We tested our method on real world data obtained from a diffusion MRI scan of a human head. The diffusion in the brain is represented by a three-dimensional second rank symmetric tensor field. Example slices of those data are shown in Fig. 3, where the 3×3 tensor in each pixel is visualized through color coding as described in the following section.

4.1   Obtaining Data and Visualization

In this experiment, we used real DTI scans of a human brain. The device used for an examination was a 3T Siemens TrioTim MR scanner using spin-echo echo-planar imaging (SE EPI) sequence. The acquisition parameters were the following: repetition time (TR) of 8300 ms, echo time (TE) of 84 ms, voxel size of $2 \times 2 \times 2$ mm, 68 axial slices, two averages, field of view (FOV) of 256 mm, number of diffusion directions 30, two b-values: 0, and $900 \, \text{s/mm}^2$. The output of the machine are 62 measurements each with different gradient settings. One measurement is a set of 68 slices in the axial plane with a of resolution 112×128 pixels resulting in 62 $112 \times 128 \times 68$ volumes, each representing the diffusion in a certain direction. These volumes can be stacked together (as described in [15]) to produce a 3D volume with of 3×3 symmetric matrices - one tensor in each voxel.

It is challenging to visualize such data [12]. We used a method, where we visualize the tensor field so that colors are used to indicate the direction of the diffusion. First, we assign the RGB colors to the world coordinates x, y, z i.e. red for the coronal, green for the sagittal and blue for the axial plane. Then for each voxel, the diffusion tensor is transformed into a diagonal matrix. Components on the diagonal - eigenvalues - represent a magnitudes of diffusion in a new coordinate system given by the corresponding eigenvectors. This efficiently transfers a tensor into a vector. Each of the new coordinates can be assigned a color based on the $x, y, z \iff R, G, B$ correspondence. Finally, the three colors can be merged into one as their linear combination with coefficients being the corresponding eigenvalues.

For clarity, only the voxels with strongly anisotropic diffusion (one prevailing direction of diffusion) were colored, while the rest was assigned gray level based on the fractional anisotropy (scalar value, the total diffusion).

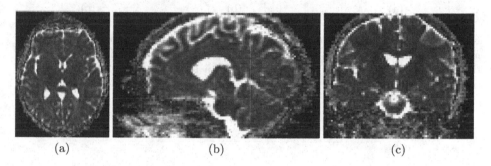

(a)   (b)   (c)

Fig. 3. Original MRI diffusion tensor, cut in (a) axial, (b) sagittal and (c) coronal plane. (Color figure online)

4.2 Invariance

We generated ten random affine transformations without translation. They were composed of two rotations (Euler angles with uniform distribution on the intervals $\langle 0, 2\pi \rangle$, $\langle 0, \pi \rangle$, and $\langle 0, 2\pi \rangle$) and a non-uniform scaling (Gauss distribution with mean one and standard deviation 0.2) between them. An example of such an affine transformation is shown in Fig. 4. The values of four representative invariants are depicted in Fig. 5. The horizontal lines show that they indeed do not change under the different affine transforms. The labels on the horizontal axis note specific affine transformations, 0 are values of the original tensor field. The average relative error over all invariants and all affine transformations is 1.043%. For comparison, we also converted the moments directly, without re-sampling the tensor field. The relative average error then decreased to $1.2604 \cdot 10^{-12}$ %. It is caused just by numerical imprecisions during computation.

4.3 Template Matching

We tested our invariants in a template matching experiment. We generated 10 random spherical templates with a diameter of 15 voxels (see Fig. 7 for an example of the template). At least 90% of the volume of all the templates is inside the patient's head (i.e. valid data) and there is no overlap between them. Then, we again generated two random affine transformations of the whole diffusion tensor field and searched the templates in them.

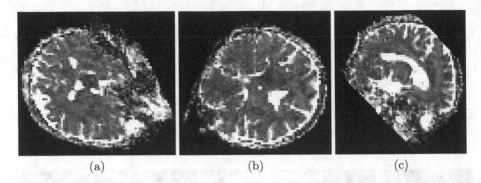

|(a)|(b)|(c)|

Fig. 4. The MRI diffusion tensor field from Fig. 3 after an example affine transformation, cut in (a) x-y plain, x-z plain, (c) y-z plain. The color coding is descibed in the main text. The artifacts are part of the data and have nothing to do with the transformation. (Color figure online)

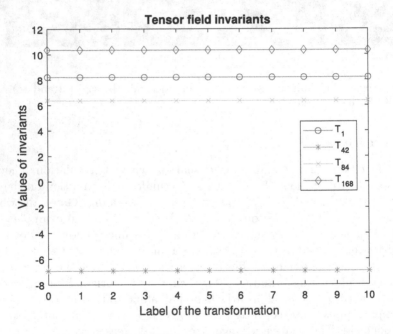

Fig. 5. Values of some TFAMI.

We searched the tensor field voxel by voxel, computed TFAMIs in each position, and tested for a match with TFAMIs of the templates. We used 205 invariants of symmetric tensor fields from the 2nd to the 6th order, their list can be downloaded from [29]. The sorted errors can be seen in Fig. 6. The errors are computed as Euclidean distances of the affinely transformed template positions and the best matches of TFAMIs.

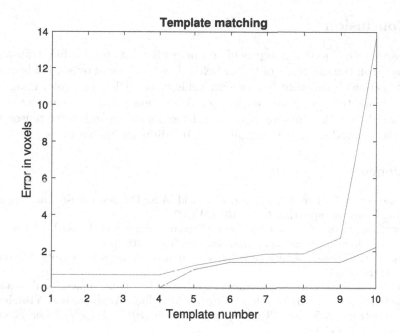

Fig. 6. Errors in the two template matching experiments. Each line corresponds to a random affine transform.

We consider errors greater than 5 voxels as failures. So, the templates in the first transformation were found successfully, while in the second attempt, we encountered one mismatch. This template size is a limit, where the mismatches are rare. The templates of a bigger size are found reliably, while the matching of smaller templates fails frequently because of a shortage of significant information contained in the template. We intentionally chose limit parameter values, where we can study the behavior of the invariants. When we use bigger templates, we obtain errorless result.

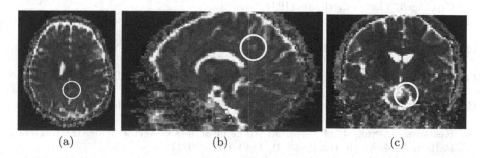

(a)   (b)   (c)

Fig. 7. MRI diffusion tensor with example of a template, (a) x-y plain, x-z plain, (c) y-z plain. The color coding is described in the main text. (Color figure online)

5   Conclusion

This paper introduced invariants of tensor fields w.r.t. total affine transformations based on the moments of tensor fields. The behavior of tensor fields in affine transformations is different from vector fields, scalar fields, and color images and the traditional techniques cannot be used. We developed a set of complete invariants that allow affine invariant pattern detection on second order tensor fields and demonstrated their performance on MRI diffusion tensor fields.

References

1. Alexander, A.L., Lee, J.E., Lazar, M., Field, A.S.: Diffusion tensor imaging of the brain. Neurotherapeutics **4**(1), 316–329 (2007)
2. Bowen, R., Wang, C.: Introduction to Vectors and Tensors. Dover books on mathematics, Dover Publications, Mineola, New York (2008)
3. Bujack, R., Flusser, J.: Flexible basis of rotation moment invariants. In: WSCG, pp. 11–20 (2017)
4. Bujack, R., Hagen, H.: Moment invariants for multi-dimensional data. In: Schultz, T., Özarslan, E., Hotz, I. (eds.) Modeling, Analysis, and Visualization of Anisotropy. MV, pp. 43–64. Springer, Cham (2017). https://doi.org/10.1007/978-3-319-61358-1_3
5. Bujack, R., Hlawitschka, M., Scheuermann, G., Hitzer, E.: Customized TRS invariants for 2D vector fields via moment normalization. Pattern Recogn. Lett. **46**(1), 46–59 (2014)
6. Bujack, R., Zhang, X., Suk, T., Rogers, D.: Systematic generation of moment invariant bases for 2D and 3D tensor fields. Pattern Recogn. **123**, 108313 (2022)
7. Gong, M., Hao, Y., Mo, H., Li, H.: Naturally combined shape-color moment invariants under affine transformations. Comput. Vis. Image Underst. **162**, 46–56 (2017)
8. Grinfeld, P.: Introduction to Tensor Analysis and the Calculus of Moving Surfaces. Springer, New York (2013)
9. de Groot, M., et al.: White matter degeneration with aging: longitudinal diffusion MR imaging analysis. Radiology **279**(2), 532–541 (2015)
10. Gurevich, G.B.: Osnovy teorii algebraicheskikh invariantov. Moskva, The Union of Soviet Socialist Republics, OGIZ (1937)
11. Gurevich, G.B.: Foundations of the theory of algebraic invariants. Nordhoff, Groningen, The Netherlands (1964)
12. Hergl, C., et al.: Visualization of tensor fields in mechanics. Comput. Graph. Forum **40**(6), 135–161 (2021)
13. Hickman, M.S.: Geometric moments and their invariants. J. Math. Imaging Vis. **44**(3), 223–235 (2012)
14. Hu, M.K.: Visual pattern recognition by moment invariants. IRE Trans. Inf. Theory **8**(2), 179–187 (1962)
15. Kingsley, P.B.: Introduction to diffusion tensor imaging mathematics: Part i–iii. Concepts Magn. Reson. Part A **28**(2), 101–179 (2006)
16. Kostková, J., Suk, T., Flusser, J.: Affine invariants of vector fields. IEEE Trans. Pattern Anal. Mach. Intell. **43**(4), 1140–1155 (2021)
17. Langbein, M., Hagen, H.: A generalization of moment invariants on 2D vector fields to tensor fields of arbitrary order and dimension. In: Bebis, G., et al. (eds.) ISVC 2009. LNCS, vol. 5876, pp. 1151–1160. Springer, Heidelberg (2009). https://doi.org/10.1007/978-3-642-10520-3_110

18. Liu, M., Yap, P.T.: Invariant representation of orientation fields for fingerprint indexing. Pattern Recogn. **45**(7), 2532–2542 (2012)
19. Liu, W., Ribeiro, E.: Detecting singular patterns in 2-D vector fields using weighted Laurent polynomial. Pattern Recogn. **45**(11), 3912–3925 (2012)
20. Mindru, F., Tuytelaars, T., Gool, L.V., Moons, T.: Moment invariants for recognition under changing viewpoint and illumination. Comput. Vis. Image Underst. **94**(1–3), 3–27 (2004)
21. Reiss, T.H.: Recognizing Planar Objects Using Invariant Image Features. LNCS, vol. 676. Springer, Berlin, Germany (1993). https://doi.org/10.1007/BFb0017553
22. Rothe, I., Süsse, H., Voss, K.: The method of normalization to determine invariants. IEEE Trans. Pattern Anal. Mach. Intell. **18**(4), 366–376 (1996)
23. Schlemmer, M., et al.: Moment invariants for the analysis of 2D flow fields. IEEE Trans. Vis. Comput. Graph. **13**(6), 1743–1750 (2007)
24. Suk, T., Flusser, J.: Graph method for generating affine moment invariants. In: Proceedings of the 17th International Conference on Pattern Recognition ICPR'04, pp. 192–195. IEEE Computer Society (2004)
25. Suk, T., Flusser, J.: Affine moment invariants generated by automated solution of the equations. In: Proceedings of the 19th International Conference on Pattern Recognition ICPR'08. IEEE Computer Society (2008)
26. Suk, T., Flusser, J.: Affine moment invariants of color images. In: Jiang, X., Petkov, N. (eds.) CAIP 2009. LNCS, vol. 5702, pp. 334–341. Springer, Heidelberg (2009). https://doi.org/10.1007/978-3-642-03767-2_41
27. Suk, T., Flusser, J.: Affine moment invariants generated by graph method. Pattern Recogn. **44**(9), 2047–2056 (2011)
28. Suk, T., Flusser, J.: Afintensors - generation of invariants of tensor fields (2019). http://zoi.utia.cas.cz/afintensors
29. Suk, T., Flusser, J.: Affine moment invariants of tensor fields (2020). http://zoi.utia.cas.cz/affine-tensor-fields
30. Wheeler-Kingshott, C.A.M.G., et al.: Investigating cervical spinal cord structure using axial diffusion tensor imaging. Neuroimage **16**(1), 93–102 (2002)
31. Wikipedia: einstein notation (last edit 2022). http://en.wikipedia.org/wiki/Einstein_notation
32. Yang, B., Kostková, J., Flusser, J., Suk, T., Bujack, R.: Rotation invariants of vector fields from orthogonal moments. Pattern Recogn. **74**, 110–121 (2018)
33. Yang, B., Kostková, J., Suk, T., Flusser, J., Bujack, R.: Recognition of patterns in vector fields by Gaussian-Hermite invariants. In: Luo, J., Zeng, W., Zhang, Y.J. (eds.) International Conference on Image Processing ICIP'17, pp. 2350–2363. IEEE (2017)

Fashion CUT: Unsupervised Domain Adaptation for Visual Pattern Classification in Clothes Using Synthetic Data and Pseudo-labels

Enric Moreu[1,2]([✉])[ID], Alex Martinelli[1], Martina Naughton[1], Philip Kelly[1], and Noel E. O'Connor[2][ID]

[1] Zalando SE, Valeska-Gert-Straße 5, 10243 Berlin, Germany
enric.moreu@zalando.ie
[2] Insight Centre for Data Analytics, Dublin City University, Dublin, Ireland

Abstract. Accurate product information is critical for e-commerce stores to allow customers to browse, filter, and search for products. Product data quality is affected by missing or incorrect information resulting in poor customer experience. While machine learning can be used to correct inaccurate or missing information, achieving high performance on fashion image classification tasks requires large amounts of annotated data, but it is expensive to generate due to labeling costs. One solution can be to generate synthetic data which requires no manual labeling. However, training a model with a dataset of solely synthetic images can lead to poor generalization when performing inference on real-world data because of the domain shift. We introduce a new unsupervised domain adaptation technique that converts images from the synthetic domain into the real-world domain. Our approach combines a generative neural network and a classifier that are jointly trained to produce realistic images while preserving the synthetic label information. We found that using real-world pseudo-labels during training helps the classifier to generalize in the real-world domain, reducing the synthetic bias. We successfully train a visual pattern classification model in the fashion domain without real-world annotations. Experiments show that our method outperforms other unsupervised domain adaptation algorithms.

Keywords: Domain adaptation · Synthetic data · Pattern classification

1 Introduction

In 2021, 75% of EU internet users bought goods or services online [1]. One of the main drivers of increased e-commerce engagement has been convenience, allowing customers to browse and purchase a wide variety of categories and brands in a single site. If important product metadata is either missing or incorrect, it becomes difficult for customers to find products as the number of available products on e-commerce sites grows. Online stores typically offer a set of filters

R. Gade et al. (Eds.): SCIA 2023, LNCS 13885, pp. 314–324, 2023.
https://doi.org/10.1007/978-3-031-31435-3_21

(e.g. pattern, color, size, or sleeve length) that make use of such metadata and help customers to find specific products. If such critical information is missing or incorrect then the product cannot be effectively merchandised. Machine learning has been used for fashion e-commerce in recent works to analyze product images, e.g. clothes retrieval [2], detecting the outline [3], or to find clothes that match an outfit [4]. In this paper, our prime interest is a visual classification task which consists of classifying patterns in catalog images of clothing. Patterns describe the decorative design of clothes, and they are important because they are widely used by customers to find products online. Figure 1 shows fashion visual pattern examples in the syntetic and real-world domains.

Fig. 1. Synthetic samples from our Zalando SDG dataset (first row) and real samples from the DeepFashion dataset [5] (second row) representing the striped, floral, and plain categories.

Fashion pattern classification is challenging. Fashion images often include models in different poses with complex backgrounds. Achieving high performance requires large annotated datasets [5] [6] [7]. However, public datasets are only available for non-commercial use or do not cover the specific attributes or diversity we require, while generating private datasets with fine-grained and balanced annotations is expensive. In addition, publicly available fashion datasets typically have underrepresented classes with only a few samples. For example, in the Deep Fashion dataset [5] there are 6633 images with the "solid" pattern while only 242 images contain the "lattice" pattern. Categories that are underrepresented during training achieve a lower performance, thus reducing the overall performance.

We address these problems by generating artificial samples using Synthetic Data Generation (SDG) techniques. Synthetic data has shown promising results in domains where few images are available for training [8] [9]. The main advantage of synthetic data is that it can generate unlimited artificial images because labels are automatically produced by the 3D engine when rendering the images.

However, synthetic images are not a precise reflection of the real-world domain in which the model will operate. Computer vision models are easily biased by the underlying distribution in which they are trained [10]. Even if the synthetic images use realistic lighting and textures that look realistic to humans, the model will tend to over-optimize against the traits of the synthetic domain and won't generalize well to real data.

We consider this problem in the context of unsupervised domain adaptation [11] by using the knowledge from another related domain where annotations are available. We assume that we have abundant annotated data on the source domain (synthetic images) and a target domain (real images) where no labels are available.

Unsupervised domain adaptation has shown excellent results when translating images to other domains [11]; nevertheless, translated images can't be readily used to train classification models because image features, such as patterns, are distorted during the translation step since the translation model doesn't have information about the features. Specifically, when complex patterns are shifted to a different domain, they can be distorted to a level that they no longer adhere to the original pattern label for the synthetic image. For example, when an image with the "camouflage" pattern is translated from the synthetic to the real domain, the pattern could be accidentally distorted to "floral".

In this paper, we introduce a new unsupervised domain adaptation approach that doesn't require groundtruth labels. First, we produce a synthetic dataset for fashion pattern classification using SDG that equally represents all the classes. Second, we jointly train a generative model and a classifier that will make synthetic images look realistic while preserving the class patterns. In the final stage of the training, real-world pseudo-labeled images are used to improve the model generalization toward real images. The contributions of this paper are as follows:

- We propose a novel architecture that performs the image translation task while jointly training a classification model.
- We outperform other state-of-the-art unsupervised domain adaptation algorithms in the visual fashion pattern classification task.

The remainder of the paper is organized as follows: Sect. 2 reviews relevant work; Sect. 3 explains our method; Sect. 4 presents our synthetic dataset and experiments, and Sect. 5 concludes the paper.

2   Related Work

Synthetic data has been used extensively in the computer vision field. Techniques to generate synthetic datasets range from simple methods generating primitive shapes [12] to photorealistic rendering using game engines [13]. Although high quality synthetic images can appear realistic to humans, they don't necessarily help the computer vision models to generalize to real-world images. Convolutional neural networks easily overfit on synthetic traits that are not present in the real-world. This is addressed by using domain adaptation techniques that reduces the disparity between the synthetic and real domains. Some works app-roach domain adaptation by simply improving the realism aspect [14], or by pushing the randomization and distribution coverage at the source [15]. These approaches imply additional modeling effort and longer generation times per image, for example by relying on physically based renderers for higher photore-alistic results, making the synthetic data better match the real data distribution.

In the context of unsupervised domain adaptation, non-adversarial approaches consist of matching feature distributions in the source and the target domain by transforming the feature space to map the target distribution [16]. Gong et al. [17] found that gradually shifting the domains during training improved the method's stability. Recent methods are based on generative adversarial networks [18] because of their unsupervised and unpaired nature. Generative domain adaptation approaches rely on a domain discriminator that distinguishes the source and target domains [19] and updates the generator to produce better images. Our approach improves existing adversarial approaches by optimizing a classifier alongside the generator, producing realistic data that retrain the source category.

3 Fashion CUT

Our approach has two components: 1) An image translation network that gen-erates realistic images. 2) A classifier that enforces the generated images to keep the class patterns. The overall architecture is shown in Fig. 2.

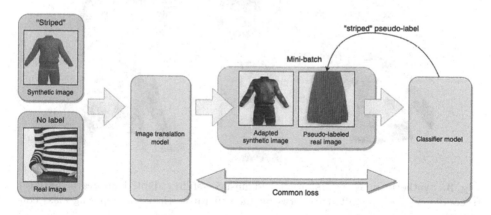

Fig. 2. The proposed architecture includes a translation model (CUT) and a classifier model (ResNet50), which are optimized together via a common loss that ensures real-istic images with reliable annotations. Pseudo-labeled real images are included in each mini-batch to improve the classifier generalization.

Acquiring paired images from both domains can be difficult to achieve in the fashion domain, resulting in high costs. As such, we use Contrastive Unpaired

Translation (CUT) [20] for the image translation module. Synthetic images don't have to match the exact position or texture of real images in the dataset because we use an unpaired translation method. CUT learns a mapping that translates unpaired images from the source domain to the target domain. It operates on patches that are mapped to a similar point in learned feature space using an infoNCE [21] contrastive loss. In addition, CUT uses less GPU memory than other two-sided image translation models (e.g. CycleGAN) because it only requires one generator and one discriminator. By reducing memory usage, the joint training of an additional classifier becomes tractable on low cost GPU setups with less than 16GB of memory.

While CUT produces realistic images, the class patterns can be lost or mixed with other classes since CUT doesn't enforce that these category features are consistent across the image translation. The generator's only objective is to produce realistic images that resemble the real-world domain, but it ignores the nature of each pattern. Any pattern distorted during the translation will impact the performance of a classifier trained on this synthetic data. Figure 3 showcases unsuccessful examples of mixed patterns by the generator, and Fig. 4 shows successful translations using Fashion CUT.

Fig. 3. Synthetic images (first row) and unsuccessfully adapted images using CUT (second row) due to shifted patterns by the generator when not imposing class constraints.

In order to enforce stability in the generated patterns, we add a ResNet50 model that predicts the category of the images generated by CUT. The classifier is optimized alongside the CUT generator to fulfill both classification and translation tasks. Figure 5 shows how the classifier preserves the pattern features in comparison to vanilla CUT. Training both models simultaneously is faster and

Fig. 4. Synthetic images (first row) and adapted domain images using Fashion CUT(second row).

provides better results than training them separately. The generator loss function is given by:

$$\lambda g * \mathcal{L}_{GAN}(G, D, X, Y) + \lambda c * \mathcal{L}_{classifier}(C) +$$
$$\lambda ncex * \mathcal{L}_{NCEx}(G, D, X) + \lambda ncey * \mathcal{L}_{NCEy}(G, D, Y) \qquad (1)$$

where $\mathcal{L}_{GAN}(G, D, X, Y)$ is the generator loss, $\mathcal{L}_{classifier}(C)$ is the cross-entropy loss on the classifier inferred from the images generated by the generator. $\mathcal{L}_{NCEx}(G, D, X)$ and $\mathcal{L}_{NCEy}(G, D, Y)$ are the contrastive losses that encourage spatial consistency for the synthetic and real images, respectively. G is the generator model, D the discriminator model, X the real image, Y the synthetic image, and C the classification model. λg, λc, $\lambda ncex$, and $\lambda ncey$ are hyperparameters that control the weight of the generator, the classifier, and both contrastive losses, respectively.

In our experiments we empirically choose to replace half of the synthetic minibatch with images from the target domain. As real-world annotations are not available for generated images, we use pseudo-labels predicted by the classifier. The model suffers from the cold start problem when introducing pseudo-labels in the early epochs because the classifier struggles to converge. We found that the classifier requires at least 1 epoch of synthetic samples in order to generate reliable pseudo-labels for real-world images. We obtained the best results when enabling pseudo-labels at the end of epoch 2.

Synthetic   CUT   Fashion CUT

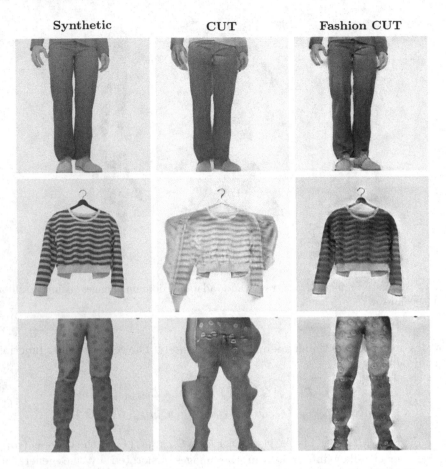

Fig. 5. Comparison of CUT and Fashion CUT image translation. Note that the annotations (gradient, striped, dotted) are preserved when using Fashion CUT.

4   Experiments

This section describes the synthetic dataset we generated and the two experiment setups used to evaluate Fashion CUT.

4.1   Zalando SDG Dataset

The Zalando SDG dataset is composed of 31,840 images of 7 classes: plain, floral, striped, dotted, camouflage, gradient, and herringbone. The dataset has been generated using Blender, an open-source 3D computer-graphic software [22]. We relied on a basic set of professionally modeled 3D objects from CGTrader representing a variety of fashion silhouettes (e.g. shirt, dress, trousers) and implemented a procedural material for each of the 7 target classes. Each procedural material is implemented as a Blender shader node, where multiple properties can

Fig. 6. For each render we start with a provided 3D object, add environment and spot lights, apply a procedural material and then randomize its properties (e.g. colors, scale).

be exposed and controlled via Blender Python API. Examples of such properties include pattern scale, color or color pairing, orientation and image-texture. This setup allows an arbitrary amount of different images for each 3D object and class pair to be generated programmatically. We randomized background, lighting, and camera position, as seen in Fig. 6. We didn't use physically based renderers as those are more resource intensive, instead we traded off rendering accuracy for speed and adopted the real-time Blender Eevee render engine [23].

The procedural materials can be applied to any new 3D objects. As such they provide a powerful generalized approach to data creation, and the generated images do not require any manual human validation as long as the procedural randomization guarantees that each possible output belongs to the expected target domain class.

4.2 Evaluation on Zalando SDG Dataset

In our experiments we train end-user pattern classification models using datasets from both 31,840 synthetic fashion imagery (the source domain, which includes groundtruth labels), and 334,165 real-world fashion imagery (the target domain, which has no groundtruth labels and is used solely to train our domain adaptation transformation). We evaluate the performance of the algorithms using a validation set and a test set composed of 41,667 annotated real images each. The metric used is accuracy and all algorithms use a ResNet50 [24] as the classifier. Fashion CUT is optimized using Adam with learning rate 10^{-5} and $\lambda g = 0.1$, and $\lambda classifier = 0.1$ for $N = 5$ epochs.

In Table 1, we compare the performance of domain adaptation algorithms trained only on our 31,840 synthetically generated dataset and evaluated on the 41,667 real fashion images.

First, we measured the performance of training without domain adaptation. In other words, the classifier was trained only on synthetic images. The performance was poor because the model didn't have information about real world images.

Second, we evaluate Zalando SDG on other domain adaptation algorithms in the fashion domain. All experiments were performed in the environment provided

Table 1. Comparison of unsupervised domain adaptation algorithms on our Zalando SDG dataset. The metric used is accuracy.

Method	Accuracy
No adaptation	0.441
BSP [25]	0.499
MDD [26]	0.540
AFN [16]	0.578
Fashion CUT (ours)	0.613
Fashion CUT with pseudo-labels (ours)	**0.628**

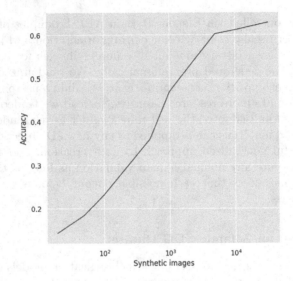

Fig. 7. Evaluation of Fashion Cut with varying amounts of the Zalando SDG dataset and 10.000 unlabeled real images. The performance is measured in accuracy.

by Jiang et al. [27]. Our approach outperforms the other algorithms for the pattern classification task. Finally, we found that using pseudo-labels improves the results with minor changes in the training.

4.3   Synthetic Dataset Size

We explore the required number of synthetic images to successfully train our unsupervised domain adaptation algorithm. For this experiment we train our model using 10,000 unlabeled real images and changing the number of synthetic images. Figure 7 shows that Fashion CUT performance benefits from large synthetic datasets. We found that at least 5,000 synthetic images are required to outperform other algorithms in visual pattern classification.

5   Conclusions

Combining synthetic data generation with unsupervised domain adaptation can successfully classify patterns in clothes without real-world annotations. Furthermore, we found that attaching a classifier to an image translation model can enforce label stability, thus improving performance. Our experiments confirm that Fashion CUT outperforms other domain adaptation algorithms in the fashion domain. In addition, pseudo-labels proved to be beneficial for domain adaptation in the advanced stages of the training. As future work, we will explore the impact of fashion synthetic data in a semi-supervised setup. We hope this study will help enforce 3D rendering as a replacement for human annotations.

References

1. Lone, S., Harboul, N., Weltevreden, J.: 2021 European e-commerce report
2. Liang, X., Lin, L., Yang, W., Luo, P., Huang, J., Yan, S.: Clothes co-parsing via joint image segmentation and labeling with application to clothing retrieval. IEEE Trans. Multimedia **18**(6), 1175–1186 (2016)
3. Liu, Z., Yan, S., Luo, P., Wang, X., Tang, X.: Fashion landmark detection in the wild. In: Leibe, B., Matas, J., Sebe, N., Welling, M. (eds.) ECCV 2016. LNCS, vol. 9906, pp. 229–245. Springer, Cham (2016). https://doi.org/10.1007/978-3-319-46475-6_15
4. Jagadeesh, V., Piramuthu, R., Bhardwaj, A., Di, W., Sundaresan, N.: Large scale visual recommendations from street fashion images. In: Proceedings of the 20th ACM SIGKDD International Conference on Knowledge Discovery and Data Mining, pp. 1925–1934 (2014)
5. Liu, Z., Luo, P., Qiu, S., Wang, X., Tang, X.: DeepFashion: powering robust clothes recognition and retrieval with rich annotations. In: Proceedings of IEEE Conference on Computer Vision and Pattern Recognition (CVPR) (June 2016)
6. Rostamzadeh, N., et al.: Fashion-Gen: the generative fashion dataset and challenge. arXiv preprint arXiv:1806.08317 (2018)
7. Wu, H., et al.: The fashion IQ dataset: retrieving images by combining side information and relative natural language feedback. CVPR (2021)
8. Sankaranarayanan, S., Balaji, Y., Jain, A., Lim, S.N., Chellappa, R.: Learning from synthetic data: addressing domain shift for semantic segmentation. In: Proceedings of the IEEE Conference on Computer Vision and Pattern Recognition, pp. 3752–3761 (2018)
9. Moreu, E., Arazo, E., McGuinness, K., O'Connor, N.E.: Joint one-sided synthetic unpaired image translation and segmentation for colorectal cancer prevention. Expert Syst., e13137 (2022)
10. Nam, H., Lee, H., Park, J., Yoon, W., Yoo, D.: Reducing domain gap by reducing style bias. In: Proceedings of the IEEE/CVF Conference on Computer Vision and Pattern Recognition, pp. 8690–8699 (2021)
11. Wang, M., Deng, W.: Deep visual domain adaptation: a survey. Neurocomputing **312**, 135–153 (2018)
12. Rahnemoonfar, M., Sheppard, C.: Deep count: fruit counting based on deep simulated learning. Sensors **17**(4), 905 (2017)

13. Wang, Q., Gao, J., Lin, W., Yuan, Y.: Learning from synthetic data for crowd counting in the wild. In: Proceedings of IEEE Conference on Computer Vision and Pattern Recognition (CVPR), pp. 8198–8207 (2019)

14. Ros, G., Sellart, L., Materzynska, J., Vazquez, D., Lopez, A.M.: The synthia dataset: a large collection of synthetic images for semantic segmentation of urban scenes. In: Proceedings of the IEEE Conference on Computer Vision and Pattern Recognition, pp. 3234–3243 (2016)

15. Moreu, E., McGuinness, K., Ortego, D., O'Connor, N.E.: Domain randomization for object counting. arXiv preprint arXiv:2202.08670 (2022)

16. Xu, R., Li, G., Yang, J., Lin, L.: Larger norm more transferable: an adaptive feature norm approach for unsupervised domain adaptation. In: The IEEE International Conference on Computer Vision (ICCV) (October 2019)

17. Gong, B., Shi, Y., Sha, F., Grauman, K.: Geodesic flow kernel for unsupervised domain adaptation. In: 2012 IEEE Conference on Computer Vision and Pattern Recognition, pp. 2066–2073. IEEE (2012)

18. Goodfellow, I., et al.: Generative adversarial networks. Commun. ACM **63**(11), 139–144 (2020)

19. Ganin, Y., et al.: Domain-adversarial training of neural networks. J. Mach. Learn. Res. **17**(1), 1–35 (2016)

20. Park, T., Efros, A.A., Zhang, R., Zhu, J.Y.: Contrastive learning for unpaired image-to-image translation. In: European Conference on Computer Vision (2020)

21. Gutmann, M., Hyvärinen, A.: Noise-contrastive estimation: a new estimation principle for unnormalized statistical models. In: Proceedings of the Thirteenth International Conference on Artificial Intelligence and Statistics, pp. 297–304. JMLR Workshop and Conference Proceedings (2010)

22. Community, B.O.: Blender - a 3D modelling and rendering package. Blender Foundation, Stichting Blender Foundation, Amsterdam (2018). http://www.blender.org

23. Guevarra, E.T.M.: Modeling and animation using blender: blender 2.80: the rise of Eevee. Apress (2019)

24. He, K., Zhang, X., Ren, S., Sun, J.: Deep residual learning for image recognition. In: Proceedings of the IEEE Conference on Computer Vision and Pattern Recognition, pp. 770–778 (2016)

25. Chen, X., Wang, S., Long, M., Wang, J.: Transferability vs. discriminability: batch spectral penalization for adversarial domain adaptation. In: International Conference on Machine Learning, pp. 1081–1090. PMLR (2019)

26. Zhang, Y., Liu, T., Long, M., Jordan, M.: Bridging theory and algorithm for domain adaptation. In: International Conference on Machine Learning, pp. 7404–7413. PMLR (2019)

27. Jiang, J., Baixu Chen, B.F.M.L.: Transfer-learning-library. https://github.com/thuml/Transfer-Learning-Library (2020)

Long Range Object-Level Monocular Depth Estimation for UAVs

David Silva[1]([✉])(iD), Nicolas Jourdan[2](iD), and Nils Gählert[1](iD)

[1] Wingcopter GmbH, Weiterstadt, Germany
{silva,gaehlert}@wingcopter.com
[2] TU Darmstadt, Darmstadt, Germany
n.jourdan@ptw.tu-darmstadt.de

Abstract. Computer vision-based object detection is a key modality for advanced Detect-And-Avoid systems that allow for autonomous flight missions of UAVs. While standard object detection frameworks do not predict the actual depth of an object, this information is crucial to avoid collisions. In this paper, we propose several novel extensions to state-of-the-art methods for monocular object detection from images at long range. Firstly, we propose Sigmoid and ReLU-like encodings when modeling depth estimation as a regression task. Secondly, we frame the depth estimation as a classification problem and introduce a Soft-Argmax function in the calculation of the training loss. The extensions are exemplarily applied to the YOLOX object detection framework. We evaluate the performance using the Amazon Airborne Object Tracking dataset. In addition, we introduce the Fitness score as a new metric that jointly assesses both object detection and depth estimation performance. Our results show that the proposed methods outperform state-of-the-art approaches w.r.t. existing, as well as the proposed metrics.

Keywords: Monocular Depth Estimation · Unmanned Aerial Vehicles · Detect-And-Avoid · Object-level · Amazon Airborne Object Tracking Dataset · Long Range Detection

1 Introduction

Within recent years, significant technological progress in Unmanned Aerial Vehicles (UAVs) was achieved. To enable autonomous flight missions and mitigate the risk of in-flight collisions, advanced Detect-And-Avoid (DAA) systems need to be deployed to the aircraft. By design, these systems shall maintain a *well-clear volume* around other airborne traffic [1]. As a result, DAA systems are required to reliably detect potential intruders and other dangerous objects at a long range to allow for sufficient time to plan and execute avoidance maneuvers. Specifically in small and lightweight UAVs, the usage of Lidar and Radar systems is challenging due to their power consumption, weight, and the required long range detection capabilities. However, computer vision approaches based on monocular images have proved their effectiveness in related use cases such as autonomous driving [5, 7, 22]. In addition, cameras can be equipped with lenses

© The Author(s), under exclusive license to Springer Nature Switzerland AG 2023
R. Gade et al. (Eds.): SCIA 2023, LNCS 13885, pp. 325–340, 2023.
https://doi.org/10.1007/978-3-031-31435-3_22

that employ different focal lengths depending on the application. Camera systems are therefore a powerful base modality for the perception stack of small and lightweight UAVs.

Depending on the actual use case and application, engineers might choose from several computer vision-related tasks such as image classification, object detection, or semantic segmentation. Those tasks are nowadays usually solved by Convolutional Neural Networks (CNNs) specifically designed for the selected use case. For vision-based DAA systems, single-stage object detection frameworks like SSD [28] or YOLO [16,31,32] are often employed to detect the objects of interest. By default, these frameworks detect objects in the two-dimensional image space by means of axis-aligned, rectangular bounding boxes. To reliably detect and avoid potentially hazardous objects, additional information about their three-dimensional position and trajectory is crucial. This capability, however, is missing in most vanilla object detection frameworks and specific extensions are needed to provide it.

In this paper, we specifically address the problem of object-level depth estimation based on monocular images for long range detections in the use case of UAVs. Several studies focusing on monocular 3D object detection have been conducted in autonomous driving [5,14,15,22,34]. For UAV-related use cases, however, object-level monocular depth estimation at long range is not yet widely researched. The two fields of application, UAVs and autonomous driving, differ in two major aspects: 1. The range of the objects. UAVs are required to keep a well clear volume of at least 2000 ft or approximately 600 m to prevent potential mid-air collisions [1]. In autonomous driving, on the other hand, objects are mostly limited to less than 200 m [5,14,17,22,34,35]. 2. Knowledge of the full 9 degrees of freedom 3D bounding box is not required to maintain the well clear volume. The distance is sufficient. In addition to simplifying the task, this aspect greatly eases the annotation process. As objects do not require a fully annotated 3D bounding box, one can can save both time and money.

Thus, we summarize our contributions as follows: 1. We propose two encodings, Sigmoid and ReLU-like, to improve long range depth estimation modeled as a regression task. 2. We frame the task of depth estimation as a classification problem and introduce Soft-Argmax based loss functions to improve the performance of monocular depth estimation. 3. We introduce a novel *Fitness Score* metric to assess the quality of depth estimation on object-level combined with the object detection metrics. 4. We demonstrate the extension of the state-of-the-art YOLOX object detection framework and benchmark the proposed approaches against existing methods applied to long range detections.

2   Related Work

The problem of depth estimation from monocular RGB images has been the subject of intense academic research in the past decade. Due to the ambiguity between an object's size and the object's distance to the camera, it is mathematically an ill-posed problem [14,24]. Thus, machine learning approaches, specifically ones that rely on CNNs, gained traction in this field. Two research streams can be identified in monocular depth estimation: 1. *Dense* or *Pixel-level* depth estimation, which estimates a dense map of distances for every pixel of a given image, and 2. *Object-level* depth estimation, which estimates distances only for detected objects of interest. While dense depth estimation is more prominent in computer vision literature, 3D object detection is gaining popularity in relevant application domains such as environment perception for autonomous driving [5,14,15,22,34]. Nevertheless, there's limited related work in the domain of 2D object-level depth estimation at long ranges [18].

In the case of depth prediction and corresponding loss functions, we distinguish between continuous regression approaches in contrast to approaches that rely on discretization of the depth estimation.

Continuous Regression. The reverse Huber loss (berHu) is employed in [23] to model the value distribution of depth predictions as a continuous regression problem for dense depth estimation. [18] uses the L2 loss for training a continuous, object-level depth regressor. The log-distance is used within the loss calculation to scale larger distance values.

Depth Discretization. [6] formulates depth estimation as a classification problem by discretizing the depth values into intervals. The cross-entropy (CE) loss is used to train a classifier that assigns a depth bin to every pixel for dense depth estimation. [25,26] use a soft-weighted sum inference strategy to compute final depth predictions based on the softmax predictions scores of the depth bins. [13] proposes an ordinal regression loss function to learn a meaningful inter-depth-class ordinal relationship. The depth intervals for discretization are growing in width for increasing distance to the camera as the uncertainty about the true depth increases as well. [8] extends on the idea of using ordinal regression for depth estimation by using a softmax-like function to encode the target vector for depth classification as a probability distribution.

3 Methodology

In this section, we give information on YOLOX as the selected base framework for 2D object detection. We outline the mathematical foundation of the different depth estimation strategies and embed our proposed methods into these paradigms. In addition, we introduce the Fitness score metric in detail.

3.1 YOLOX – Base Framework for 2D Object Detection

To tackle the problem of depth estimation at the object level, we start with a pure 2D object detection framework that outputs a confidence score, class label, and 2D bounding box for each object using axis-aligned rectangles. Given our use case, in which the trade-off between inference speed and detection performance is of high importance, we choose YOLOX Tiny [16] as the base object detection framework. YOLOX was released in 2021 and is one of the latest advances within the YOLO family [31,32].

To allow for object-level depth estimation, we create a separate new head dedicated to depth estimation. The architecture of the depth head is based on the existing classification head with the necessary adjustments to the number of output channels in the last layer. This separation between the depth head and the other heads allows for a modular combination of various outputs.

While we have selected YOLOX as the foundation for this work, the ideas presented in the following sections can be carried over to other modern 2D object detectors.

3.2 Depth Regression

The most natural way to estimate depth d is to frame it as a continuous regression problem. In this case, the model is trained to predict a single and continuous value by minimizing the distance between the model predictions $\hat{y}$ and the ground truth target y. In its simplest form, depth can be regressed directly, *i.e.* $y = d$ and $\hat{y} = \hat{d}$. This simple model, however, allows for negative distances, which are not meaningful in the context of monocular depth estimation. Thus, we can use a differentiable transfer function, $g\,(x)$, which supports us in encoding and decoding the network output given a set of constraints. To avoid negative predictions, we propose the encoding

$$g\,(x) = \frac{x - b}{a}. \tag{1}$$

Its corresponding decoding can be calculated as:

$$g\,(x)^{-1} = \max\,(d_{\min}, a \cdot x + b), \tag{2}$$

with a and b being hyperparameters that allow for better control over the domain and range of the model outputs. As g^{-1} follows the ReLU structure, we refer to this approach as the ReLU-like encoding. We argue that designing a differentiable transfer function with this constraint not only eases training but also enhances robustness against out-of-distribution objects [4] or adversarial attacks.

Besides direct regression, there are encodings based on non-linear transfer functions *e.g.*, the inverse $g\,(x) = \frac{1}{x}$ [15] and the logarithm $g\,(x) = \log x$ [3,9, 10,18,24]. All previously mentioned encodings lack an upper bound. Thus, they allow for any positive number to be predicted as the depth of the object. As a result, the calculated loss is also unbound and may cause instabilities during training. In some use cases, however, it is possible to assign an upper bound or a maximum distance to the objects of interest. For those settings, we propose a bounded transfer function that maps the domain $(d_{\min}, d_{\max})$ to the range $(-\infty, +\infty)$:

$$g\,(x) = \mathrm{logit}\left(\frac{x - d_{\min}}{d_{\max} - d_{\min}}\right). \tag{3}$$

The corresponding decoding operation, based on the sigmoid function σ, is then calculated as:

$$g(x)^{-1} = (d_{\max} - d_{\min})\,\sigma(x) - d_{\min}, \tag{4}$$

where $d_{\max}$ and $d_{\min}$ are the maximum and the minimum depth, respectively. As g^{-1} uses the sigmoid function, we refer to this approach as the Sigmoid encoding.

3.3 Depth Bin Classification

Depending on the application and the use case, a coarse depth estimation might be sufficient. In such cases, depth estimation can be framed as a multiclass classification task with K discretized depth intervals $\{d_0, d_1, ..., d_{K-1}\}$ [6]. Each depth interval links to an individual class in the classification paradigm. Relaxing the need for fine-grained and continuous depth estimation also eases the process of ground truth generation and data annotation. This, in return, can be beneficial from a business perspective.

During training, in a classification setting, the softmax function is typically used in CNNs to compute the pseudo-probability distribution and is paired with CE loss. At test time, the selected depth bin is obtained by using the argmax over the pseudo-probabilities. Reformulating depth estimation as a simple classification task is straightforward. In our experiments, we will use this approach as the baseline for classification-based approaches.

Employing CE, however, models the classes – and thus the depth bins – as being independent of each other. In particular, the default CE loss doesn't penalize predictions more if they are further away from the target bin compared to predictions that are closer to the target.

Depth bins, however, are ordered. We naturally would consider predictions that are far away from the actual depth as *more wrong* compared to closer ones. Thus, we propose to design a loss that considers the distance of the predicted depth bin to the target depth bin.

Designing a loss based on the distance between the prediction and ground truth implies the knowledge of the argmax of the predicted depth classes. Once the argmax and ground truth is known, an arbitrary distance loss function *e.g.*, Smooth L1 or MSE, can easily be computed. The implementation of this approach, however, renders a challenge as the default argmax function is not differentiable. Thus, we replace it with the Soft-Argmax [12, 20]

$$\text{Soft-Argmax}(\hat{y}, \beta) = \sum_{i=0}^{K-1} i \cdot \text{softmax}(\beta\hat{y})_i \tag{5}$$

where $\beta > 0$ is a parameter that scales the model predictions $\hat{y}$. The larger the β, the more it approximates a one-hot encoded argmax. In our experiments, we found $\beta = 3$ to be a good choice. Soft-Argmax provides an approximated bin index that is used to compute a distance loss between it and the target bin.

During inference, we can naturally obtain the predicted depth bin, $\hat{d}_i$, by applying the argmax function to the model output, $\hat{y}$, and set the depth value, $\hat{d}$, to its center.

3.4   Fitness Score

As described previously, depth estimation can be formulated as a regression or a classification task. A natural choice for a metric capable of assessing the quality of depth estimation is the mean absolute localization error [14].

If depth estimation is framed as a classification task, the predicted depth is by default not a continuous number and depends on the *real* depth assigned to this bin *e.g.*, its center. As a result, predictions might cause a large absolute localization error despite being assigned to the proper depth bin. This effect makes it difficult to compare both regression and classification models.

To solve this challenge, we suggest to also discretize the network prediction in the case of a regression model into K bins and apply a metric suitable for classification tasks. By doing so, we are able to compare both regression and classification models. Finally, this approach also simplifies the proper model selection by a single metric across the different depth estimation paradigms.

As the network predicts confidence, class label, bounding box parameters as well as the depth, we effectively have set up a multitask network. Thus, we need to be able to assess both depth estimation as well as standard 2D object detection performance. Assessing the performance of a multitask network is challenging as all included tasks might perform and be weighted differently. We, however, favor a single number summarizing the model performance in all tasks.

In typical object detection benchmarks, mean Average Precision (mAP) is commonly used [11]. mAP measures both the classification performance as well as the bounding box localization quality by utilizing the Intersection-over-Union (IoU) of the predicted and the ground truth bounding box as an auxiliary metric. In addition, F1 jointly assesses both precision and recall in a single number. Note that because several properties of the object – *e.g.* its size and full 3D bounding box – are unknown and thus not needed for our use case, metrics commonly used in 3D object detection – *e.g.* AP_{3D} [17] and DS [14] – are not suitable.

As we propose to calculate the performance of depth estimation as a classification task, we suggest employing a scheme similar to F1 for depth estimation as well.

Eventually, we calculate the joint measure as the harmonic mean between the mean F_1-Score of the object detector, mF_1^{OD}, and the mean F_1-Score of the depth estimation, mF_1^{DE}, given the detected objects. As the harmonic mean between two numbers is the same as the F_1-Score, we refer to this metric as F_1^{Comb}.

The mean F_1-Scores for both object detection as well as for depth estimation are dependent on the confidence threshold t_c as well as on the minimum required IoU threshold t_{IoU}. It is $t_c \in \{0.00, 0.01, ..., 0.99, 1.00\}$. All predictions with confidence below this threshold will be discarded. For t_{IoU}, we obtain the values according to [27] such that $t_{IoU} \in \{0.50, 0.55, ..., 0.90, 0.95\}$. Predictions with an IoU $\geq t_{IoU}$ will be treated as TP. Predictions with an IoU $< t_{IoU}$ are assumed to be FP.

Finally, it is

$$mF_1^{OD} = mF_1^{OD}(t_c, t_{IoU}) \tag{6}$$

$$mF_1^{DE} = mF_1^{DE}(t_c, t_{IoU}) \tag{7}$$

$$F_1^{Comb} = F_1^{Comb}(t_c, t_{IoU}) \tag{8}$$

$$= \frac{2 \cdot mF_1^{OD} \cdot mF_1^{DE}}{mF_1^{OD} + mF_1^{DE}}. \tag{9}$$

The domain of the combined score F_1^{Comb} is $[0, 1]$ with higher values representing a better combined performance. As the combined score still depends on both t_c and t_{IoU}, we distill it into a single value, which we refer to as *Fitness* score. We define it as the maximum of the combined score over all confidence and IoU thresholds,

$$Fitness = \max_{t_c, t_{IoU}} F_1^{Comb}. \tag{10}$$

By doing so, we are able to assess the model performance *as is* when productively deployed.

4   Experiments

To demonstrate the effectiveness of the proposed methods for long range object-level monocular depth estimation we design several experiments and compare them using the Amazon Airborne Object Tracking (AOT) dataset.

We split the experiments into three major groups: regression, bin classification, and ordinal regression. Each group formulates the depth estimation task differently, implying different network architectures and loss functions. Eventually, we evaluate the performance of each experiment. We use 2D mAP as well as the mean absolute localization error (MALE) as individual metrics for object detection and depth estimation, respectively. In addition, we assess the quality of each tested approach using the joint Fitness Score.

4.1   Dataset

The Amazon AOT dataset was introduced in 2021 as part of the Airborne Object Tracking Challenge [2]. It contains a collection of in-flight images with other aircraft flying by as planned encounters. Planned objects are annotated with a 2D bounding box (in pixels), the object label, and the distance (in meters) from the camera to a specific object. As the metadata only contains the euclidean distance from the camera without splitting it into x, y, and z, we use the terms *distance* and *depth* interchangeably within this study. Additionally, the sequences may contain encounters with unplanned objects. Those objects are annotated with bounding box parameters and their specific class label – the distance, however, is unknown.

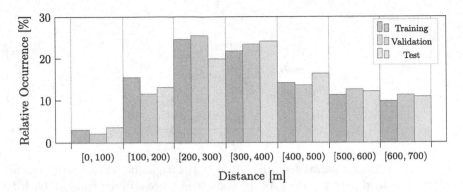

Fig. 1. Distance distribution for all objects within *train*, *val*, and *test* training set.

While most other datasets that feature object-level depth annotations mostly focus on autonomous driving and only contain objects up to 200 m [5,14,17,22, 34,35], the AOT dataset features objects up to several hundreds of meters. In our experiments, we use the *partial* dataset as provided by the authors. This subset contains objects up to 700 m away from the camera. We have observed that some objects, specifically with a range below 10 m, are mislabeled with respect to the object's distance. Thus, we removed the range annotation for these objects but kept the bounding box and classification labels so that they can still be used for training the object detector. The images in the flight sequences are collected at a rate of 10 Hz. As such, many of the images tend to be quite similar. With the goal of shortening training time without significant degradation of performance, we use only every 5th image of this dataset. This subset is equivalent to 2 Hz or 20 % of the initial dataset. We further split the flight sequences in the dataset into dedicated sets for training (60 %), validation (20 %), and testing (20 %). By splitting the entire dataset on a sequence level, we ensure that no cross-correlations between training, validation, and test set occur. Our selected split also provides similar distance as well as class distributions as depicted in Figs. 1 and 2. Sample images of the dataset including typical objects are shown in Fig. 3.

4.2 Experimental Setup

Our models are trained on 2464 × 2464 px images. We upsample and slightly stretch the original images with a resolution of 2448 × 2048 px to the target resolution as the network requires squared input images with dimensions multiple of 32. We use Stochastic Gradient Descent (SGD) as our optimizer and combine it with a cosine learning rate scheduler with warm-up [16]. In total, we train for 15 epochs with a batch size of 14 using 2 Nvidia RTX3090 GPUs.

As described previously, our network architecture features a de-facto multi-task setting. Thus, we calculate our overall multitask loss function $\mathcal{L}$ as:

$$\mathcal{L} = \mathcal{L}_{OD} + w_{DE}\mathcal{L}_{DE}. \tag{11}$$

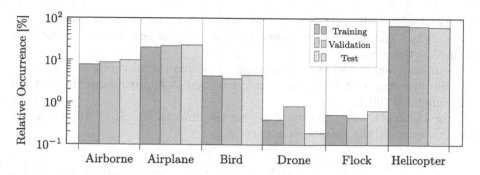

Fig. 2. Class distribution for all objects within *train*, *val*, and *test* training set, log-scaled to improve visibility.

Fig. 3. Sample images of the Amazon AOT dataset [2].

Accordingly, the detector loss function, $\mathcal{L}_{\mathrm{OD}}$, is defined as:

$$\mathcal{L}_{\mathrm{OD}} = w_{\mathrm{obj}}\mathcal{L}_{\mathrm{obj}} + w_{\mathrm{loc}}\mathcal{L}_{\mathrm{loc}} + w_{\mathrm{class}}\mathcal{L}_{\mathrm{class}}, \tag{12}$$

with $\mathcal{L}_{\mathrm{obj}}$ being the objectness, $\mathcal{L}_{\mathrm{loc}}$ the localization, and $\mathcal{L}_{\mathrm{class}}$ the classification loss. w_{obj}, w_{loc}, and w_{class} refer to the corresponding balancing weights. We leave the detector loss function from YOLOX [16] unchanged. Thus, it is $w_{\mathrm{obj}} = 1$, $w_{\mathrm{loc}} = 5$, and $w_{\mathrm{class}} = 1$. We conduct experiments with different depth loss functions, $\mathcal{L}_{\mathrm{DE}}$. At the same time, the depth weight w_{DE} is a hyperparameter.

Regression. Our first set of experiments frames depth estimation as a regression task. As such, we set the number of output channels for the last convolutional layer of our depth estimation head to 1. As described in Sect. 3.2, there are different methods of encoding depth information. Moreover, each encoding can be combined with different distance-based loss functions.

As mentioned in Sect. 4.1, the distance of the objects to the camera is at most 700 m. Therefore, we parameterize the Sigmoid encoding such that it is defined in the domain $(d_{\min}, d_{\max}) \to (0, 700)$.

Similarly, for the ReLU-like encoding, we obtain the best results when defining the hyperparameters a and b in a way that it approximates the Sigmoid encoding: $a = 100$ and $b = \frac{700}{2}$.

For the depth loss function, $\mathcal{L}_{\text{DE}}$, we use Smooth L1 (SL1) [19] and mean squared error (MSE) loss for each encoding:

$$
\text{SL1}(y, \hat{y}) = \begin{cases} \frac{1}{2N} \sum_{i=1}^{N} (y_i - \hat{y}_i)^2, & \text{if } |y_i - \hat{y}_i| \leq 1 \\ \frac{1}{N} \sum_{i=1}^{N} |y_i - \hat{y}_i| - 0.5, & \text{otherwise} \end{cases} \tag{13}
$$

$$
\text{MSE}(y, \hat{y}) = \frac{1}{N} \sum_{i=1}^{N} (y_i - \hat{y}_i)^2. \tag{14}
$$

In addition, we follow [23] and combine direct depth regression with the reverse Huber (berHu) loss:

$$
\text{berHu}(y, \hat{y}, c) = \begin{cases} \frac{1}{N} \sum_{i=1}^{N} |y_i - \hat{y}_i|, & \text{if } |y_i - \hat{y}_i| \leq c \\ \frac{1}{N} \sum_{i=1}^{N} \frac{(y_i - \hat{y}_i)^2 + c^2}{2c}, & \text{otherwise.} \end{cases} \tag{15}
$$

$\hat{y}$ refers the model prediction and y is the target. c is a pseudo-constant that is originally calculated as a function, $c(y, \hat{y}) = \frac{1}{5} \max_i (|y_i - \hat{y}_i|)$ [23]. N refers to the overall number of predictions.

Bin Classification. The second set of experiments models depth estimation as a classification task. The depth interval $(d_{\min}, d_{\max}) \to (0, 700)$ is uniformly discretized into $K = 7$ bins with a uniform bin width of 100 m. Choosing the proper bin size is rather subjective and highly dependent on the use case. For our use case, we find that 100 m is suitable since the environment is less cluttered and objects are found at larger distances when compared to other similar applications, *e.g.* autonomous driving, where smaller bin sizes might be desired. Similarly, and in agreement with [29], we choose uniform discretization over a log-space discretization strategy because the latter increases bin sizes at larger distances where most objects are found. Moreover, for our use case, early detections are beneficial as we want to avoid entering other objects' airspace.

To allow the model to predict K depth bins, we change the number of output channels in the last convolutional layer of our depth estimation head to K.

Our baseline experiment in this group uses softmax (*c.f.* Sect. 3.3) as the final activation and CE as the loss function. In total, we design two experiments that employ the proposed Soft-Argmax (SA) with Smooth L1 and MSE loss.

Ordinal Regression. In our last set of experiments, we follow the guidelines of [13], framing depth estimation as an ordinal regression problem. First, we uniformly discretize the depth into 7 bins, as previously described. The number of output channels in the last convolution layer is set to $2 \cdot (K - 1)$, where the number of bins, K, equals 7. Finally, we reimplement the proposed loss function, applying it to objects instead of pixels.

Metrics. We evaluate the performance of the experiments based on different metrics. The Fitness score proposed in Sect. 3.4 is our primary metric. To compute it, depth is once again uniformly discretized into 7 bins with a width of 100 m. During training, we search for hyperparameters that maximize the Fitness score on the validation dataset. Once optimal hyperparameters are found, we evaluate on the *test* set and report the Fitness score as our primary metric.

Additionally, we report secondary metrics including 2D mAP with 10 IoU thresholds $t_{\mathrm{IoU}} \in \{0.50, 0.55, ..., 0.90, 0.95\}$ and the mean absolute localization error. We furthermore evaluate the performance w.r.t. the number of parameters, GFLOPs, inference, and post-processing times, allowing us to compare the methods in terms of computational constraints.

4.3   Results

Table 1 summarizes the experiment results on the *test* set. Within the depth regression methods, the proposed Sigmoid encoding outperforms all other encodings. The ReLU-like encoding performs worse compared to the Sigmoid encoding but is still competitive with the best encoding from the state-of-the-art, the logarithm. The combination Sigmoid/SL1 performs best within this group.

Within the classification methods, we observe that the proposed loss functions based on Soft-Argmax perform better than the baseline with CE loss. We obtain the best results w.r.t. Fitness by combining Soft-Argmax with Smooth L1 loss. Ordinal regression also outperforms the classification with CE loss. Our results are consistent with the results of [13]. Overall though, it is outperformed by our proposed loss functions based on Soft-Argmax.

Table 1 also shows that in most experiments, extending YOLOX with an additional depth head slightly degrades the base 2D performance by means of 2D mAP. There are notable exceptions, the combination SA/SL1 is one of them.

While the combination of Soft-Argmax and Smooth L1 loss performs best w.r.t. the Fitness score and 2D mAP, it doesn't yield the lowest absolute localization error. This can easily be understood as we select the middle point of the predicted bin as the actual distance of the object, *c.f.* Sect. 3.3. In particular, Table 1 shows that models using the Sigmoid and the ReLU-like encoding regression perform better in this aspect.

We attempt to further improve absolute localization in the classification setting by using bin interpolation, as a postprocessing step, instead of simply choosing the center of the bin. Following [29], we define the interpolation function f as:

$$f \left(p \left(d_{i-1} \right), p \left(d_i \right), p \left(d_{i+1} \right) \right) = f \left(\frac{p \left(d_i \right) - p \left(d_{i-1} \right)}{p \left(d_i \right) - p \left(d_{i+1} \right)} \right). \tag{16}$$

Table 1. Experiment results obtained on the *test* set. Object detection and depth estimation are jointly evaluated on the proposed Fitness score. 2D mAP and mean absolute localization error (MALE) individually evaluate object detection and depth estimation, respectively.

Method	Loss function	Fitness	2D mAP	MALE
2D Only		–	27.7%	–
Direct	SL1	42.2%	26.7%	52.7 m
	MSE	43.0%	26.9%	50.3 m
	berHu [23]	43.4%	24.7%	49.6 m
Inverse	SL1	39.9%	26.4%	92.8 m
	MSE	35.0%	25.5%	94.7 m
Log	SL1	48.2%	27.0%	35.5 m
	MSE	46.7%	26.6%	38.0 m
Sigmoid	SL1 (Ours)	51.6%	25.7%	**28.9 m**
	MSE (Ours)	50.4%	25.3%	32.4 m
ReLU-like	SL1 (Ours)	48.3%	27.9%	33.3 m
	MSE (Ours)	47.7%	28.0%	35.5 m
Classification	CE	50.9%	24.9%	37.9 m
	SA/SL1 (Ours)	**53.6%**	**28.5%**	37.9 m
	SA/MSE (Ours)	52.8%	26.9%	38.5 m
Ordinal Regression		52.7%	27.0%	37.9 m

$p(d_i)$ refers to the probability of the predicted bin, $p(d_{i-1})$, and $p(d_{i+1})$ are the probabilities of the neighboring bins. The predicted depth bin is refined using:

$$\hat{d} = \begin{cases} \hat{d} - \frac{s_i}{2} \cdot (1 - f(x)), & \text{if } p(d_{i-1}) > p(d_{i+1}) \\ \hat{d} + \frac{s_i}{2} \cdot \left(1 - f\left(\frac{1}{x}\right)\right), & \text{otherwise} \end{cases} \tag{17}$$

where s_i is the bin size *i.e.*, the width, of the predicted bin i.

Any function f must shift the predicted depth towards the previous bin if $p(d_{i-1}) > p(d_{i+1})$, shift towards the next bin if $p(d_{i-1}) < p(d_{i+1})$, and leave it unchanged if $p(d_{i-1}) = p(d_{i+1})$. We then select the following strictly monotone functions $f : [0,1] \rightarrow [0,1]$ depicted in Fig. 4:[1]

Equiangular [33]	$f(x) = x,$	(18)
Parabola [33]	$f(x) = \frac{2x}{x+1},$	(19)
SinFit [21]	$f(x) = \sin\left(\frac{\pi}{2}(x-1)\right) + 1,$	(20)
MaxFit [30]	$f(x) = \max\left(\frac{1}{2}\left(x^4 + x\right), 1 - \cos\left(\frac{\pi x}{2}\right)\right),$	(21)
SinAtanFit[29]	$f(x) = \sin\left(\frac{\pi}{2}\arctan\left(\frac{\pi x}{2}\right)\right).$	(22)

---

[1] The authors of [29] did not name the function. We refer to it as *SinAtanFit* to easily identify it.

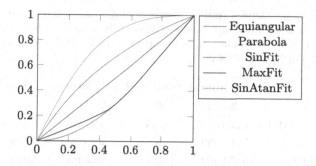

Fig. 4. Different bin interpolation functions.

Table 2. Results of the different bin interpolation functions evaluated on mean absolute localization error (MALE).

Function	MALE
None (baseline)	37.9 m
Equiangular	31.1 m
Parabola	32.5 m
SinFit	**30.1 m**
MaxFit	**30.1 m**
SinAtanFit	34.6 m

As shown in Table 2, all interpolation functions show improvements over the baseline. SinFit and MaxFit obtain the same results and perform the best out of our selection. Despite the improvements, it is not able to surpass the Sigmoid-encoded model. As the interpolation is part of the postprocessing and does not change the network architecture or the predicted depth bin, both Fitness score and 2D mAP remain unchanged.

Runtime Comparison. Besides the quality of the predictions, another important aspect is how the different models compare at runtime. In Table 3, representative models are benchmarked and compared.

Compared to pure 2D object detection, the inference time increases by approx. 4 ms for all proposed methods. This is mainly caused by the increased GFLOPs coming from the additional prediction head.

Amongst the proposed methods, GFLOPs, number of parameters, and inference speed do not vary meaningfully. Looking at postprocessing though, we observe that the classification and ordinal regression models are slower than the regression models. This result is expected as there are more steps involved for both in order to transform the model output into the depth value. Moreover, classification and ordinal regression models grow in complexity with an increasing number of depth bins. Lastly, we conclude that the cost of bin interpolation is negligible.

Table 3. Inference benchmark results on representative models for regression, classification, classification with bin interpolation, and ordinal regression. Results measured with 2464×2464 px image resolution, batch size 1 and FP16 using PyTorch on an Intel Core i9-10920X and Nvidia RTX3090.

Method	Parameters	GFLOPs	Inference	Postprocessing
2D Only	5 034 321	56.1	21.5 ms	1.5 ms
Sigmoid & Smooth L1	5 200 690	66.5	25.9 ms	**1.7 ms**
SA/SL1	5 201 369	66.5	**25.4 ms**	2.9 ms
SA/SL1 & SinFit	5 201 369	66.5	25.5 ms	3.0 ms
Ordinal Regression	5 201 951	66.6	25.7 ms	3.0 ms

5   Conclusion

In this work, we addressed the problem of long range object-level monocular depth estimation and exemplarily extended the YOLOX object detection framework. We modeled the depth estimation task as a regression, classification, and ordinal regression problem. To jointly assess object detection and depth estimation performance, we introduced the Fitness score as a novel metric. We proposed two novel encodings for regression, Sigmoid and ReLU-like. The former outperforms other state-of-the-art encodings w.r.t. Fitness score and absolute localization error, while the latter is competitive with the best encoding from the state-of-the-art. Moreover, for classification, we proposed a novel loss function based on the Soft-Argmax operation that minimizes the distance between the predicted and target depth bins. In conjunction with the Smooth L1 loss, it outperforms all other models, including ordinal regression, w.r.t. Fitness score. Furthermore, its 2D mAP performance even surpasses the baseline 2D model. However, it doesn't reach the same accuracy by means of absolute localization error compared to the proposed Sigmoid encoding – even when with bin interpolation functions. In general, regression based models have a slight advantage in postprocessing which lead to an overall faster runtime. Based on the conducted experiments, we find that our proposed methods provide great extensions to standard 2D object detection frameworks, enabling object-level depth estimation at long range.

References

1. Standard specification for detect and avoid system performance requirements ASTM f3442/ASTM f3442m (2020). https://www.astm.org/f3442_f3442m-20.html
2. Aicrowd, airborne object tracking challenge (2021). https://www.aicrowd.com/challenges/airborne-object-tracking-challenge
3. Bhat, S.F., Alhashim, I., Wonka, P.: Adabins: Depth estimation using adaptive bins. In: CVPR (2021)
4. Blei, Y., Jourdan, N., Gählert, N.: Identifying out-of-distribution samples in real-time for safety-critical 2D object detection with margin entropy loss. In: ECCVW (2022)
5. Caesar, H., et al.: nuScenes: a multimodal dataset for autonomous driving. In: CVPR (2020)
6. Cao, Y., Wu, Z., Shen, C.: Estimating depth from monocular images as classification using deep fully convolutional residual networks. In: TCSVT (2017)
7. Cordts, M., et al.: The cityscapes dataset for semantic urban scene understanding. In: CVPR (2016)
8. Diaz, R., Marathe, A.: Soft labels for ordinal regression. In: CVPR (2019)
9. Eigen, D., Fergus, R.: Predicting depth, surface normals and semantic labels with a common multi-scale convolutional architecture. In: ICCV (2015)
10. Eigen, D., Puhrsch, C., Fergus, R.: Depth map prediction from a single image using a multi-scale deep network. In: NeurIPS (2014)
11. Everingham, M., Van Gool, L., Williams, C.K., Winn, J., Zisserman, A.: The pascal visual object classes (voc) challenge. In: IJCV (2010)
12. Finn, C., Tan, X.Y., Duan, Y., Darrell, T., Levine, S., Abbeel, P.: Deep spatial autoencoders for visuomotor learning. In: ICRA (2016)
13. Fu, H., Gong, M., Wang, C., Batmanghelich, K., Tao, D.: Deep ordinal regression network for monocular depth estimation. In: CVPR (2018)
14. Gählert, N., Jourdan, N., Cordts, M., Franke, U., Denzler, J.: Cityscapes 3D: dataset and benchmark for 9 DoF vehicle detection. In: CVPRW (2020)
15. Gählert, N., Wan, J.J., Jourdan, N., Finkbeiner, J., Franke, U., Denzler, J.: Single-shot 3D detection of vehicles from monocular RGB images via geometrically constrained keypoints in real-time. In: IV (2020)
16. Ge, Z., Liu, S., Wang, F., Li, Z., Sun, J.: YOLOX: exceeding yolo series in 2021. arXiv preprint arXiv:2107.08430 (2021)
17. Geiger, A., Lenz, P., Urtasun, R.: Are we ready for autonomous driving? the KITTI vision benchmark suite. In: CVPR (2012)
18. Ghosh, S., Patrikar, J., Moon, B., Hamidi, M.M., et al.: Airtrack: onboard deep learning framework for long-range aircraft detection and tracking. arXiv preprint arXiv:2209.12849 (2022)
19. Girshick, R.: Fast R-CNN. In: ICCV (2015)
20. Goroshin, R., Mathieu, M.F., LeCun, Y.: Learning to linearize under uncertainty. In: NeurIPS (2015)
21. Haller, I., Pantilie, C., Mariţa, T., Nedevschi, S.: Statistical method for sub-pixel interpolation function estimation. In: ITSC (2010)
22. Janai, J., Güney, F., Behl, A., Geiger, A., et al.: Computer vision for autonomous vehicles: Problems, datasets and state of the art. In: FTCGV (2020)
23. Laina, I., Rupprecht, C., Belagiannis, V., Tombari, F., Navab, N.: Deeper depth prediction with fully convolutional residual networks. In: 3DV (2016)

24. Lee, J.H., Han, M.K., Ko, D.W., Suh, I.H.: From big to small: multi-scale local planar guidance for monocular depth estimation. arXiv preprint arXiv:1907.10326 (2019)
25. Li, B., Dai, Y., He, M.: Monocular depth estimation with hierarchical fusion of dilated cnns and soft-weighted-sum inference. Pattern Recogn. **83**, 328–339 (2018)
26. Li, R., Xian, K., Shen, C., Cao, Z., Lu, H., Hang, L.: Deep attention-based classification network for robust depth prediction. In: ACCV (2018)
27. Lin, T.-Y., et al.: Microsoft COCO: common objects in context. In: Fleet, D., Pajdla, T., Schiele, B., Tuytelaars, T. (eds.) ECCV 2014. LNCS, vol. 8693, pp. 740–755. Springer, Cham (2014). https://doi.org/10.1007/978-3-319-10602-1_48
28. Liu, W., et al.: SSD: single shot multibox detector. In: Leibe, B., Matas, J., Sebe, N., Welling, M. (eds.) ECCV 2016. LNCS, vol. 9905, pp. 21–37. Springer, Cham (2016). https://doi.org/10.1007/978-3-319-46448-0_2
29. Miclea, V.C., Nedevschi, S.: Monocular depth estimation with improved long-range accuracy for UAV environment perception. In: TGRS (2021)
30. Miclea, V.C., Vancea, C.C., Nedevschi, S.: New sub-pixel interpolation functions for accurate real-time stereo-matching algorithms. In: ICCP (2015)
31. Redmon, J., Divvala, S., Girshick, R., Farhadi, A.: You only look once: unified, real-time object detection. In: CVPR (2016)
32. Redmon, J., Farhadi, A.: YOLO9000: better, faster, stronger. In: CVPR (2017)
33. Shimizu, M., Okutomi, M.: Sub-pixel estimation error cancellation on area-based matching. In: IJCV (2005)
34. Sun, P., et al.: Scalability in perception for autonomous driving: Waymo open dataset. In: CVPR (2020)
35. Ye, X., et al.: Rope3D: the roadside perception dataset for autonomous driving and monocular 3d object detection task. In: CVPR (2022)

RadarFormer: Lightweight and Accurate Real-Time Radar Object Detection Model

Yahia Dalbah[ID], Jean Lahoud[ID], and Hisham Cholakkal[(✉)][ID]

Mohamed Bin Zayed University of Artificial Intelligence, Abu Dhabi, UAE
{yahia.dalbah,jean.lahoud,hisham.cholakkal}@mbzuai.ac.ae

Abstract. The performance of perception systems developed for autonomous driving vehicles has seen significant improvements over the last few years. This improvement was associated with the increasing use of LiDAR sensors and point cloud data to facilitate the task of object detection and recognition in autonomous driving. However, LiDAR and camera systems show deteriorating performances when used in unfavorable conditions like dusty and rainy weather. Radars on the other hand operate on relatively longer wavelengths which allows for much more robust measurements in these conditions. Despite that, radar-centric data sets do not get a lot of attention in the development of deep learning techniques for radar perception. In this work, we consider the radar object detection problem, in which the radar frequency data is the only input into the detection framework. We further investigate the challenges of using radar-only data in deep learning models. We propose a transformers-based model, named RadarFormer, that utilizes state-of-the-art developments in vision deep learning. Our model also introduces a channel-chirp-time merging module that reduces the size and complexity of our models by more than 10 times without compromising accuracy. Comprehensive experiments on the CRUW radar dataset demonstrate the advantages of the proposed method. Our RadarFormer performs favorably against the state-of-the-art methods while being 2x faster during inference and requiring only one-tenth of their model parameters.

Keywords: Radar · Object detection · Autonomous driving

1 Introduction

Autonomous driving technology heavily relies on a combination of cameras and LiDAR sensors, mostly due to the complementary benefits that LiDAR sensors bring to most detection pipelines. LiDAR sensors provide dense and detailed point cloud maps using rotating sensors with spherical/semi-spherical coverage of the surrounding area. These sensors' pre-processed data features can be easily integrated with images from camera sensors in autonomous driving. However, LiDAR sensors have shorter wavelengths, causing the following limitations in LiDAR object detectors [9]. (i) Signals are highly prone to errors under poor weather conditions and occlusion (ii) they have a relatively shorter sensing range.

R. Gade et al. (Eds.): SCIA 2023, LNCS 13885, pp. 341–358, 2023.
https://doi.org/10.1007/978-3-031-31435-3_23

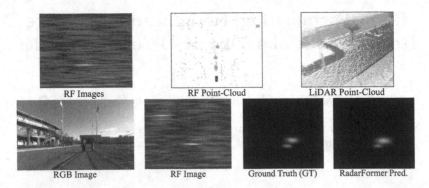

Fig. 1. The first row shows an example of different data samples from a radar-frequency image-like data point (from [35]), a point-cloud radar-frequency data sample (from [28]), and a point-cloud LiDAR sample (from [3]). The RF image heatmap signifies the magnitude of the echoed radar signal from the radar, with blue being the minimum. The second row shows a sample RGB image and its corresponding RF image. Our model, RadarFormer, takes in the RF image only and produces a heatmap prediction with the object class (shown in different colors), illustrated in the rightmost image. (Color figure online)

In contrast to LiDAR signals, radar's Frequency Modulated Continuous Wave (FMCW) signals operate at the millimeter-wave (mmW) band, or in the frequency band between 30 to 300 GHz. The mmW band is much lower than visible light, which allows radar signals to go through occlusion particles such as smoke and dust, enabling radars to function more robustly in extreme weather conditions. Furthermore, radar signals' longer wavelength (mmW) provides a larger range for detection with acquisition capabilities reaching up to 3000 m. Radars are also more accessible and cheaper to introduce to dynamic systems compared to LiDARs [18].

It is possible to extract point cloud data from raw radar signals, however, it is more common to extract them as radar frequency (RF) image-like data. Radar signals are sent and received through a multi-input multi-output (MIMO) antenna array, which is then passed through a series of fast Fourier Transform (FFT) to extract range-angle-doppler maps. These maps compactly describe the 3D space in the range plane (distance to detection), azimuth plane (angle of arrival), and doppler information (relative velocity) [32]. An example of RF data and LiDAR data in comparison can be seen in Fig. 1. In the RF image, we see an RF range-azimuth (RA) map shown in an image-like processed format. The second image shows RF point cloud data, which can be compared to the LiDAR point cloud data, where we see that LiDAR sensors provide more detailed object descriptions compared to radar data. However, radar data can provide readings from a longer range and contain velocity information, as discussed earlier. Multiple works demonstrated the use of radar data as a feasible alternative to cameras and LiDARs in object classification [1,4,5]. Recently, [2,35] explored radars as an opportunity to be fused with other sensors such as camera-radar fusion to produce more accurate predictions.

The increasing availability of radar frequency data now has opened the path to explore more complex approaches for radar perception. Common radar datasets provide a variety of RF data as input, for instance, [35] provides only the RA map, while [24] provides RA, Range-Doppler (RD), and Range-Azimuth-Doppler (RAD) maps. Some datasets [24,42] provide the original radar tensors in addition to the maps, while others [18] provide the digitized output after the Analog to Digital Converter (ADC) stage. Radar data can also be provided as point cloud data, as was shown earlier and provided by [28] with range-azimuth information in a 3D space point cloud data. The previous works for radar object detection used computationally expensive models that use large 3D convolutions and might be impractical in generating rapid real-time predictions.

In this work, we propose a transformer-based deep-learning model that operates on radar frequency data exclusively and produces state-of-the-art results in object detection and classification. The proposed model is lightweight in size and generates inferences in real-time, making it suitable for the task of autonomous detection. Figure 1 (bottom row) shows a camera RGB image, a radar RF image, and the corresponding ground truth annotation from CRUW dataset [35]. The proposed RadarFormer takes only the RF image as input (without the RGB image) and produces a heatmap prediction of the localized object class shown in the rightmost image. The key contributions of the proposed approach are:

- We explore the effectiveness of vision transformers in radar perception and introduce a novel architecture, RadarFormer, for real-time radar object detection. To the best of our knowledge, we are the first to introduce a transformer-based architecture for RF maps data for the task of object detection.
- We propose a channel-chirp-time merging module that contributes to reducing the size of radar perception models and using less computationally expensive modules.
- Our proposed method, RadarFormer, achieves state-of-the-art performance with one-tenth the model size of the previous state-of-the-art model and a two-times faster inference speed.

2   Object Detection on Radar Data

Radar data is usually visualized as two-dimensional or three-dimensional maps with two sets of channel dimensions. The first is the real and complex part of the RF signal, and the other is the chirps of the echoed signal. This data form makes them suitable for multiple deep learning models utilizing Convolutional Neural Networks (CNN). While conventional CNNs do not work with point-cloud data, multiple models were still explored through the voxelization of input data, such as LidarMultiNet [39], PanopticPolarNet [43], or works that encourage interaction between the model CNNs and voxelized input like JS3C-Net [38]. For 2D image-like data, multiple works detail the process of generating RA, RD, RAD, and RAMaps [24,32,35]. The output of said maps is then passed through a Constant False Alarm Rate (CFAR) algorithm, which checks the amplitude of pixels to determine their magnitude relative to the average noise level in surrounding pixels, and classifies pixels as 'object' and 'non-object'. After CFAR, object classes are then determined following different techniques. For example, CRUW data set uses a 3D object localization and class recognition to generate ground truth for the data [33]. It is common to also use temporally and spatially aligned cameras and LiDars to generate annotations for object detection purposes without classifications [18]. In a similar fashion, the work in [17] generated the annotations using radar LiDAR fusion with the latter being the ground truth. Other works [24] use clustering techniques that rely on the rich doppler information properties of radars to create clustered annotations, with some manual annotations as a quality check [28].

Following the generation of data sets, most of the works in the literature heavily rely on CNN-based deep learning models. RADDet [42] uses a radar-tailored ResNet backbone followed by a YOLO-inspired [27] dual detection head to produce object detections and classifications. Encoder-decoder style models are very popular and were adapted differently to different datasets. RODNet [35] uses a stacked-hourglass model to generate predictions, while TMVA-Net [23] uses a temporal-multi-map encoder decoder in their CARRADA data set. Other works like RadSegNet [2] and LidarMultiNet [39] introduces the encoder-decoder block after the input voxelization step to point cloud data. Following this discussion, we notice a pattern in the over-reliance on CNN-based deep learning models. While they perform adequately in most detection-based models, our proposed model extends beyond CNNs to include more developed deep learning techniques for radar perception systems.

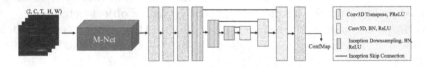

Fig. 2. Original RODNet hourglass with inception model as per [35].

Given the benefits of radar frequency data for object detection and classification, multiple radar datasets have been collected for this purpose. The Camera-Radar of University of Washington (CRUW) data set [35] contains RF images collected through radar sensors with synchronized cameras connected. The FMCW mmW radars along with the camera collect synchronized radar maps and images at 30 frames per second (FPS) with 255 chirps per frame and a range and azimuth resolutions of 0.23 m and 15°, respectively. The data set contains three classes: pedestrians, cyclists, and cars, with around 5 objects per frame on average. RODNet was proposed as a 'student' module that learns alongside a camera-radar fusion (CRF) cross-modal approach. RODNet alone takes only the RF images and produces confidence maps (ConfMaps) which are passed later into a location-based non-maximum suppression (L-NMS). Figure 2 illustrates the RODNet model architecture, which consists of a chirp merging module (M-Net) that downsamples chirps into one layer, followed by the stacked-hourglass architecture featuring the temporal inception convolutional layers. The model reported an average precision (AP) and average recall (AR) of 77.40% and

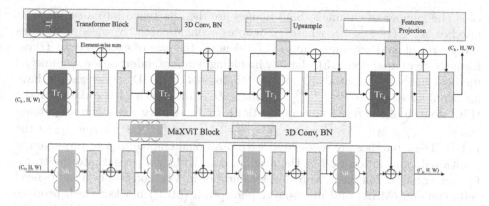

Fig. 3. Our transformer-based proposed models, 2D transformers (top) and Radar-Former (bottom). The input to both models is the output from the output of the channel-chirp-time merging module, which will be discussed in Sect. 3.3. The output of the ViT transformers downsamples the input, requiring an upsample block to retrieve the original resolution. The same flow is followed in the MaXViT-based model, without any resolution changes to the inputs/outputs.

80.80%, respectively, without the CRF and using only a camera-only supervision method, which will function as our baseline in terms of the accuracy of predictions.

3 The Proposed RadarFormer

While CNNs have been a dominant architectural design block for a lot of tasks in both image recognition and radar-based recognition [12,30], we propose a transformer-based architecture that utilizes recent developments in deep learning transformer techniques, shown in Fig. 3. Our model introduces a hybrid model between CNNs, transformers, and multi-axis attention following the work done in [31]. We also introduce an updated channel-chirp merging module that includes the temporal domain. Our module extends the merging of extra channels to include residuals connecting the temporal domain in the downsampling/upsampling stream and is shown in Fig. 4.

3.1 Transformers

In recent years, vision transformers (ViTs) [8] were introduced as a new paradigm that is fully free of convolutional networks and produced state-of-the-art results for image recognition tasks. Various approaches and versions of self-attention modules were proposed with some works reporting state-of-the-art results in object detection without the use of CNNs [19,20]. Transformers were originally developed for Natural Language Processing (NLP) tasks using 1D sequence inputs. Transformers were then repurposed for images in ViTs, taking in a 2D input image of size $x \in \mathbb{R}^{H \times W \times C}$, where (H, W) is the image resolution and

C is the number of channels. Said images are then flattened into a sequence of patches, $x_p \in \mathbb{R}^{N \times (P^2 \cdot C)}$, where x_p denotes a single patch, $N = \frac{H \times W}{P^2}$ denotes the length of said sequence, and (P, P) denotes the size of every patch. Transformers were expanded to 3D data by including temporal information, volumetric medical images, or spatial 3D images [15,26,36,41]. In 3D sequences, we have an input of size $x \in \mathbb{R}^{D \times H \times W \times C}$, where D is the depth of the data (or the third domain of the respective application). Here, we have a sequence of patches shaped as $x_p \in \mathbb{R}^{N \times (P^3 \cdot C)}$ where $N = \frac{H \times W \times D}{P^3}$ and (P, P, P) is the size of the patch. These patches are associated with positional embeddings to preserve the positional information tied to the original data and then passed into the transformer encoder. The basic transformer encoder in ViT consists of a multi-head self-attention (MSA) and multi-layer perceptron (MLP) blocks, which produce an output $x_z \in \mathbb{R}^{N \times S}$, where S is the projection following the MLP blocks. The MSA block learns a mapping between a query (q), a corresponding key (k), and a value (v) representation of the encoder output, measured by $[\mathbf{q},\mathbf{k},\mathbf{v}] = x_z \mathbf{U_{qkv}}$ where U_{qkv} is the projection MLP weights. The $\mathbf{q}$ and $\mathbf{k}$ representations are then used to find the attention weights A through

$$\mathbf{A} = \mathrm{Softmax}(\frac{\mathbf{qk}^\top}{\sqrt{S_l}}) \tag{1}$$

S_l a scaled version of S by a factor l and is set to $\frac{S}{l}$, keeping the total number of parameters constant given a variation in the number of key values $\mathbf{k}$. Using the attention weights, we then measure the self-attention SA with $\mathrm{SA}(x_z) = A\mathbf{v}$. MSA scales the previous expression up to a sequence of self-attention heads in the form of a vector with each head having its own unique set of weights in U_{msa}, described by

$$\mathrm{MSA}(x_z) = [\mathrm{SA}_1(\mathbf{x_z}); \mathrm{SA}_2(\mathbf{x_z}); ...; SA_m(\mathbf{x_z})]\mathbf{U_{msa}} \tag{2}$$

The data set we operate on can be adjusted to work in both 3D and 2D (including the temporal domain or with a downsampled temporal domain), as will be shown in Sect. 3.3. RadarFormer's early stages used expensive 3D transformers that are based on ViT [8] and improved by [11]. While this provided good results, the computational complexity was still high. We can use 2D transformers by passing our input data through our proposed channel-chirp-time merging stream (Sect. 3.3), providing a much lighter model and higher accuracy as well. Using 2D data allows us to explore a computationally inexpensive 2D variation of attention as we will see in Sect. 3.2.

3.2   Attention Variation

Following ViT, multiple works have been developed that aim at more robust vision transformer models. Some works introduced many features to image transformers such as hybrid models between transformers and CNNs [6,7,16], with many surveys evaluating and contrasting these models and their variations [14]. In addition to introducing hybrid models, variations in attention modules were also explored by many works [10]. Our module uses a transformer block, called MaXViT (Multi-Axis Vision Transformer) [31], which consists of an inverted residual block, MBConv [13], followed by an attention block and grid attention. The MBConv consists of 3 convolutional layers with a wide-narrow-wide style of channels, and a small-large-small style of kernel sizes, with a residual connecting the first convolution to the last one. The multi-axis attention block creates windowed partitions of the input and performs self-attention on these partitions, whose shape is $(\frac{H}{P}, \frac{W}{P}, P \times P, C)$, creating non-overlapping windows of size $P \times P$ following the notation used earlier. Similarly, a grid partitioning module uses a $G \times G$ uniform grid to partition the input with adaptive size $\frac{H}{G} \times \frac{W}{G}$, resulting in a dilated mix of tokens that provide global information. In the module proposition, stacking both window and grid attention provides local and global contexts in transformer operations, hence the name multi-axis attention.

As an attempt to improve the transformers models, we explored multiple variations of transformer models and attentions along with convolutional combinations. The first is a high-low attention (HiLo) [25] approach that splits attention into a high-resolution branch and a low-resolution (downsampled) branch. This approach did provide more consistency in training but did not perform adequately, where it capped at 74.2% AP. This led us to explore the attention and resolution variation of the models more thoroughly. To this end, we explored a high-resolution transformer (HRFormer)-like architecture, following the work proposed in [40]. HRFormer was very computationally expensive both in its 3D and 2D variants and did not perform well. The number of trainable parameters, 872.9 million, was too large to justify its use, and too large for the data set to train and performed poorly as expected. In our proposition, MaxViT blocks performed consistently better with a lot of variations when compared to other baseline models. Other architectures like UNETR [11] and UNETR++ [29] were computationally expensive in their 3D format and did not provide an accuracy high enough to justify their use. However, downsampling to 2D and using a UNETR/ViT-inspired transformer design provided a good baseline for transformer architectures which we referred to as '2D Transformer' in Fig. 3. We discuss the quantitative results of relevant models in Sect. 4.2.

3.3   Use of 2D Information

In the original model of RODNet, the input has a shape of $(B, 2, T, C, H, W)$, where B is the batch size, C is the number of chirps, T is the window size (temporal), 2 is the number of RF channels (being a constant referring to the real and imaginary magnitudes of RF signals), and H, W are the height and width of the RF image, respectively. The RF channels and chirps are merged into one dimension using M-Net, a channel-chirp merging module suggested in [35]. We extend M-Net to go beyond channel-chirp merging and propose a module that includes temporal merging as well whose inputs and outputs are shown in Fig. 4. The chirps and RF channels dimensions are compressed first into one channel with dimension C_h, shown as (B, C_h, T, H, W), and then downsampled in the temporal domain to be passed into the models. However, we preserve temporal context through a temporal residual connection towards the upsampling stream at the end before generating the ConfMaps.

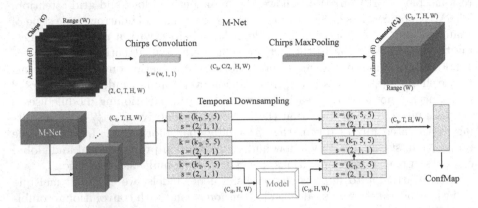

Fig. 4. The M-Net module (top) reported in the original RODNet architecture and used by us and proposed temporal downsampling/upsampling flow (bottom). The M-Net module takes in the full RF image with shape $(2, T, C, H, W)$ and merges the channels (2) and the chirps (C) into one dimension, C_h. The second module downsamples the temporal domain to receive a 2D tensor of shape (C_h, H, W). The tensors are connected via element-wise addition to their upsampling counterpart after the models forward the input. Both modules are cascaded and output a 3D output of shape (C_h, T, H, W).

The motive for downsampling comes from two observations. First, 3D convolutions and temporal 3D convolutions are extremely computationally expensive and their weights have a large size, especially when using many filters. This is undesirable in the context of autonomous vehicles that have controllers with limited computing and storage capabilities. Second, we noticed that downsampling 3D inputs into 2D, performing all deep learning operations after downsampling the frames into one frame, and upsampling before inference with residuals did not compromise the accuracy to a noticeable level. The downsampling however reduced the inference time to one-fifth of the original value and the size of the weights to one-tenth of the original value. We can conclude that the effect of the time domain is less pronounced when treated as separate channels and employing expensive 3D convolutions or 3D transformer modules like UNETR is not necessary. Other models in radar perception used temporal downsampling techniques like TMVA-NET [23]. The reduced base RODNet which has a temporal downsampling/upsampling stream added to is referred to as '2D CNN' from this point forward. It is worth mentioning that downsampling is more useful than distributing channels. We noticed that when we skip the 3D downsampling module and instead stack our model input to the shape $(T \times B, C_h, H, W)$ we lose a lot of contextual temporal information. This information was retained through the convolutional filters and the residual connections between the input and output.

We noticed that RODNet's performance degraded by a negligible margin by downsampling the temporal domain to a single channel and performing most of the operation with reduced numbers of channels with 2D convolutions. In addition, the model did produce heavily fluctuating results as the model trains, which initially was attributed to the configuration of the learning rate, however, altering the learning rate value, the scheduling, and the optimizer [21] did not make the model converge to a consistent value. Furthermore, the reproducibility of networks trained on this data set was inconsistent and sometimes inaccurate. The same model could be trained using the same configurations and yield completely different prediction accuracies. Replicating the results of reported values of RODNet was also not feasible despite attempting to train the same model multiple times. Introducing residual connections between the input downsampling convolutions and the output upsample convolutions helped remedy this issue. The difficulty of learning consistent weight sets is one of the challenging aspects of this data set and application, but we noticed better reproducibility of transformer-based architectures compared to the CNN-only approaches.

3.4 Effect of Receptive Fields and Residuals

Transformers did introduce a large reduction in required computational complexity at the early stages of our model design, providing a large margin for us to utilize more complex architectures and modules. To take advantage of both transformers and CNNs, we noticed that introducing CNNs before and after the transformers along with residual connections following our design in Fig. 3 improved the model's accuracy. This was in line with the suggested conclusions of [37], which was encouraging to explore two main ideas in our experiments. The first is exploring the effects of varying the number of convolution layers, input/output channels, and residual connections before and after the transformers. The second is to assess the effect of varying the receptive fields and the size of the convolutional layers' kernels. This discussion excludes the first transformer layer of both models due to it being preceded by 3 convolutional downsampling layers. For the first point, using two convolutional layers before the transformer and one after the network provided the best results without residuals. Introducing residuals with element-wise addition and a convolutional layer that equalizes the number of channels, followed by a convolution after the element-wise addition, provided similar performance with a slight decrement in required computational time. We noticed that MaXViTs perform the best without downsampling and the convolution in the residual connection is not necessary and does not improve the performance either with MaXViTs, providing a lighter model in this aspect by removing it. For the second point, varying the kernel sizes of said layers did contribute to better learning of data. Increasing the size of the kernel as we go deeper in the network allows for learning of a more general receptive field that gets improved by the transformer's global/local attention dynamic. In a similar fashion, establishing residual connections between the input 3D downsampling to the output 3D upsampling compensates for the information loss in propagating through the network and improves the upsampling stream.

Implementation Details. We train our models on a single NVIDIA A100 GPU with Adam optimizer and an initial learning rate of 10^{-4} with step decay.

4 Experiments

Dataset Details. We build and test our model on the CRUW radar data set [35]. CRUW consists of roughly 400k frames of recorded driving sequences. The data is processed to be represented as Range-Azimuth Heatmaps (RAMap)s, which describe a bird's-eye view of the scene seen from the ego-vehicle. The x-axis depicts the azimuth plane, describing the angle, and the y-axis depicts the range plane, describing the distance to the object, with the intensity describing the magnitude of the RF signal. These can be described as an image with a resolution of 128×128 each, with a sample shown in Fig. 1. The acquisition setup consists of two 77 GHz FMCW antennas that collect 256 chirps every frame (30 frames a second), and only 4 are chosen out of these chirps (0, 64, 128, 192). The details of the data we use and inputs will strictly follow the work done in [35] which most of our work will use as an evaluation baseline. This work will also focus on the behavior of deep learning modules with radar-based data, which has not been studied extensively in most works regarding radar data. Unlike the model associated with the CRUW data set, RODNet, our aim was to create a deep learning model that uses radar data without any sort of sensor infusion, which was illustrated in the model discussed in Sect. 3. The input to the model is a (B, C_h, H, W) input tensor, with the batch size B, the number of channels C_h, and (H, W) is the resolution of the RF image. We note that this is the output of the downsampling/upsampling module discussed earlier in Sect. 3.3. The RF image resolution (H, W) is fixed at 128. There are 40 sequences reserved for training and 10 sequences for testing. Testing annotations and images are not publicly shared and evaluation of the testing is done on a private evaluation server for the RODNet2021 challenge [34].

Evaluation Metrics. We use the same evaluation metrics as in [35] throughout all of our evaluations. RODNet uses an object location similarity (OLS) metric that takes the role of Intersection over Union (IoU). OLS is then passed to a location-based non-maximum suppression (L-NMS) algorithm to generate confidence maps (ConfMaps). ConfMaps in the range-azimuth represent predicted locations for the objects, with multiple channels attributing the location to a class. Similar to previous work for pose estimation [22], the output is a gaussian heatmap-like prediction with a mean equal to object location and variance attributed to the object class and scale information. Our main evaluation metric will be the average precision (AP) and average recall (AR) calculated through the variation of OLS threshold between 0.5 to 0.9 with steps of 0.05.

4.1 Baselines

The baseline to the CRUW data set is the RODNet model associated with it, reporting an AP of 77.40% and AR of 80.80% using camera-only (CO) supervision. Instead of using the reported inference times, we instead retrain the model and report inference speed using our equipment to provide a consistent and scalable assessment and comparison. The model reports 61.2 million total trainable parameters (single stack hourglass). The time it takes for a single backpropagation iteration with a window size of 16 for this model on our setup is 1920 ms, and the average inference time is 148 ms. This is associated with 47664 GMAC ($\times 10^9$ multiply-accumulate) operations. The listed parameters will be the main comparison points to our model.

4.2   Quantitative Results

Due to the recency of emerging RF-based data, and the lack of other works to compare to, we compare our proposed models, 2D transformer and Radar-Former, to RODNet considering the latter the state-of-the-art method. Table 1 shows the performance comparisons of mentioned models. The evaluation metric here is AP and AR, each is split into four categories: PL, CR, CS, and HW, referring to parking lot, campus road, city street, and highway data categories, respectively. Both AP and AR are then averaged into AP_{total} and AR_{total}. We can see that RadarFormer achieved an AP almost on par with RODNet, with a higher AR. They varied in their performance in different situations, where we noticed that RadarFormer had slightly better performance in the parking lot scenarios, while RODNet performed noticeably better on campus roads. In city streets, RadarFormer had a higher AR than RODNet, but a lower AP, and on highways, RadarFormer outperformed RODNet significantly. Looking at total AP, the models perform comparably the same, but RadarFormer had a tendency to have a higher recall, on average, implying that RadarFormer has a tendency to produce fewer predictions, but an inclination to have higher confidence for the produced predictions. We also note that 2D CNN, which is a 2D version of RODNet that utilized our channel-chirp-time merging module did not compromise the accuracy greatly when compared to the retrained version of RODNet. However, we will see in Sect. 4.2 how this module reduced the computational cost and size of RODNet to a large degree.

Table 1. Quantitative comparisons of AP and AR on the CRUW data set with CO supervision, categories are described in Sect. 4.2. RadarFormer performed comparable to RODNet, being the accessible and replicable state-of-the-art model for this data set. The exact values of RODNet were not replicable, so we use the values reported in the RODNet challenge server [34] and report the values trained of the publicly available model (*). We discuss the discrepancy in the results in Sect. 3.3.

Method	Total		PL		CR		CS		HW	
	AP	AR	AP	AR	AP	AR	AP	AR	AP	AR
RODNet*	72.32	79.62	94.52	95.59	68.12	72.50	52.30	72.13	67.48	71.63
RODNet [34]	**77.40**	80.80	95.50	96.40	**75.30**	**78.40**	**66.10**	71.70	68.00	72.20
2D CNN	71.58	81.52	94.93	96.04	66.52	73.56	52.15	75.62	69.23	74.13
2D Transformer	75.03	81.99	**96.68**	**97.55**	68.51	77.04	58.67	75.08	70.16	72.46
RadarFormer	77.18	**83.45**	95.88	96.99	71.74	77.66	61.19	**76.46**	**74.37**	**77.30**

Model Comparisons. We further compare the models' computational cost and inference times using Table 2. We note that we use MAC operations instead of floating-point operations (FLOPs) due to PyTorch's inclination to use MAC operations. The M-Net module used earlier remains constant and its performance is reported separately. The word 'size' is used interchangeably with the number of parameters. Despite RODNet having reported a marginally higher accuracy in CO supervision predictions, the model requires almost twenty times more MAC operations than our proposed model, while also being ten times as big in size. This is significant in regards to employing said models on devices that don't have much computing power. Similarly in the same table, the inference and backpropagation (BP) times are aligned with the MAC discussion, where RadarFormer does provide inferences at roughly half the time of RODNet. We would like to note that these numbers are relative, and they scale up and down based on the used GPU for training and inference. Furthermore, we normalize and take into account the used window sizes and batch sizes for the training and testing. We also point out that reporting the BP time for a single iteration instead of training time per epoch is a more accurate measurement for this data set to factor out the loading time from the storage devices to the CUDA GPU. 2D CNN is a very lightweight and fast model that compromises the accuracy marginally compared to RODNet, while 2D Transformers provides a middle-ground between training and inference time and accuracy, but at the cost of having a large model size and the number of parameters.

Table 2. Comparisons of (10^9) multiply-accumulate (GMAC) operations, the number of parameters (No. Parm.) in millions (m), back-propagation (BP) time per iteration, and inference times between the different models (inclusive of M-Net). All time units are in milliseconds (ms). M-Net has no stand-alone inference or BP time since all models use it. The downsampling/upsampling streams are counted in our proposed models. All predictions and inferences were adjusted and normalized to take into account the batch size, window size, and test stride for equal comparisons.

Model	GMACs	No. Parm. (m)	BP Time (ms)	Infer. Time (ms)
M-Net	0.805	0.224	–	–
RODNet [35]	47664	61.22	1920	148.17
2D CNN	388.6	2.71	330	51.92
2D Transformer (ours)	1950	20.88	380	60.38
RadarFormer (ours)	2123	6.42	700	84.35

4.3   Qualitative Results

We refer to Fig. 5 for samples of the predictions from the mentioned models. It shows the original RGB image, RF image, ground truth (GT), then the prediction heatmap of the models for three samples. We also note that since we do not have access to the images of the test set, the models were re-trained on a 90% split of the training data, and these predictions were generated on the unseen

10% of the data. As discussed earlier and is evident by the relatively high AR of RadarFormer, the model produces predictions with relatively higher confidence when it decides on an object and a class, depicted through the accurate class label and higher brightness of the heatmap. We also notice that RadarFormer performs better in scenes with far objects as can be seen in the third row.

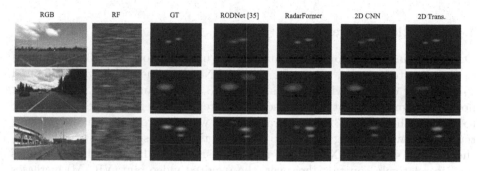

Fig. 5. Test cases for all models mentioned. Pedestrians, cyclists, and cars are referred to with red, green, and blue, respectively in the prediction and ground truth map. We see that RadarFormer provides predictions with acceptably high confidence when they are presented, in contrast to other models that are prone to occasional false labeling. (Color figure online)

We further explore the incorrect prediction cases of RadarFormer and list some of the model's limitations. Failure is most common due to difficult cases or due to inaccurate annotation of the ground truth, and we show examples of

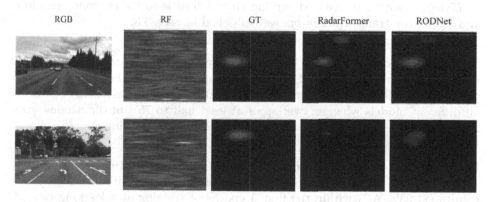

Fig. 6. Incorrect predictions generated by our proposed model, compared with predictions from RODNet. We note that despite there inaccurate annotations cases like this, the model attempts to predict objects that were not annotated. We can see this evident in the top row where we see predictable objects in the RGB images and the RF images, but not present in the ground truth. The second row is a difficult case where RODNet generated a better prediction than RadarFormer.

this in Fig. 6. First, the model tends to generate predictions of objects that it perceives in the RF images but are not present in the ground truth, which can be seen in the first row. We can see a second car, however, the ground truth doesn't show its presence and is then determined as an incorrect prediction case. We also notice that the model does not generate predictions of objects that are close to other objects within sensor's view, as can be seen in the second row.

4.4 Ablation Studies

Data Format and Model Variations. We use a window size of 32 (32 frames) in the reported results, except for the RODNet baseline which was configured to 16. Using different window sizes between 8, 16, 24, and 48 did not yield any noticeable improvement to the model and reported degraded performance (1 to 2% less than using 32). This was evident when operating with and without temporal downsampling. Similarly, we varied the patch size of the transformers, the window size of MaXViT, and the MLP sizes. The BP and inference times remain relatively constant, however, the required video ram (VRAM) increases exponentially with larger attention windows without any improvement to the accuracy. Similarly, the size of the window attention for MaXViT should remain a fraction of the image resolution (window size of 7 for an RF image of resolution 128). Using the base window size of 7 yielded consistently higher AP than other counterparts. Increasing the number of neurons in transformer MLP caused the model to not converge to a high enough accuracy compared with a lower number of neurons, so we use an MLP ratio range between 20 and 150, depending on how deep the model goes (smaller ratio for deeper models). Any value higher than 150 yielded degraded performance, and the same applies to values less than 20 (AP maximum of 66%). Lastly, we observed an increase in accuracy from 75.10% to 77.18% when we introduced varying spatial kernels to the convolutions after and before the transformer layers as mentioned in Sect. 3.3.

Training Stride. The stride controls how overlapped the data input is. The model's performance degraded noticeably when we remove the stride and take every unique window size. A jump of around 10% in accuracy is usually observed on different models when an overlap of at least half to 75% of the window size is introduced (e.g. overlap of 24 frames in a 32 window size input).

Learning Rate & Scheduling. The model trains on multiple iterations per epoch, and it is possible to converge to a set of weights that produce acceptable results (71.23% AP) within the first 3 epochs of training at a learning rate of 10^{-4}. This learning rate is too large, however, starting with a learning rate of 10^{-5} proved to be too small. We used multiple scheduling techniques and starting/end points [21]. Starting with a learning rate of 10^{-4} and ending at 10^{-6} using both step scheduling and cosine annealing yielded the final models.

5 Conclusion

We introduce a novel transformers-based architecture for deep learning applications on radar frequency images, named RadarFormer. The main novelty lies in the reduction of total computing complexity and training/inference times of the original model by using transformers, a lightweight and efficient deep learning module. We also introduce a channel-chirp-time merging block that contributes to the reduction in computation complexity without compromising accuracy. Our multi-axis attention-based model produces state-of-the-art results while having significantly fewer parameters and inference time compared to the previous state-of-the-art method. The proposed models can pave the way for transformer-based research in radar frequency deep learning research.

References

1. Angelov, A., Robertson, A., Murray-Smith, R., Fioranelli, F.: Practical classification of different moving targets using automotive radar and deep neural networks. IET Radar Sonar Navig. **12**(10), 1082–1089 (2018). https://doi.org/10.1049/iet-rsn.2018.0103. https://ietresearch.onlinelibrary.wiley.com/doi/abs/10.1049/iet-rsn.2018.0103
2. Bansal, K., Rungta, K., Bharadia, D.: RadSegNet: a reliable approach to radar camera fusion (2022). https://doi.org/10.48550/ARXIV.2208.03849. https://arxiv.org/abs/2208.03849
3. Behley, J., et al.: Towards 3D LiDAR-based semantic scene understanding of 3D point cloud sequences: the SemanticKITTI Dataset. Int. J. Robot. Res. **40**(8–9), 959–967 (2021). https://doi.org/10.1177/02783649211006735
4. Cao, P., Xia, W., Ye, M., Zhang, J., Zhou, J.: Radar-ID: human identification based on radar micro-doppler signatures using deep convolutional neural networks. IET Radar Sonar Navig. **12**(7), 729–734 (2018). https://doi.org/10.1049/iet-rsn.2017.0511. https://ietresearch.onlinelibrary.wiley.com/doi/abs/10.1049/iet-rsn.2017.0511
5. Capobianco, S., Facheris, L., Cuccoli, F., Marinai, S.: Vehicle classification based on convolutional networks applied to FMCW radar signals. In: Leuzzi, F., Ferilli, S. (eds.) TRAP 2017. AISC, vol. 728, pp. 115–128. Springer, Cham (2018). https://doi.org/10.1007/978-3-319-75608-0_9
6. Dai, Z., Liu, H., Le, Q.V., Tan, M.: CoatNet: marrying convolution and attention for all data sizes. In: Beygelzimer, A., Dauphin, Y., Liang, P., Vaughan, J.W. (eds.) Advances in Neural Information Processing Systems (2021). https://openreview.net/forum?id=dUk5Foj5CLf
7. D'Ascoli, S., Touvron, H., Leavitt, M.L., Morcos, A.S., Biroli, G., Sagun, L.: ConViT: improving vision transformers with soft convolutional inductive biases. In: Internation Conference on Machine Learning, pp. 2286–2296 (2021)
8. Dosovitskiy, A., et al.: An image is worth 16x16 words: transformers for image recognition at scale (2020). https://doi.org/10.48550/ARXIV.2010.11929. https://arxiv.org/abs/2010.11929
9. Feng, D., et al.: Deep multi-modal object detection and semantic segmentation for autonomous driving: datasets, methods, and challenges. IEEE Trans. Intell. Transp. Syst. **22**(3), 1341–1360 (2021). https://doi.org/10.1109/TITS.2020.2972974

10. Hassanin, M., Anwar, S., Radwan, I., Khan, F.S., Mian, A.: Visual attention methods in deep learning: An in-depth survey (2022). https://doi.org/10.48550/ARXIV.2204.07756. https://arxiv.org/abs/2204.07756

11. Hatamizadeh, A., et al.: UNETR: transformers for 3D medical image segmentation. In: Proceedings of the IEEE/CVF Winter Conference on Applications of Computer Vision, pp. 574–584 (2022)

12. He, K., Zhang, X., Ren, S., Sun, J.: Deep residual learning for image recognition. In: 2016 IEEE Conference on Computer Vision and Pattern Recognition (CVPR), pp. 770–778 (2016). https://doi.org/10.1109/CVPR.2016.90

13. Howard, A.G., et al.: MobileNets: efficient convolutional neural networks for mobile vision applications. CoRR abs/1704.04861 (2017). http://arxiv.org/abs/1704.04861

14. Khan, S., Naseer, M., Hayat, M., Zamir, S.W., Khan, F.S., Shah, M.: Transformers in vision: a survey. ACM Comput. Surv. **54**(10s), 1–41 (2022). https://doi.org/10.1145/3505244. https://doi.org/10.1145/3505244

15. Lahoud, J., et al.: 3D vision with transformers: a survey. arXiv preprint arXiv:2208.04309 (2022)

16. Li, Y., et al.: MViTv 2: improved multiscale vision transformers for classification and detection. In: CVPR (2022)

17. Lim, T.Y., et al.: Radar and camera early fusion for vehicle detection in advanced driver assistance systems. In: NeurIPS Machine Learning for Autonomous Driving Workshop (2019)

18. Lim, T.Y., Markowitz, S.A., Do, M.N.: Radical: A synchronized FMCW radar, depth, IMU and RGB camera data dataset with low-level FMCW radar signals. IEEE J. Select. Top. Sig. Process. **15**(4), 941–953 (2021). https://doi.org/10.1109/JSTSP.2021.3061270

19. Liu, Z., et al.: Swin transformer v2: Scaling up capacity and resolution. In: International Conference on Computer Vision and Pattern Recognition (CVPR) (2022)

20. Liu, Z., et al.: Swin transformer: hierarchical vision transformer using shifted windows. In: Proceedings of the IEEE/CVF International Conference on Computer Vision (ICCV) (2021)

21. Loshchilov, I., Hutter, F.: SGDR: Stochastic gradient descent with warm restarts. In: International Conference on Learning Representations (2017). https://openreview.net/forum?id=Skq89Scxx

22. Newell, A., Yang, K., Deng, J.: Stacked Hourglass Networks for Human Pose Estimation. In: Leibe, B., Matas, J., Sebe, N., Welling, M. (eds.) ECCV 2016. LNCS, vol. 9912, pp. 483–499. Springer, Cham (2016). https://doi.org/10.1007/978-3-319-46484-8_29

23. Ouaknine, A., Newson, A., Pérez, P., Tupin, F., Rebut, J.: Multi-view radar semantic segmentation. In: Proceedings of the IEEE/CVF International Conference on Computer Vision (ICCV), pp. 15671–15680 (2021)

24. Ouaknine, A., Newson, A., Rebut, J., Tupin, F., Pérez, P.: Carrada dataset: camera and automotive radar with range- angle- doppler annotations. In: 2020 25th International Conference on Pattern Recognition (ICPR), pp. 5068–5075 (2021). https://doi.org/10.1109/ICPR48806.2021.9413181

25. Pan, Z., Cai, J., Zhuang, B.: Fast vision transformers with HiLo attention. In: NeurIPS (2022)

26. Peiris, H., Hayat, M., Chen, Z., Egan, G., Harandi, M.: A robust volumetric transformer for accurate 3D tumor segmentation. In: Wang, L., Dou, Q., Fletcher, P.T., Speidel, S., Li, S. (eds.) Medical Image Computing and Computer Assisted Intervention–MICCAI 2022. MICCAI 2022. LNCS, vol. 13435, pp. 162–172. Springer, Cham (2022). https://doi.org/10.1007/978-3-031-16443-9_16

27. Redmon, J., Divvala, S.K., Girshick, R.B., Farhadi, A.: You only look once: unified, real-time object detection. CoRR abs/1506.02640 (2015). http://arxiv.org/abs/1506.02640

28. Schumann, O., et al.: RadarScenes: a real-world radar point cloud data set for automotive applications. CoRR abs/2104.02493 (2021). https://arxiv.org/abs/2104.02493

29. Shaker, A., Maaz, M., Rasheed, H., Khan, S., Yang, M.H., Khan, F.S.: UNETR++: delving into efficient and accurate 3D medical image segmentation. arXiv:2212.04497 (2022)

30. Szegedy, C., et al.: Going deeper with convolutions. In: 2015 IEEE Conference on Computer Vision and Pattern Recognition (CVPR), pp. 1–9 (2015). https://doi.org/10.1109/CVPR.2015.7298594

31. Tu, Z., et al.: MaxViT: Multi-axis vision transformer. In: Avidan, S., Brostow, G., Farinella, G.M., Hassner, T. (eds) Computer Vision–ECCV 2022. ECCV 2022. LNCS, vol. 13684. Springer, Cham (2022). https://doi.org/10.1007/978-3-031-20053-3_27

32. Vogginger, B., et al.: Automotive radar processing with spiking neural networks: Concepts and challenges. Front. Neurosci. **16**, 851774 (2022). https://doi.org/10.3389/fnins.2022.851774. https://www.frontiersin.org/articles/10.3389/fnins.2022.851774

33. Wang, Y., Huang, Y.T., Hwang, J.N.: Monocular visual object 3D localization in road scenes. In: Proceedings of the 27th ACM International Conference on Multimedia, pp. 917–925. ACM (2019)

34. Wang, Y., et al.: Rod 2021 challenge: a summary for radar object detection challenge for autonomous driving applications. In: Proceedings of the 2021 International Conference on Multimedia Retrieval, pp. 553–559 (2021)

35. Wang, Y., Jiang, Z., Gao, X., Hwang, J.N., Xing, G., Liu, H.: RODNet: radar object detection using cross-modal supervision. In: 2021 IEEE Winter Conference on Applications of Computer Vision (WACV), pp. 504–513 (2021). https://doi.org/10.1109/WACV48630.2021.00055

36. Wang, Y., Guizilini, V., Zhang, T., Wang, Y., Zhao, H., Solomon, J.M.: DETR3D: 3D object detection from multi-view images via 3D-to-2D queries. In: The Conference on Robot Learning (CoRL) (2021)

37. Xiao, T., Singh, M., Mintun, E., Darrell, T., Dollár, P., Girshick, R.: Early convolutions help transformers see better. Adv. Neural. Inf. Process. Syst. **34**, 30392–30400 (2021)

38. Yan, X., et al.: Sparse single sweep lidar point cloud segmentation via learning contextual shape priors from scene completion. Proceed. AAAI Conf. Artif. Intell. **35**(4), 3101–3109 (2021). https://doi.org/10.1609/aaai.v35i4.16419. https://ojs.aaai.org/index.php/AAAI/article/view/16419

39. Ye, D., et al.: LidarMultiNet: unifying lidar semantic segmentation, 3D object detection, and panoptic segmentation in a single multi-task network (2022). https://doi.org/10.48550/ARXIV.2206.11428. https://arxiv.org/abs/2206.11428

40. Yuan, Y., et al.: HRFormer: high-resolution transformer for dense prediction. In: NeurIPS (2021)

41. Yuan, Z., Song, X., Bai, L., Wang, Z., Ouyang, W.: Temporal-channel transformer for 3D lidar-based video object detection for autonomous driving. IEEE Trans. Circuits Syst. Video Technol. **32**(4), 2068–2078 (2022). https://doi.org/10.1109/TCSVT.2021.3082763

42. Zhang, A., Nowruzi, F.E., Laganiere, R.: RADDet: range-azimuth-doppler based radar object detection for dynamic road users. In: 2021 18th Conference on Robots and Vision (CRV), pp. 95–102 (2021). https://doi.org/10.1109/CRV52889.2021.00021
43. Zhou, Z., Zhang, Y., Foroosh, H.: Panoptic-PolarNet: proposal-free lidar point cloud panoptic segmentation. In: Proceedings of the IEEE/CVF Conference on Computer Vision and Pattern Recognition (CVPR) (2021)

Drawing and Analysis of Bounding Boxes for Object Detection with Anchor-Based Models

Manav Madan[1]([envelope]) [ID], Christoph Reich[1] [ID], and Frank Hassenpflug[2]

[1] Institute for Data Science, Cloud Computing, and IT Security, Furtwangen University, Furtwangen, Germany
{manav.madan,christoph.reich}@hs-furtwangen.de
[2] C.R.S. iiMotion GmbH, Peterzeller Str. 8, 78048 Villingen-Schwenningen, Germany
frank.hassenpflug@crs-iimotion.com

Abstract. Supervised object detection models are trained to recognize certain objects. These models are classified into two types: single-stage detectors and two-stage detectors. The single-stage detectors just need one pass through the model to anticipate all the bounding boxes, whereas the two-stage detectors require to first estimate the image portions where the object could be located. Due to their speed and simplicity, single-stage anchor-based models are used in many industrial settings. Training such models require bounding boxes that describe the spatial location of an object, which are usually drawn by an expert. However, the question remains, how much area should be considered when drawing the bounding boxes? In this paper, we demonstrate the effects that the size and placement of a rectangular bounding box can have on the performance of the anchor-based models. For this, we first perform experiments on a synthetically generated binary dataset and then on a real-world object detection dataset. Our results show that fixing the size of the bounding boxes can help in improving the performance of the model in the case of single class object detection (approximately 50% improvement in mAP@[.5:.95] for real world dataset). Furthermore, we also demonstrate how freely available tools can be combined for obtaining the best possible semi automated object labeling pipeline.

Keywords: Object detection · Yolo · Bounding box labelling

1 Introduction

Object detection is the task of classifying an object of interest in an image, in the addition to predicting the spatial location through some sets of coordinates. After the progress made in the field of image classification, deep learning (DL) is being extensively used for object detection also. There are numerous applications of this such as in autonomous driving, visual quality inspection, etc., which

Supported by the Ministry of Science, Research and the Arts of the State of Baden-Württemberg (MWK BW).

© The Author(s), under exclusive license to Springer Nature Switzerland AG 2023
R. Gade et al. (Eds.): SCIA 2023, LNCS 13885, pp. 359–373, 2023.
https://doi.org/10.1007/978-3-031-31435-3_24

are currently being implemented using DL [1,2]. Such techniques have been already translated into different industries. Object detection is divided into two categories mostly, i.e., single-stage detectors and Region-Based two-stage detectors [2]. The industrial requirement is often performance in terms of both time and accuracy. Therefore, the faster object detectors such as single-stage anchor-based models which detect important regions and classify objects in a single step being used for many different applications [10,11]. Their counterparts are the two or multi-stage models which require more than one iteration over the input for object detection. Many models (both single and two-stage detectors) use simplification techniques such as pre defined fixed sized anchor boxes, which can help the model to converge fast. An anchor box is a pre-defined bounding box with a specific height and width that is given to the model during training. These are often chosen based on the object sizes encountered in the dataset before the training starts to capture the scale and aspect ratio of the different object classes to detect. However, the ideal size and the effects of the drawing strategy of bounding boxes have not been studied extensively, which is the focus of this work.

1.1   Common Object Detection Framework with DL

The success of DL for object detection has been largely due to the availability of large labeled datasets with annotations [8,9]. This is not always true in industrial settings, where labeled data is scarce, and it needs to be curated and labeled by different domain experts [6]. In Fig. 1 the traditional workflow for object detection is portrayed (adapted from [7]). It is also common to improve data annotations once the model performance is observed in production, as shown in the figure. It can be seen that data curation and data labeling are not mostly included in the machine learning operations (MLOps) pipeline and hence are not fully automated [6]. One reason for this is the lack of fast quality estimation metrics for high dimensional data such as images and videos. Numerous iterations are required for acquiring high-quality labeled data which is labeled by multiple experts. Some major causes of model failure in production are label inconsistency, evaluation bias, and low-quality data used during model training [19].

There has been a lot of effort put in for improving the object detection models, but labeling still remains a costly procedure, with the success of such models dependent on abundant labelled data [9]. To acquire this labelled data, a group of domain experts are required. The process involves iterating over each sample manually, wherein, each expert needs to be confident about the size and location of the object of interest in the sample. Once the expert is confident, a bounding box (also referred to as annotations and labels) is drawn to the region of the object but there is no information regarding how much size should be incorporated into the box. For providing the labeling expert with some guidance, this paper focuses on estimating how the bounding boxes should be drawn for real-time object detectors i.e., mainly single stage detectors which use prior anchor boxes for training the detector. Additionally, a pipeline for semi-automated object labeling is also presented, which is built using tools Labelbox [27] and Voxel FiftyOne [28]. FiftyOne is a data visualization tool whereas Labelbox is a labeling tool that can be integrated with FiftyOne therefore these two were chosen. In Table 4, a comparison of some tools for data labeling is provided. All of these tools offer a free version if they are not fully open-sourced. Another contribution of this paper is the two separate datasets which are open-sourced with three different annotation sets containing bounding boxes of different sizes.

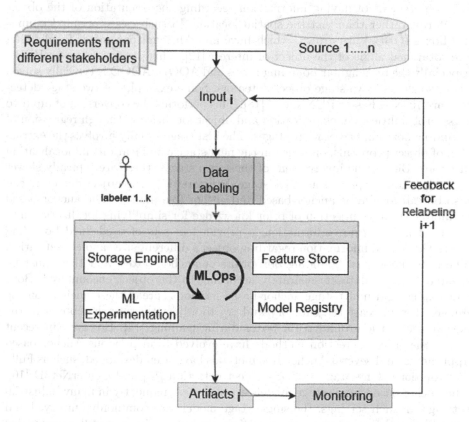

Fig. 1. Overview of an MLOps workflow, with data curation and labeling included as external processes.

2 Related Work

It is known that the annotations i.e., Ground Truth (GT) bounding boxes available for many publicly licensed object detection datasets contain only approximations of the actual labels. These discrepancies in GT arise due to various factors such as human bias (labeler), occlusion, the presence of multiple objects, and errors due to the generation of sub-optimal bounding boxes from the segmentation masks [3]. The bounding boxes are metaphorically not just boxes, but can be designed in different shapes. Some common type of bounding boxes is polygons, square, and rectangle. Furthermore, they can be categorized as horizontal bounding boxes (HBBs) if they are used just to capture the object's position in the images. In such cases, the problem of object detection is referred to as Arbitrary-Oriented Object Detection (AOOD) [13]. However, these are still limited to encompassing the object of interest. In some applications, there is a requirement of having information regarding the orientation of the object of interest rather than just the spatial location. In such cases, oriented bounding boxes (OBBs) are drawn, which have an additional parameter for defining the rotational angle of the object of interest [12]. This paper focuses on HBBs, especially the rectangular bounding boxes and AOOD. AOOD is typically solved using single and two-stage object detectors. Some examples of two-stage detectors are RCNN, Faster RCNN, etc. [18]. This performs the conversion of input to class probabilities, i.e. classification and object localization though regression of bounding boxes in two or more stages. The first stage usually involves the extraction of object proposals, on which in the next stage classification and localization are done. Due to the involvement of these two stages, these are typically slower than their counterparts the single-stage detectors [12]. The object detectors can also be categorized as anchor-based and anchor-free models. The anchor-based technique employs injection of prior knowledge for simplifying localization into the model. This prior information is provided in sets of predefined bounding boxes that entail information regarding object's different sizes and aspect ratios. These are known as bounding box priors or anchor boxes, and they must be created for each dataset separately according to the object encountered. Both single-stage and multi-stage anchor-based methods spread these anchors on the image, then forecast the category and try to adjust these anchor boxes to the actual size of the ground truth boxes during training [14]. However, in recent years, the object detection methods have evolved from previous Anchor-based approaches, and several Anchor-free methods have been developed, such as Fully Convolutional One-Stage (FCOS) object detection [15] and CenterNet3D [16]. The anchor based single stage models are still used frequently in many industrial settings. In such settings, the single-stage models are commonly employed and are called real-time detectors. Their size and complexity can still increase due to factors such as high input dimension or complex network design formations. If the structure and complexity are compromised, then the model can lose accuracy while gaining some inference speed. For solving this issue, various techniques such as model scaling, and multi-scale inference have been introduced. One such network that employs such technique is the YOLO Scaled v4 [5]. This particular anchor based model has allowed DL-based single-stage detectors to attain high accuracy while maintaining near real-time inference speed on various kinds of computing platforms [5]. The Real-time object detectors such as YOLO Scaled v4 operating on standard Graphics Processing Units (GPU) enables widespread usage while reducing overall cost. Generally, because of its great mix of speed and accuracy, the YOLO series has been the most preferred detection frameworks in industrial applications [4].

For any supervised object detector, the data with correct annotations is the key to having a well-trained model. However, drawing bounding boxes over objects correctly and consistently requires competent annotators, and quality control of annotation is a difficult undertaking [17]. Further challenges with bounding box annotations and label inconsistency are well defined in the work [19, 20]. Some techniques use active learning, unsupervised learning, and transfer learning for automated labeling [23]. Other labeling techniques have also emerged that use a trained model first acquired using human-based labels, which can then propose labels on unlabelled images. This is commonly referred to as AI-assisted or model-assisted labeling. However, its performance is still debated. Even for professionals, there is little information accessible about the importance of the consistency of drawn annotations in labeling tools. In this work, we show that fixing the size of the anchor boxes can give a performance boost irrespective of the object size encountered in the dataset. The selected object detector i.e., YOLO Scaled v4 and its functioning are not discussed in detail in this work and for that the original paper from the authors [5] provides a clear guideline on the working and its sub-parts such as anchor box refinement, scaling of the model, etc.

3 Experimentation

As mentioned in the previous section, the anchor boxes define the scale and aspect ratio of specified object classes in the dataset at hand. If there are multiple objects present of contrasting shapes and sizes, then there is a vast number of possible priors out of which the ideal ones are mostly learned using some clustering algorithm. The structure of the single-stage detectors in the YOLO series (from YOLO v3 onwards) includes specific network hierarchy as structural choice which forces the model to make predictions at different scales, i.e., 9 anchor sets in total divided over 3 different scales. Once the different sets of anchor boxes according to the scale are tilted over the image, using non-maximum suppression (NMS) a significant amount of learning capacity is spent in learning the offsets to the chosen box such that there is the best fit to the GT box using the intersection over union (IoU) metric. Based on a certain threshold of IoU, the mean average precision (mAP) metric is calculated for evaluating the accuracy of any object detector over a set of images. The most common forms of mAP are at 50% threshold (mAP@0.5) and an average over different threshold values, i.e., mAP@[.5:.95]. The models learn to refine the different anchor boxes in terms of the position and size, such that localization can be learned with the identification of the object within the best fitting box.

Motivation behind experiments: In many industrial cases, including the targeted application of this work, deal with single class object detection. Thereby the same object might occur in different sizes in the images at different locations. Such cases are quite commonly implemented for quality control or anomaly detection scenarios. So the question arises, do we need a set of different anchor boxes? or can we fix the size by restricting the anchor box size to be greater than the maximum size of the shape of the object? If the size of the box can be fixed then the labeling expert does not have to spend time drawing an ideal boundary around the object for labels, rather he just has to locate the object in the image. Furthermore, the model would not have to refine sets of bounding boxes at different scales and this should ease learning and faster convergence. We tested this hypothesis on two different datasets presented in Fig. 2. One was synthetically generated where the object of interest was selected as a circle, and the other was from a real-world use case. Both the datasets and corresponding annotations are made available on the GitHub (https://github.com/MadanMl/Drawing-and-analysis-of-bounding-boxes-for-object-detection-with-anchor-based-models-).

3.1   Datasets Used

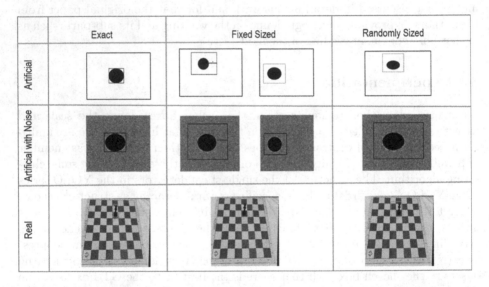

Fig. 2. Overview of the dataset used for experimentation. The first two rows illustrate the synthetically generated dataset and the last row indicates the modified version of real chess-board dataset with rectangular bounding boxes drawn over the pieces as a single class detection problem. In the second column "Fixed Sized" for synthetic datasets, two sub-sets are created where in one the radius is kept constant (displayed with a line of length 2*radius marked) and in the other where radius is varied.

Binary images dataset (synthetically generated): This dataset helps to test the hypothesis in a controlled setting. The data was generated using the OpenCV package [25] with the resolution being set to 416 by 416 and the single class object chosen to be a circle. The circle is similar to the industrial use case which the motivation is also from where similarly shaped anomalies are encountered in lenses of microscopes and endoscopes. As illustrated in Fig. 2, the dataset had two divisions (row 1 and row 2) and three sub-sets depicting the following scenarios:

1. Exact (Fig. 2, column 1): the bounding box is drawn at the circumference of the circle.
2. Fixed Sized (Fig. 2, column 2): the bounding boxes are drawn at a fixed distance from the center i.e. 2 * radius resulting in a square. It is further subdivided into two parts, one where the radius is constant for all images and so is the bounding box size and the circle is always in the center of the box. The second subset depicts the scenario where the radius is randomly changed, but the circle's area never exceeds the size of the box as the maximum radius is set to half of the length of the box.
3. Randomly sized (Fig. 2, column 3): No restrictions on the size of the box. The box dimensions change for each image, depicting how in the real world the domain experts draw the bounding boxes.

The total number of generated images were 500 where a train, valid and test spilt was made of 90%, 5% and 5%. For the second division gaussian blurring was done with the kernel size being varied randomly for each sample and rest everything remains as the original setting.

The original chessboard dataset is a collection of chessboard images where all images were taken from the same angle, with a tripod to the left of the board. The dataset has single and multiple objects in certain images. All pieces' are annotated with a total of 12 classifications as individual classes with rectangular bounding boxes. After augmentation, the accessible roboflow dataset has 693 photos and over 2000 labels which is available publicly [29].

Modified Chess board dataset: We used a modified version of the original chessboard dataset. The total number of images used was 306 where 276 were used in the training and 30 were used for the validation set. The first modification made to the dataset was a conversion from a multi-class dataset to a single-class dataset. For this, all the images with multiple objects were removed. In the next step, three different sets of labels were generated for the same training and test images. For the exact version, the original labels from the dataset were used where the class information in the annotations was set to a single number to adapt the class label to a single class. In the original labels, the bounding boxes are drawn very close to the chess pieces, which represent the exact settings from the first set of the synthetic dataset. Next, the fixed-size relabeling of the whole subset was done with the focus on drawing a box in a 3 by 3 chessboard cell irrespective of the size of the piece. Therefore, in this set, the size of the bounding boxes is bigger than the original and represents the second synthetic set with fixed bounding boxes. For the final set, the bounding boxes were drawn randomly during the relabeling process. The width and height of the bounding boxes varied approximately.

3.2 Configuration

The implementation of scaled YOLO v4 is available freely by the authors and can be easily implemented on custom datasets [26]. Three model configurations are available according to the hardware of choice for inference, i.e., large, CSP, and tiny. The YOLOv4-CSP configuration provided higher accuracy than the YOLOv4-tiny and better inference than the YOLOv4-large, that's why that was chosen. The model's hyperparameters were chosen from those tuned on the COCO dataset (Common Objects in Context) [22], which contains approximately 120,000 photos for training and testing and 80 object class labels. These were fixed for all the experiments, and the only change between the synthetic and the chessboard dataset were the number of epochs (100 for synthetic and 300 for chessboard dataset). Further parameters regarding the training of the model are attached in the appendix (see Table 5).

4   Results

We observed that the object detector converged faster when there was less randomness in the size of the bounding boxes, and this resulted in better localization of the object as the detector spent less effort in optimizing the shifts to the anchor boxes to fit the ground truth. This is helpful in a setting where only one object of some defined shape is involved, as then we do not need to define different sets of anchor boxes. Usually, these are defined to capture objects of all sizes.

In Fig. 3, 4, 5 the normalized bounding boxes' width and height are plotted on the x and y-axis for the synthetically generated binary datasets (training set). It can be seen that for the labels drawn at random, the maximum variance is

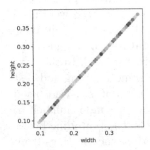
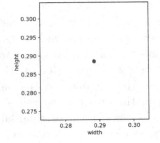
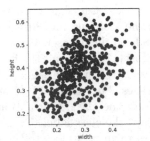

Fig. 3. Label distribution for dataset with bounding box at the boundary of circle.

Fig. 4. Label distribution for dataset with fixed sized bounding boxes around the circle.

Fig. 5. Label distribution for dataset with randomly drawn bounding boxes around the circle.

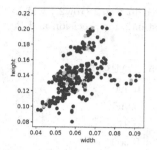 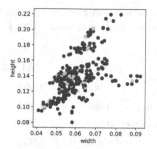 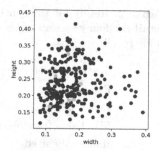

Fig. 6. Label distribution for chessboard dataset with exact bounding boxes.

Fig. 7. Label distribution for chessboard dataset with fixed sized bounding boxes around the chess piece.

Fig. 8. Label distribution for dataset with randomly drawn bounding boxes around the chess piece.

present, whereas for the boxes with the fixed sized (random and fixed radius) we have a single point. Figure 3 shows the case in which labels were drawn exactly at the circumference, due to which we have a diagonal of points representing square boxes of different sizes according to the size of the radius. Whereas the label distribution plots (Fig. 6, 7, 8) for the chessboard dataset are similarly categorized but still they are not labeled in a controlled setting. Therefore, there is not a single diagonal of points but a distribution for the exact category in Fig. 6 like the one observed in Fig. 3.

Table 1. Result for the synthetic dataset (testset) presented in Fig. 2 (row1) with fixed seed value. The P stands for precision and R is for recall.

Synthetic original	P	R	MAP@0.5	MAP@[.5:.95]
Exact	0.956	1	0.995	0.995
Fixed size (fixed radius)	0.965	1	0.995	0.995
Fixed size (random radius)	0.965	1	0.995	0.962
Randomly sized	0.886	0.92	0.965	0.520

The results in Table. 1, 2, and 3 demonstrate that the overall performance in terms of mAP is higher in both of the datasets where the bounding box size is restricted or kept constant. The mAP@[.5:.95] is an absolute metric in which different precision over different IoU thresholds are averaged, whereas the precision and recall displayed are calculated at a pre-defined IoU threshold. There is a significant improvement in the mAP metric for both the synthetic and real datasets when there is less randomness in annotations size.

Table 2. Result for the synthetic dataset (testset) presented in Fig. 2 (row2.) which entail random gaussian noise with fixed seed value. The P stands for precision and R is for recall.

Synthetic with noise	P	R	MAP@0.5	MAP@[.5:.95]
Exact	0.965	1	0.995	0.990
Fixed size (fixed radius)	0.965	1	0.995	0.995
Fixed size (random radius)	0.965	1	0.995	0.971
Randomly sized	0.813	0.92	0.936	0.509

Table 3. Result for the modified chessboard dataset (validation set) presented in Fig. 2 (row 3) with experiments repeated thrice with different seed values. The mAP@[.5:.95] is averaged and presented with deviation, while the other values define the iteration where best mAP@0.5 was achieved. The P stands for precision and R is for recall.

Modified Chessboard	P	R	MAP@0.5	Average mAP@[.5:.95]
Exact	0.892	1	0.995	0.791 ±0.0058
Fixed size (3 by 3 grid)	0.970	1	0.995	0.744 ±0.0123
Randomly sized	0.436	1	0.977	0.289 ±0.0029

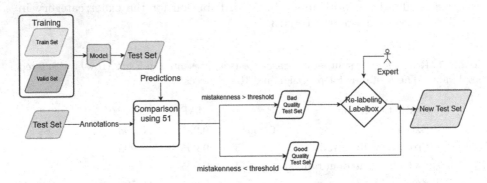

Fig. 9. The relabeling workflow constructed with Labelbox and FiftyOne tools for fast and simplification of the data labeling process. The diagram showcases that the most difficult data points are only relabeled by the expert, where the initial model predicts the highest number of annotation errors.

5   Conclusion

Our results show that fixing the size of the bounding boxes improves the ability of the model to learn the position of the object with high precision for single stage anchor-based detection. However, as shown in Fig. 1 the data relabeling is still constructed as a manual task. Data is an important part of any deep learning application, since models can only extract those important patterns in supervised AOOD that are present in the training set. Typically, data is subjected to several sources of inaccuracy at various phases of creation, and it needs to undergo multiple rounds of improvements. This could also occur due to challenges such as data or concept drift [21]. We need workflows where data can be relabeled automatically and quickly as possible and with the knowledge learned from our experiments showcasing the importance of fixing the size of the bounding boxes. Therefore, we combined two publicly available tools i.e. Labelbox and FiftyOne to develop a semi-automated data relabeling workflow inside a google colab notebook. It can be visualized in Fig. 9. The workflow utilizes a YOLO v5 trained model. The FiftyOne offers integration for Labelbox which offers a free subscription to the tool with limited capabilities. The workflow utilizes the initial labels by the labeler through which an initial detector (YOLO v5 model) is acquired. Then we utilize the predictions through these models and metrics from FiftyOne such as the mistakenness to select a subset of images that have maximum errors. The mistakenness metric in FiftyOne gives a float value after comparing the ground truth and predicted labels, which tell the likelihood that ground truth labels drawn by the labeler are wrong. These few problematic samples can be relabelled again if the labeler has missed them. If they were already labeled correctly, then this key information provides an insight into the performance of the initial detector, which then can be improved. An implementation of the workflow can be visualized in Fig. 10 attached in appendix where an example is shown with a public dataset of blood cells [24] wherein 7 samples are relabeled which were sorted by the metric from the FiftyOne tool.

In the currently implemented relabelling workflow, there is no option for fixing the size of the drawn bounding boxes, but this is planned for the next release. The workflow is open sourced (https://github.com/MadanMl/Fiftyone-yolov5-tutorial) and can adapted for different single-stage models belonging to the YOLO family. In the future, we also plan to propose a fixed bounding box size according to the object size in the workflow with fully automated relabeling using reinforcement learning. This will result in a higher quality of annotations and faster convergence in terms of training the detector, as shown by our experiments in this work.

Acknowledgements. The contents of this publication are taken from the research project "(QAMeLiA) Quality Assurance of Machine Learning Applications", funded by the Ministry of Science, Research and the Arts of the State of Baden-Württemberg (MWK BW) under reference number 32–7547.223–6/12/4. The authors also wish to thank Mr. Rainer Zwing, Mr. Tobias Martin, Mr. Michel Kronenthaler, and Mr. Fritz Griese from C.R.S. iiMotion GmbH for their feedback, support, and guidance.

6   Appendix

Table 4. Comparison of some open source tools for labeling with the feature Model Assisted Labeling (MAL) compared for each tool.

Name	Features and data modality	Extensions
Labelbox	Web based, MAL = No (in free version), support for images, videos, text, documents, maps, audio, and HTML	FiftyOne, SDK
Label Studio	Need to be setup, MAL = Yes, support for audio, time series, images and text	FiftyOne, SDK
make-sense	Web based, MAL = Yes, support for images	No SDK or extn.
CVAT	web + local, MAL = Yes based solution, for images and videos	FiftyOne, SDK and CLI
OpenLabeling	Need to be setup, MAL = NO, support for images and videos	No SDK or extn.
PixelAnnotationTool	need to be setup, MAL = No support for images	No SDK or extn.
SuperAnnotate	Web based, MAL = No (in free version) support for images, video and text	SDK and CLI

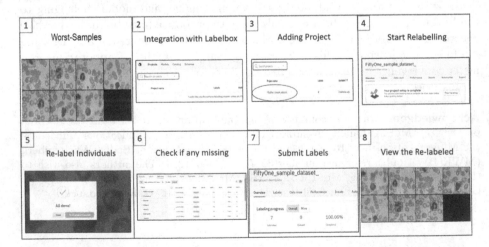

Fig. 10. A sequential portrayal of the workflow showcases how selected subsamples are pushed to the Labelbox tool from the python notebook which is then relabelled and available to be loaded back. The left corner displays the different stages marked which show the whole sequential process.

Table 5. Configuration for training and validation.

Parameters	Binary dataset	Chessboard dataset
epochs	100	300
initial lr	0.01	0.01
momentum	0.937	0.937
weight decay	0.0005	0.0005
giou	0.05	0.05
cls	0.5	0.5
cls pw	1	1
obj	1	1
obj pw	1	1
iou t	0.2	0.2
anchor t	4	4
fl gamma	0.0	0.0
hsv h	0.015	0.015
hsv s	0.7	0.7
hsv v	0.4	0.4
degrees	0.0	0.0
translate	0.5	0.5
scale	0.5	0.5
shear	0.0	0.0
perspective	0.0	0.0
flipud	0.0	0.0
fliplr	0.5	0.5
mixup	0.0	0.0
batch size	6	6
resolution	416*416	416*416

References

1. Xiongwei, W., Doyen, S., Steven, C.H.: Recent advances in deep learning for object detection Neurocomputing Elsevier **396**, 39–64 (2020)
2. Liu, L., et al.: Deep learning for generic object detection: a survey. Int. J. Comput. Vis. **128**, 261–318 (2020)
3. Jeffri, M.-L., Kirsten, L.N., Jung, C.R.: Can We Trust Bounding Box Annotations for Object Detection? In: Proceedings of the IEEE/CVF Conference on Computer Vision and Pattern Recognition (CVPR) Workshops, pp. 4813–4822 (2022)
4. Chuyi, L., et al.: YOLOv6: a single-stage object detection framework for industrial applications. In: (2022) arXiv preprint arXiv:2209.02976
5. Wang, C.-Y., Alexey, B., Mark, L.H.-Y.: In: Proceedings of the IEEE/CVF conference on computer vision and pattern recognition 2021, pp. 13029–13038 (2021)

6. Ruf, P., Madan, M., Reich, C., Ould-Abdeslam, D.: Demystifying MLOps and presenting a recipe for the selection of open-source tools. Appl. Sci. **11**, 8861 (2021)

7. Melde, A., et al.: Tackling key challenges of AI development-insights from an industry-academia collaboration. the upper-Rhine artificial intelligence symposium UR-AI 2022: AI Applications in Medicine And Manufacturing, 19 Oct 2022, Villingen-Schwenningen, Germany, pp. 112–121 (2022)

8. Li, X., et al.: A free lunch for unsupervised domain adaptive object detection without source data. Proc. OAAAI Conf. Artif. Intell. **35**, 8474–8481 (2021)

9. Perez-Rua, J., Zhu, X., Hospedales, T., Xiang, T.: Incremental few-shot object detection. In: Proceedings Of The IEEE/CVF Conference On Computer Vision And Pattern Recognition (CVPR), 6(2020)

10. Luo, R., Yu, Z.: Ai enhanced visual inspection of post-polished workpieces using you only look once vision system for intelligent robotics applications. In: 2022 International Conference on Advanced Robotics and Intelligent Systems (ARIS), pp. 1–6 (2022)

11. Adibhatla, V., Chih, H., Hsu, C., Cheng, J., Abbod, M., Shieh, J.: Applying deep learning to defect detection in printed circuit boards via a newest model of you-only-look-once. American Institute of Mathematical Sciences (AIMS) (2021)

12. Yi, J., Wu, P., Liu, B., Huang, Q., Qu, H., Metaxas, D.: Oriented object detection in aerial images with box boundary-aware vectors. In: Proceedings Of The IEEE/CVF Winter Conference On Applications Of Computer Vision, pp. 2150–2159 (2021)

13. Wang, H., Huang, Z., Chen, Z., Song, Y., Li, W.: Multi grained Angle Representation for Remote-Sensing Object Detection. IEEE Trans. Geosci. Remote Sens. **60**, 1–13 (2022)

14. Zhang, S., Chi, C., Yao, Y., Lei, Z., Li, S.: Bridging the gap between anchor-based and anchor-free detection via adaptive training sample selection. In: Proceedings Of The IEEE/CVF Conference on Computer Vision and Pattern Recognition, pp. 9759–9768 (2020)

15. Tian, Z., Shen, C., Chen, H., He, T.: Fcos: fully convolutional one-stage object detection. In: Proceedings Of The IEEE/CVF International Conference on Computer Vision, pp. 9627–9636 (2019)

16. Wang, G., Tian, B., Ai, Y., Xu, T., Chen, L., Cao, D.: Centernet3D: an anchor free object detector for autonomous driving (2020) ArXiv Preprint ArXiv:2007.07214

17. Ma, J., Ushiku, Y., Sagara, M.: The Effect of Improving Annotation Quality on Object Detection Datasets: a Preliminary Study. In: Proceedings Of The IEEE/CVF Conference on Computer Vision and Pattern Recognition, pp. 4850–4859 (2022)

18. Cheng, B., Wei, Y., Shi, H., Feris, R., Xiong, J., Huang, T.: Revisiting RCNN: On Awakening the Classification Power of Faster RCNN. In: Ferrari, V., Hebert, M., Sminchisescu, C., Weiss, Y. (eds.) ECCV 2018. LNCS, vol. 11219, pp. 473–490. Springer, Cham (2018). https://doi.org/10.1007/978-3-030-01267-0_28

19. Nassar, J., Pavon-Harr, V., Bosch, M.: McCulloh, I. Assessing data quality of annotations with Krippendorff alpha for applications in computer vision (2019) ArXiv PreprintArXiv:1912.10107

20. Vorontsov, E., Kadoury, S.: Label noise in segmentation networks: mitigation must deal with bias, pp. 251–258. Deep Generative Models, And Data Augmentation, Labelling, And Imperfections (2021)

21. Webb, G.I., Lee, L.K., Goethals, B., Petitjean, F.: Analyzing concept drift and shift from sample data. Data Min. Know. Discov. **32**(5), 1179–1199 (2018). https://doi.org/10.1007/s10618-018-0554-1

22. Lin, T., et al.: Microsoft COCO: common objects in context, CoRR, abs/1405.0312, (2014) http://arxiv.org/abs/1405.0312
23. Shikun, Z., Omid, J., Parth, N.: A survey on machine learning techniques for auto labeling of video, audio, and text data (2021) arXiv preprint arXiv:2109.03784
24. BCCD Dataset. https://public.roboflow.com/object-detection/bccd. Last 4 Oct 2022
25. OpenCV. https://github.com/opencv/opencv-python. Accessed 4 Oct 2022
26. ScaledYOLOv4 code. https://github.com/WongKinYiu/ScaledYOLOv4 Accessed 4 Oct 2022
27. Label box. https://labelbox.com/ Accessed 4 Oct 2022
28. voxel51 51. https://voxel51.com/docs/fiftyone/ Accessed 4 Oct 2022
29. Chess Pieces dataset. https://public.roboflow.ai/object-detection/chess-full Accessed 4 Oct 2022

Raw or Cooked? Object Detection on RAW Images

William Ljungbergh[1,2](✉) [ID], Joakim Johnander[1,2] [ID], Christoffer Petersson[2] [ID], and Michael Felsberg[1] [ID]

[1] Computer Vision Laboratory, Linköping University, 581 83 Linköping, Sweden
{william.ljungbergh,michael.felsberg}@liu.se
[2] Zenseact, Lindholmspiren 2, 417 56 Gothenburg, Sweden
{joakim.johnander,christoffer.petersson}@zenseact.com

Abstract. Images fed to a deep neural network have in general undergone several handcrafted image signal processing (ISP) operations, all of which have been optimized to produce visually pleasing images. In this work, we investigate the hypothesis that the intermediate representation of visually pleasing images is sub-optimal for downstream computer vision tasks compared to the RAW image representation. We suggest that the operations of the ISP instead should be optimized towards the end task, by learning the parameters of the operations jointly during training. We extend previous works on this topic and propose a new learnable operation that enables an object detector to achieve superior performance when compared to both previous works and traditional RGB images. In experiments on the open PASCALRAW dataset, we empirically confirm our hypothesis.

Keywords: Object Detection · Image Signal Processing · Machine Learning · Deep Learning

1 Introduction

Image sensors commonly collect RAW data in a one-channel Bayer pattern [2,22], *RAW images*, that are converted into three-channel RGB images via a camera Image Signal Processing (ISP) pipeline. This pipeline comprises a number of low-level vision functions – such as decompanding [18], demosaicing [16] (or *debayering* [22]), denoising, white balancing, and tone-mapping [31,40]. Each function is designed to tackle some particular phenomenon and the final pipeline is aimed at producing a visually pleasing image.

In recent years, image-based computer vision tasks have seen a leap in performance due to the advent of neural networks. Most computer vision tasks – such as image classification or object detection – are based on RGB image inputs. However, some recent works [33,49] have considered the possibility of removing the camera ISP and instead directly feeding the RAW image into the neural network. The intuition is that the high flexibility of the neural network should

R. Gade et al. (Eds.): SCIA 2023, LNCS 13885, pp. 374–385, 2023.
https://doi.org/10.1007/978-3-031-31435-3_25

Fig. 1. Three qualitative examples from the PASCALRAW dataset. We show the ground-truth (top), the RGB baseline detector (center), and the RAW RGGB detector with a learnable Yeo-Johnson operation (bottom). Compared to the RGB baseline, our proposed RAW RGGB detector manages to detect objects subject to poor light conditions.

enable it to approximate the camera ISP if that is the optimal way to transform the RAW data. It is important to note that the camera ISP is in general not optimized for the downstream task, and the neural network might by itself be able to learn a more suitable transformation of the RAW data during the training. One possibility is that the ISP might remove information that could be crucial in adverse conditions, such as low light. Moreover, the camera ISP adds image data according to image priors, which might result in spurious network responses [21].

In this work we investigate object detection on RAW data, following the hypothesis that RAW input images lead to superior detection performance, with the aim to identify the minimal set of operations on the RAW data that results in performance that exceeds the traditional RGB detectors. Our main contributions are the following:

1. We show that naïvely feeding RAW data into an object detector leads to poor performance.
2. We propose three simple yet effective strategies to mitigate the performance drop. The outputs of the best performing strategy – a learnable version of the Yeo-Johnson transformation – are visualized in Fig. 1.
3. We provide an empirical study on the publicly available PASCALRAW dataset.

2   Related Work

Object Detection: Object detection has been an active area of research for many years, and has been approached in many different ways. It is common to divide object detectors into two categories: (i) two-stage methods [11,24,37] that first generate proposals and then localize and classify objects each proposal; and (ii) one-stage detectors that either make use of a predefined set of anchors [25,35] or make a dense (anchor-free) [42,51] prediction across the entire image. Carion et al. [5] observed that both these categories of detectors rely on hand-crafted post-processing steps, such as non-maximum suppression, and proposed an end-to-end trainable object detector, DETR, that directly outputs a set of objects. One drawback of DETR is that convergence is slow and several follow-up works [27,29,41,43,48,52] have proposed schemes to alleviate this issue. All the work above shares one property: they rely on RGB image data.

RAW Image Data: RAW image data is traditionally fed through a *camera ISP* that produces an RGB image. Substantial research efforts have been devoted into the design of this ISP, usually with the aim to produce visually pleasing RGB images. A large number of works have studied the different sub-tasks, e.g., demosaicing [9,16,23,28], denoising [3,7,10], and tone mapping [20,34,36]. Several recent works propose to replace the camera ISP with deep neural networks [8,19,39,50]. More precisely, these works aim to find a mapping between RAW images and high-quality RGB images produced by a digital single-lens reflex camera (DSLR).

Object Detection Using RAW Image Data: In this work, we aim to train an object detector that takes RAW images as input. We are not the first to explore this direction. Buckler et al. [4] found that for processing RAW data, only demosaicing and gamma correction are crucial operations. In contrast to their work, we find that also these two can be avoided. Yoshimura et al. [46], Yoshimura et al. [47], and Morawski et al. [30] strive to construct a learnable ISP that, together with an object detector, is trained for the object detection task. Based on our experiments, we argue that also the learnable ISP can be replaced with very simple operations. Most closely related to our work is the work of Hong et al. [17], which proposes to only demosaic RAW images before feeding them into an object detector. In contrast to their work, we do not find the need for an auxiliary image construction loss nor for demosaicing.

3   Method

In this section, we first introduce a strategy for downsampling RAW Bayer images (Sect. 3.1). This enables us to downsample high-resolution images to be more suitable for standard computer vision pipelines while maintaining the Bayer pattern in the RAW image. In Sect. 3.2, we introduce the three *learnable* operations.

Fig. 2. Downsampling method for Bayer-pattern RAW data. Each of the colors in the filter array of the downsampled RAW image (right) is the average over all cells in the corresponding region in the original image with the same color (left and center). The figure illustrates the downsampling of an original image patch of size $2d \times 2d$ (with $d = 5$ in this example), down to a patch of size 2×2, i.e. with a downsampling factor d in each dimension.

3.1 Downsampling RAW Images

When working with high-resolution images, it is sometimes necessary to downsample the images to make them compatible with existing computer vision pipelines. However, standard downsampling schemes, such as bilinear or nearest neighbor, do not preserve the Bayer pattern that was present in the original image. To remedy this, we adopt a simple Bayer-pattern-preserving downsampling method, shown in Fig. 2. Given an original RAW image $\mathbf{x}^{\mathrm{orig}} \in \mathbb{R}^{H \times W}$ and an uneven downsampling factor $d \in 2\mathbb{N} + 1$, we divide our original image into patches $x^{\mathrm{orig}} \in \mathbb{R}^{2d \times 2d}$ with a stride $s = 2d$. Each patch is then downsampled by a factor d in each dimension, yielding a downsampled patch $x \in \mathbb{R}^{2 \times 2}$, by averaging over the elements with the correct color in that sub-array. To clarify, all elements that correspond to a red filter in the upper left sub-array of the patch x^{orig} are averaged to produce the red output element $x_{0,0}$. The downsampling operation over the entire patch x^{orig} can be described as

$$x_{i,j} = \frac{1}{N} \sum_{m=0}^{(d-1)/2} \sum_{n=0}^{(d-1)/2} x^{\mathrm{orig}}_{di+2m,dj+2n} , \tag{1}$$

where $x \in \mathbb{R}^{2 \times 2}$ is the downsampled patch, $x^{\mathrm{orig}} \in \mathbb{R}^{2d \times 2d}$ is the original patch, d is the downsampling factor, $N = (d+1)^2/4$ is the number of elements averaged over, and $i, j \in 0, 1$. All downsampled patches are then concatenated to form the downsampled RAW image $\mathbf{x} \in \mathbb{R}^{H/d \times W/d}$.

It would be possible to feed the downsampled RAW image, $\mathbf{x}$, directly into an object detector. There is however one thing to note about the first layer of the image encoder. In the standard RGB image setting, each weight in this layer is only applied to one modality – red, green, or blue. This enables the first layer to capture color-specific information, such as gradients from one color to another. When fed with RAW images, as described above, we can assert the same property by ensuring that the stride of the first layer is an even number. Luckily, this is the case with the standard ResNet [14] architecture.

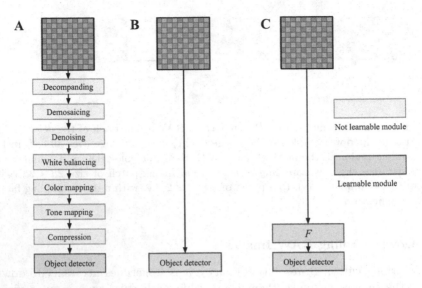

Fig. 3. Traditional (A), naïve (B), and proposed (C) detection pipelines. The traditional pipeline uses a set of common image signal processing operations, such as *Demosaicing*, *Denoising*, and *Tonemapping*, and then feeds the object detector with the processed RGB images. The naïve pipeline feeds the RAW image directly into the detector while our proposed pipeline first feeds the RAW image through a *learnable* non-linear operation, F, which can be viewed as being part of the end-to-end trainable object detection network.

3.2   Learnable ISP Operations

A standard ISP pipeline usually consists of a large collection of handcrafted operations. These operations are in general parameterized and optimized to produce visually pleasing images for the human eye. Although these pipelines can produce satisfying results with respect to their objective, there is no guarantee that this – visually pleasing – representation is optimal for computer vision. In fact, there are results indicating that only a handful of operations in classical ISP pipelines actually increase the performance of downstream computer vision systems [4,32].

Many of these handcrafted operations can be defined as learnable operations in a neural network and subsequently be optimized towards other objectives than producing visually pleasing images. Inspired by this we investigate a set of *learnable* operations that are applied to the RAW image input and optimized end-to-end with respect to the downstream computer vision tasks. Inspired by the works in [1,4,32,45], we define *Learnable Gamma Correction*, *Learnable Error Function*, and *Learnable Yeo-Johnson*, which are described in detail below.

Learnable Gamma Correction: Prior work [4,32] has shown that the most essential operations in standard ISP pipelines are demosaicing and tonemapping. In both works, they make use of a bilinear demosaicing algorithm together with a gamma correction method. We also implement a *learnable* gamma correction defined as

$$F_\gamma(\mathbf{x}) = \mathbf{x}_d^\gamma \;, \tag{2}$$

where $\gamma \in \mathbb{R}$ is the learnable parameter that is trained jointly with the downstream network, and $\mathbf{x}_d$ is the input image $\mathbf{x}$ after bilinear demosaicing. Conveniently, we can model the demosaicing operation as a 2D convolution over the entire image. By using two 3×3 kernels,

$$K_g = \begin{bmatrix} 0.0 & 0.25 & 0.0 \\ 0.25 & 1.0 & 0.25 \\ 0.0 & 0.25 & 0.0 \end{bmatrix}, \quad K_{rb} = \begin{bmatrix} 0.25 & 0.5 & 0.25 \\ 0.5 & 1.0 & 0.5 \\ 0.25 & 0.5 & 0.25 \end{bmatrix}, \tag{3}$$

we can effectively achieve bilinear demosaicing by convolving the filters over their respective masked input. To further clarify, we convolve K_g over the RAW Bayer image, where all cells that do not have the green filter are set to zero. Similarly, we convolve R_{rb} over the RAW Bayer image where we only keep the red and blue cells, respectively, thus obtaining a 3-channel bilinearly interpolated RAW image.

Learnable Error Function: An even simpler approach is to feed the RAW input data through a single non-linear function. To this end, we adopt the Gauss error function. This function has been used in prior works to model disease cases [6], as an activation function in neural networks [15], and for diffusion-based image enhancement [1]. Formally, we define

$$F_{\mathrm{erf}}(\mathbf{x}) = \mathrm{erf}\left(\frac{\mathbf{x} - \mu}{\sqrt{2}\sigma}\right) \;, \tag{4}$$

where $\mu \in \mathbb{R}$ and $\sigma \in \mathbb{R}_+$ are learnable parameters optimized jointly with the encoder and detector head parameters during training. Note that the erf function saturates quickly and we found it necessary to normalize the data to be in the range of 0 to 1.

Learnable Yeo-Johnson Transformation: A common preprocessing step in deep learning pipelines is to normalize the input data, as it has shown to improve the performance and stability of deep neural networks [12,13]. In object detection pipelines, this is commonly achieved by normalizing with the mean and variance of each RGB input channel across the entire dataset. While the same approach can easily be adopted to each of the colors in the Bayer pattern, this naïve approach does not yield satisfactory results. One thing to note is that work on weight initialization [12,13] typically assume the input to have a standard normal distribution. We observed that the RGGB data distribution was highly non-Gaussian, motivating us to find a transformation that improves the normality of the data.

Yeo and Johnson proposed a new family of power transformations that aims to improve the symmetry and normality of the transformed data [45]. These transformations are parameterized by λ, which is usually optimized offline by maximizing the log-likelihood between the input data and a Gaussian distribution. However, analogously to the ISP operations that should be optimized towards the end task, we can optimize the Yeo-Johnson transformation with respect to the end goal, rather than towards a Gaussian distribution. Inspired by this, we define the *Learnable Yeo-Johnson* transformation as a point-wise non-linear operation

$$F_{\mathrm{YJ}}(\mathbf{x}) = \frac{(\mathbf{x}+1)^{\lambda} - 1}{\lambda} \ , \tag{5}$$

where $\lambda \in \mathbb{R}_+$ is the learnable parameter.

3.3   Our Raw Object Detector

Given RAW RGGB images, we downsample as described in Sect. 3.1 to obtain $\mathbf{x}$. Then, we apply one of the learnable ISP operations, F, as described in (2), (4), or (5). Finally, we apply the object detector, D,

$$\mathcal{O} = D(F(\mathbf{x})) \ , \tag{6}$$

giving us a set of predicted objects $\mathcal{O}$. We train F and D jointly.

4   Experiments

In this section, we introduce the dataset on which we evaluate the different methods (Sect. 4.1), along with some of the prominent implementation details (Sect. 4.2) used during training and evaluation. Next, we present the results, both quantitative (Sect. 4.3) and qualitative (Sect. 4.4) for all the learnable operations proposed in Sect. 3.2. Lastly, we present how the learnable parameters in each of the proposed operations evolve during training in Sect. 4.5.

4.1   Dataset

To evaluate our learnable operations, we make use of the PASCALRAW dataset [33]. This dataset contains 4259 high-resolution (6034 × 4012) RAW 12bit RGGB images, all captured with a Nikon D3200 DSLR camera during daylight conditions in Palo Alto and San Francisco. We downsample all RAW images to a resolution more compatible with standard object detection pipelines (1206 × 802) according to the Bayer-pattern-preserving downsampling described in Sect. 3.1. Note that we crop away the last four rows and two columns (0.1% of the image) to obtain an integer downsampling factor. Subsequently, we generate the corresponding RGB images (used by the RGB Baseline) from the downsampled RAW images using a standard ISP pipeline implemented in the RAW image processing library RawPy [38]. For each image, the authors provide dense annotations in the form of class-bounding-box-pairs for three different classes: pedestrian, car, and bicycle. In total, the dataset contains 6550 annotated instances, divided into 4077 pedestrians, 1765 cars, and 708 bicycles.

Table 1. Object detection results on the PASCALRAW dataset. The results are presented in terms of AP (higher is better) and we report the mean and standard deviation over 3 separate runs.

Components	AP	AP_{50}	AP_{75}	AP_{car}	AP_{ped}	AP_{bic}
RGB Baseline	50.5 ± 0.5	84.8 ± 0.3	55.2 ± 1.6	61.8 ± 0.1	48.5 ± 0.7	41.4 ± 0.8
RAW RGGB Baseline	31.3 ± 1.2	64.7 ± 1.6	25.2 ± 2.0	42.4 ± 1.8	30.5 ± 0.5	20.9 ± 1.5
RAW + Learnable Gamma	51.4 ± 0.3	85.8 ± 0.6	56.3 ± 0.7	62.5 ± 0.4	49.0 ± 0.2	42.7 ± 1.1
RAW + Learnable Error Function	49.3 ± 0.2	84.0 ± 0.4	52.8 ± 0.5	60.1 ± 0.6	46.3 ± 0.5	41.3 ± 0.8
RAW + Learnable Yeo-Johnson	$\mathbf{52.6 \pm 0.4}$	$\mathbf{86.7 \pm 0.3}$	$\mathbf{57.9 \pm 0.6}$	$\mathbf{63.6 \pm 0.5}$	$\mathbf{49.9 \pm 0.4}$	$\mathbf{44.2 \pm 0.6}$

4.2 Implementation Details

We use a standard object detection pipeline, namely a Faster-RCNN [37], with a Feature Pyramid Network [24], and a ResNet-50 [14] backbone. All models were implemented, trained, and evaluated in the Detectron2 framework [44]. We use a batch size of $B = 16$, a learning rate of $l_r = 3 \cdot 10^{-4}$, a learning-rate scheduler with 5000 warm-up iterations, and a learning-rate drop by a factor $\alpha = 0.1$ after 100k iterations. We train for 150k iterations using an SGD optimizer. The learnable parameters in the ISP pipeline, λ, γ, μ, and σ, were initialized (when used) to 0.35, 1.0, 1.0, and 1.0 respectively.

4.3 Quantitative Results

In Table 1 we present the results when training and evaluating our different learnable functions on the PASCALRAW dataset. The results are presented in terms of *mean average precision* (AP), following the COCO detection benchmark [26]. We also provide average precision for different IoU-thresholds (AP_{50} and AP_{75}) and AP for each class. We report the mean and standard deviation over three separate runs.

From the results in Table 1, we can conclude that simply feeding the RAW RGGB image (i.e., removing all ISP operations) into a standard object detection network, corresponding to the RAW RGGB Baseline in Fig. 3(B), performs substantially worse than the traditional RGB Baseline in Fig. 3(A). Further, we can corroborate the results of [4,32] and observe that the method RAW + *Learnable Gamma*, which comprises the two operations *demosaicing* and *gamma correction*, by a slight margin surpasses the performance of the RGB Baseline. Lastly, we also observe that our method RAW +*Learnable Yeo-Johnson* in Fig. 3(C) outperforms all other methods by a statistically significant margin.

4.4 Qualitative Results

From Table 1 it is evident that our *Learnable Yeo-Johnson* operation outperforms the RGB baseline. We hypothesize that this is partly because our learnable ISP can better handle poor (low) light conditions. In Fig. 1, we present three examples from the PASCALRAW test set that further support this hypothesis. Our RAW image pipeline can more accurately detect objects in the darker parts of the images, whereas the RGB Baseline fails in the same situations.

4.5 Parameter Evolution

To further analyze the behavior of our *Learnable Yeo-Johnson* operation, we show the evolution of its trainable parameter, λ, along with the functional form of the operation, in Fig. 4. We observe that the training converges to a relatively low value of λ, which, as can be seen from the functional form of the operation, implies that low-valued/dark pixels are better differentiated than high-valued/bright pixels. This characteristic suggests that the RAW object detector is able to better distinguish features in low-light regions of the image, compared to the RGB detector, thus achieving better detection performance.

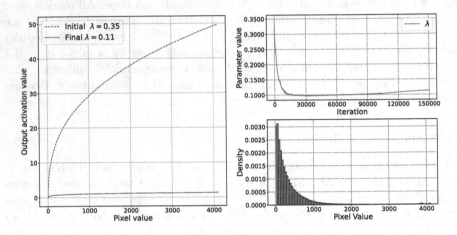

Fig. 4. Evolution of the learnable parameter λ during the entire training (top-right), the distribution of the RAW pixel values in PASCAL RAW (bottom-right), and the functional form – before and after training – of the *Learnable Yeo-Johnson* operation (left). In the left plot, the output activation values are shown across the full input range $[0, 2^{12} - 1]$.

5   Conclusion

Motivated by the observation that camera ISP pipelines are typically optimized towards producing visually pleasing images for the human eye, we have in this work experimented with object detection on RAW images. While naïvely feeding RAW images directly into the object detection backbone led to poor performance, we proposed three simple, learnable operations that all led to good performance. Two of these operators, the *Learnable Gamma* and *Learnable Yeo-Johnson*, led to superior performance compared to the RGB baseline detector. Based on qualitative comparison, the RAW detector performs better in low-light conditions compared to the RGB detector.

Acknowledgements. This work was partially supported by the Wallenberg AI, Autonomous Systems and Software Program (WASP) funded by the Knut and Alice Wallenberg Foundation.

References

1. Åström, F., Zografos, V., Felsberg, M.: Density driven diffusion. In: Kämäräinen, J.-K., Koskela, M. (eds.) SCIA 2013. LNCS, vol. 7944, pp. 718–730. Springer, Heidelberg (2013). https://doi.org/10.1007/978-3-642-38886-6_67
2. Bayer, B.E.: Color imaging array. United States Patent 3,971,065 (1976)
3. Buades, A., Coll, B., Morel, J.M.: A non-local algorithm for image denoising. In: 2005 IEEE Computer Society Conference on Computer Vision and Pattern Recognition (CVPR 2005), vol. 2, pp. 60–65. IEEE (2005)
4. Buckler, M., Jayasuriya, S., Sampson, A.: Reconfiguring the imaging pipeline for computer vision. In: Proceedings of the IEEE International Conference on Computer Vision, pp. 975–984 (2017)
5. Carion, N., Massa, F., Synnaeve, G., Usunier, N., Kirillov, A., Zagoruyko, S.: End-to-end object detection with transformers. In: Vedaldi, A., Bischof, H., Brox, T., Frahm, J.-M. (eds.) ECCV 2020. LNCS, vol. 12346, pp. 213–229. Springer, Cham (2020). https://doi.org/10.1007/978-3-030-58452-8_13
6. Ciufolini, I., Paolozzi, A.: Mathematical prediction of the time evolution of the COVID-19 pandemic in Italy by a gauss error function and monte Carlo simulations. Eur. Phys. J. Plus **135**(4), 355 (2020)
7. Condat, L.: A simple, fast and efficient approach to denoisaicking: Joint demosaicking and denoising. In: 2010 IEEE International Conference on Image Processing, pp. 905–908. IEEE (2010)
8. Dai, L., Liu, X., Li, C., Chen, J.: AWNet: attentive wavelet network for image ISP. In: Bartoli, A., Fusiello, A. (eds.) ECCV 2020. LNCS, vol. 12537, pp. 185–201. Springer, Cham (2020). https://doi.org/10.1007/978-3-030-67070-2_11
9. Dubois, E.: Filter design for adaptive frequency-domain Bayer demosaicking. In: 2006 International Conference on Image Processing, pp. 2705–2708. IEEE (2006)
10. Foi, A., Trimeche, M., Katkovnik, V., Egiazarian, K.: Practical poissonian-gaussian noise modeling and fitting for single-image raw-data. IEEE Trans. Image Process. **17**(10), 1737–1754 (2008)
11. Girshick, R., Donahue, J., Darrell, T., Malik, J.: Rich feature hierarchies for accurate object detection and semantic segmentation. In: Proceedings of the IEEE Conference on Computer Vision and Pattern Recognition, pp. 580–587 (2014)
12. Glorot, X., Bengio, Y.: Understanding the difficulty of training deep feedforward neural networks. In: Proceedings of the Thirteenth International Conference on Artificial Intelligence and Statistics, pp. 249–256. JMLR Workshop and Conference Proceedings (2010)
13. He, K., Zhang, X., Ren, S., Sun, J.: Delving deep into rectifiers: Surpassing human-level performance on imagenet classification. In: Proceedings of the IEEE International Conference on Computer Vision, pp. 1026–1034 (2015)
14. He, K., Zhang, X., Ren, S., Sun, J.: Deep residual learning for image recognition. In: Proceedings of the IEEE Conference on Computer Vision and Pattern Recognition, pp. 770–778 (2016)
15. Hendrycks, D., Gimpel, K.: Gaussian error linear units (gelus). arXiv preprint arXiv:1606.08415 (2016)

16. Hirakawa, K., Parks, T.W.: Adaptive homogeneity-directed demosaicing algorithm. IEEE Trans. Image Process. **14**(3), 360–369 (2005)

17. Hong, Y., Wei, K., Chen, L., Fu, Y.: Crafting object detection in very low light. In: BMVC, vol. 1, p. 3 (2021)

18. HP, A.W., Prasetyo, H., Guo, J.M.: Autoencoder-based image companding. In: 2020 IEEE International Conference on Consumer Electronics-Taiwan (ICCE-Taiwan), pp. 1–2. IEEE (2020)

19. Ignatov, A., Van Gool, L., Timofte, R.: Replacing mobile camera ISP with a single deep learning model. In: Proceedings of the IEEE/CVF Conference on Computer Vision and Pattern Recognition Workshops, pp. 536–537 (2020)

20. Krawczyk, G., Myszkowski, K., Seidel, H.P.: Lightness perception in tone reproduction for high dynamic range images. In: Computer Graphics Forum, vol. 24, pp. 635–646. Amsterdam: North Holland, 1982- (2005)

21. Kriesel, D.: Traue keinem scan, den du nicht selbst gefälscht hast. Mitteilungen der Deutschen Mathematiker-Vereinigung **22**(1), 30–34 (2014)

22. Langseth, R., Gaddam, V.R., Stensland, H.K., Griwodz, C., Halvorsen, P.: An evaluation of debayering algorithms on GPU for real-time panoramic video recording. In: 2014 IEEE International Symposium on Multimedia, pp. 110–115. IEEE (2014)

23. Li, X., Gunturk, B., Zhang, L.: Image demosaicing: a systematic survey. In: Visual Communications and Image Processing 2008, vol. 6822, pp. 489–503. SPIE (2008)

24. Lin, T.Y., Dollár, P., Girshick, R., He, K., Hariharan, B., Belongie, S.: Feature pyramid networks for object detection. In: Proceedings of the IEEE Conference on Computer Vision and Pattern Recognition, pp. 2117–2125 (2017)

25. Lin, T.Y., Goyal, P., Girshick, R., He, K., Dollár, P.: Focal loss for dense object detection. In: Proceedings of the IEEE International Conference on Computer Vision, pp. 2980–2988 (2017)

26. Lin, T.-Y., et al.: Microsoft COCO: common objects in context. In: Fleet, D., Pajdla, T., Schiele, B., Tuytelaars, T. (eds.) ECCV 2014. LNCS, vol. 8693, pp. 740–755. Springer, Cham (2014). https://doi.org/10.1007/978-3-319-10602-1_48

27. Liu, Z., et al.: SWIN transformer: Hierarchical vision transformer using shifted windows. In: Proceedings of the IEEE/CVF International Conference on Computer Vision, pp. 10012–10022 (2021)

28. Malvar, H.S., He, L.W., Cutler, R.: High-quality linear interpolation for demosaicing of bayer-patterned color images. In: 2004 IEEE International Conference on Acoustics, Speech, and Signal Processing, vol. 3, pp. iii–485. IEEE (2004)

29. Meng, D., et al.: Conditional DETR for fast training convergence. In: Proceedings of the IEEE/CVF International Conference on Computer Vision, pp. 3651–3660 (2021)

30. Morawski, I., Chen, Y.A., Lin, Y.S., Dangi, S., He, K., Hsu, W.H.: GENISP: neural ISP for low-light machine cognition. In: Proceedings of the IEEE/CVF Conference on Computer Vision and Pattern Recognition, pp. 630–639 (2022)

31. Mujtaba, N., Khan, I.R., Khan, N.A., Altaf, M.A.B.: Efficient flicker-free tone mapping of HDR videos. In: 2022 IEEE 24th International Workshop on Multimedia Signal Processing (MMSP), pp. 01–06. IEEE (2022)

32. Olli Blom, M., Johansen, T.: End-to-end object detection on raw camera data (2021)

33. Omid-Zohoor, A., Ta, D., Murmann, B.: Pascalraw: raw image database for object detection (2014)

34. Poynton, C.: Digital video and HD: Algorithms and Interfaces. Elsevier (2012)

35. Redmon, J., Farhadi, A.: Yolov3: an incremental improvement. arXiv preprint arXiv:1804.02767 (2018)
36. Reinhard, E., Stark, M., Shirley, P., Ferwerda, J.: Photographic tone reproduction for digital images. In: Proceedings of the 29th Annual Conference on Computer Graphics and Interactive Techniques, pp. 267–276 (2002)
37. Ren, S., He, K., Girshick, R., Sun, J.: Faster R-CNN: towards real-time object detection with region proposal networks. In: Advances in Neural Information Processing Systems, vol. 28 (2015)
38. Riechert, M.: Rawpy (2022). https://github.com/letmaik/rawpy
39. Shekhar Tripathi, A., Danelljan, M., Shukla, S., Timofte, R., Van Gool, L.: Transform your smartphone into a DSLR camera: Learning the ISP in the wild. In: Avidan, S., Brostow, G., Cissé, M., Farinella, G.M., Hassner, T. (eds) Computer Vision. ECCV 2022. ECCV 2022. LNCS, pp. 625–641. Springer, Cham (2022). https://doi.org/10.1007/978-3-031-20068-7_36
40. Suma, R., Stavropoulou, G., Stathopoulou, E.K., Van Gool, L., Georgopoulos, A., Chalmers, A.: Evaluation of the effectiveness of HDR tone-mapping operators for photogrammetric applications. Virtual Archaeol. Rev. **7**(15), 54–66 (2016)
41. Sun, Z., Cao, S., Yang, Y., Kitani, K.M.: Rethinking transformer-based set prediction for object detection. In: Proceedings of the IEEE/CVF International Conference on Computer Vision, pp. 3611–3620 (2021)
42. Tian, Z., Shen, C., Chen, H., He, T.: FCOS: fully convolutional one-stage object detection. In: Proceedings of the IEEE/CVF International Conference on Computer Vision, pp. 9627–9636 (2019)
43. Wang, Y., Zhang, X., Yang, T., Sun, J.: Anchor DETR: query design for transformer-based detector. In: Proceedings of the AAAI Conference on Artificial Intelligence, vol. 36, pp. 2567–2575 (2022)
44. Wu, Y., Kirillov, A., Massa, F., Lo, W.Y., Girshick, R.: Detectron2 (2019). https://github.com/facebookresearch/detectron2
45. Yeo, I.K., Johnson, R.A.: A new family of power transformations to improve normality or symmetry. Biometrika **87**(4), 954–959 (2000)
46. Yoshimura, M., Otsuka, J., Irie, A., Ohashi, T.: Dynamicisp: dynamically controlled image signal processor for image recognition. arXiv preprint arXiv:2211.01146 (2022)
47. Yoshimura, M., Otsuka, J., Irie, A., Ohashi, T.: Rawgment: noise-accounted raw augmentation enables recognition in a wide variety of environments. arXiv preprint arXiv:2210.16046 (2022)
48. Zhang, H., et al.: Dino: DETR with improved denoising anchor boxes for end-to-end object detection. arXiv preprint arXiv:2203.03605 (2022)
49. Zhang, X., Zhang, L., Lou, X.: A raw image-based end-to-end object detection accelerator using hog features. IEEE Trans. Circuits Syst. I: Regular Papers **69**(1), 322–333 (2021)
50. Zhang, Z., Wang, H., Liu, M., Wang, R., Zhang, J., Zuo, W.: Learning raw-to-srgb mappings with inaccurately aligned supervision. In: Proceedings of the IEEE/CVF International Conference on Computer Vision, pp. 4348–4358 (2021)
51. Zhou, X., Wang, D., Krähenbühl, P.: Objects as points. arXiv preprint arXiv:1904.07850 (2019)
52. Zhu, X., Su, W., Lu, L., Li, B., Wang, X., Dai, J.: Deformable DETR: deformable transformers for end-to-end object detection. arXiv preprint arXiv:2010.04159 (2020)

Local Neighborhood Features for 3D Classification

Shivanand Venkanna Sheshappanavar$^{(\boxtimes)}$ [ID] and Chandra Kambhamettu [ID]

Video/Image Modeling and Synthesis (VIMS) Lab, University of Delaware,
212 Smith Hall, 18 Amstel Ave, Newark, DE 19711, USA
{ssheshap,chandrak}@udel.edu
http://bigdatavision.org/

Abstract. With advances in deep learning training strategies, the training of Point cloud classification methods is significantly improving. For example, Point-NeXt, which adopts prominent training techniques and InvResNet layers into PointNet++, achieves over 7% improvement on the real-world ScanObjectNN dataset. A typical set abstraction layer in these models maps point coordinate features of neighborhood points to higher dimensional space and passes point coordinates to the next set abstraction layer. However, these models ignore already computed neighborhood point features as additional neighborhood features. In this paper, we revisit the PointNeXt model to study the usage and benefit of neighborhood point distance and directional vectors as additional neighborhood features. We train and evaluate PointNeXt on ModelNet40 (synthetic), ScanObjectNN (real-world), and a recent large-scale, real-world grocery dataset, i.e., 3DGrocery100. In addition, we provide an additional inference strategy of weight averaging the top two checkpoints of PointNeXt to improve classification accuracy. Together with the above-mentioned ideas, we gain **0.5%**, **1%**, **4.8%**, **3.4%**, **1.6%**, and **2.8%** overall accuracy on the PointNeXt model with real-world datasets, ScanObjectNN (hardest variant), 3DGrocery100's Apple10, Fruits, Vegetables, Packages, and Full subsets, respectively. We also achieve a comparable **0.2%** accuracy gain on ModelNet40. Finally, we provide detailed ablation studies discussing the trade-offs of using additional neighborhood features. Code is available at https://github.com/VimsLab/Local3DFeatures.

Keywords: Point Cloud · 3D Object Classification · Local Features

1 Introduction

3D Computer Vision is a vibrant research domain with broad applications in augmented/virtual reality and autonomous-driving vehicles. Over the past five years, the application of deep neural networks to 3D computer vision, especially for 3D point cloud processing, has progressed tremendously. This progress has led to state-of-the-art results on several computer vision tasks, such as 3D Object Classification, 3D Semantic Segmentation, 3D Scene Understanding, and 3D Shape retrieval.

Point cloud Object Classification has gained traction since the pioneering work of PointNet [1] that processed raw point sets through Multi-Layer Perceptrons (MLPs). However, PointNet, while aggregating features at the global level using max-pooling

© The Author(s), under exclusive license to Springer Nature Switzerland AG 2023
R. Gade et al. (Eds.): SCIA 2023, LNCS 13885, pp. 386–395, 2023.
https://doi.org/10.1007/978-3-031-31435-3_26

operation, lost valuable local geometric information. Overcoming this limitation, Point-Net++ [2] employed ball querying and/or k-Nearest Neighbor (k-NN) querying to query local neighborhoods to extract local semantic information.

Although PointNet++ [2] was effective in capturing local geometric information, it still lost contextual information due to the max-pooling operation. Several methods [2–4,8–10,13] since PointNet++ have used raw x, y, z point coordinates as input. First, sample farthest points (FPS) from the input points, and at each FPS point, query neighborhood points using a fixed radius ball, i.e., compute the distance to each point from each of the FPS/anchor points to check if the point is within the neighborhood of the FPS point. Secondly, group a small sample of neighborhood points at each FPS point based on this distance. From the queried neighborhood points, subtracting the anchor FPS point gives directional vectors of the neighborhood. Finally, these directional vectors are mapped to higher dimensions in each set abstraction layer (SA) to get local features. After this step, the directional vectors and the computed distance (in the case of ball querying) are ignored.

The most recent model PointNeXt [20], builds upon the PointNet++ model, which uses ball querying for neighborhood querying by computing the distance from the anchor point to the neighborhood point to check if the point is within a ball of radius r. After querying, PointNeXt computes the directional vectors from the grouped neighborhood points, normalizes the vectors using the radius, and encodes them into local neighborhood features but does not use the distance and the directional vectors as additional local features. In this paper, we emphasize using the radius-normalized distance and the directional vectors as additional local neighborhood features with minimal additional memory or computational costs. The contributions of our work are four-fold:

- We use radius r-normalized neighborhood point distance as an additional neighborhood feature to improve the classification accuracy.
- We show radius r-normalized directional vectors as additional neighborhood features benefit several models such as PointNeXt [20].
- We demonstrate that averaging the weights of two best model checkpoints (models saved in the same training session) benefits test/inference accuracy.
- We train and evaluate PointNeXt with our combined approaches on one synthetic, i.e., ModelNet40 [15], and two real-world datasets, i.e., ScanObjectNN [16] and 3DGrocery100.

2 Related Works

PointNet++ [2] is composed of Set Abstraction (SA) Layers, and each SA constitutes three layers. A Farthest Point Sampling (FPS) layer to sample a uniformly distributed subset of points in the input point cloud. A grouping layer partitions the input point cloud into common hierarchical structures. The third layer, a pointnet layer, learns contextual representation from the grouped points.

Using Kernel Correlation, KCNet [5] mines local points and measures point affinity with similar geometric structures. Different convolution operations [6] capture different levels of geometric information from the surface deformations. Using the Edge-Conv technique, Dynamic Graph Convolutional Neural Network (DGCNN) [19] captures local geometric features (kNN is used for neighborhood querying). PointGrid [7] proposed an integrated point and grid hybrid 3D CNN model for a better representation of the local geometry. DensePoint [9] recursively concatenates features from MLPs to learn sufficient contextual semantic information.

Relation Shape Convolutional Neural Network (RS-CNN) [3] proposed a relation-shape convolution to explicitly encode the geometric relation of points for better aware-ness of the underlying shape. Although RS-CNN uses the Euclidean distance as one of the relations, it does not normalize the distance with the ball radius. The critical differ-ence between our approach and RS-CNN is that our approach normalizes the distance (between the neighborhood point and its anchor point) with the radius r.

Although novel neighborhood querying methods [11, 12] have demonstrated the use of Eigenvalues of the neighborhood as additional features, they devise a two-pass query-ing of the neighborhood, first with a ball and then using an oriented (and scaled) ellip-soid resulting in extra training and inference time. While PointNeXt only normalizes directional vectors, it does not include the vectors as additional features after mapping them to higher dimensions. Our approach uses these normalized directional vectors as neighborhood features along with the radius r-normalized distance.

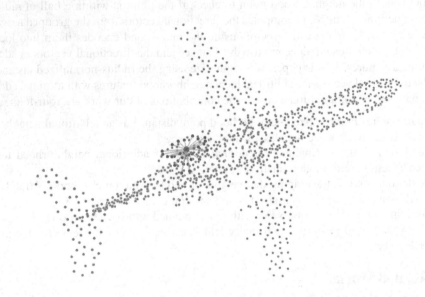

Fig. 1. Airplane: Green points - input point cloud, Red point - anchor point for a neighborhood obtained from the farthest point sampling step, Blue points - points queried from the neighbor-hood of anchor point, Red Vectors - Directional Vectors from the anchor point to the neighbor points. The magnitude of each directional vector represents the distance of the neighbor point to its anchor point. The direction of these vectors is from the anchor point to the neighbor point. (Color figure online)

3 Method

In this section, we elaborate on the usage of distances and directional vectors from the anchor point to neighborhood points already computed in the grouping stage of classification models from PointNet++ [2] to the most recent PointNeXt [20] method.

3.1 Neighborhood Point Distance

With the inception of PointNet++ [2], which introduced neighborhood querying using a fixed radius ball (as shown in Fig. 1) or k-nearest neighbor(kNN), several methods have continued to take advantage of these querying techniques. In the case of ball-queried neighborhood points, the distance between the anchor point and a neighborhood point is computed to determine if the given point is within the ball of radius r. i.e., a point (x_i, y_i, z_i) is selected from the input point set xyz_1, if it lies within the ball of radius r centered at (x_j, y_j, z_j) (anchor points are obtained as xyz_2 using the farthest point sampling method on the input xyz_1 set). Euclidean distance d_i of the point from the anchor point/center of the ball is calculated, and the distance is normalized with the radius r as shown in Eq. 1. When d_i is less than 1, the point is within the ball; if the distance is equal to 1, then the point is on the ball. And if the distance is greater than 1, the point is outside the ball. A point (x_i, y_i, z_i) with distance d_i less than 1 is selected. We save this distance and use it as an additional neighborhood feature.

$$d_i = \frac{\sqrt{(x_i - x_j)^2 + (y_i - y_j)^2 + (z_i - z_j)^2}}{r} \qquad (1)$$

3.2 Directional Vectors as Neighborhood Features

Fig. 2. Left: Input point cloud (green points), Farthest Point Sampled points (red points), Right: Directional vectors at each of the red points. (Color figure online)

After querying, the neighborhood points are centered at the anchor point by subtracting the anchor point from the neighborhood points. By doing so, the directional vectors (dv_j) at the grouping level are computed in models that adopted the grouping layer of PointNet++ [2] (check Fig 2b). PointNeXt [20] normalizes directional vectors using the radius r as shown in Eq. 2. These radius normalized-directional vectors are mapped to relatively higher dimensions but are not directly included as features while hierarchically learning the discriminative features. Instead, we include these directional vectors as neighborhood-level features by concatenating these vectors to the relatively-higher dimensional features in a given neighborhood. Figure 3 shows visualization of directional vectors.

$$dv_j = \frac{[(x_i - x_j), (y_i - y_j), (z_i - z_j)]}{r} \tag{2}$$

Fig. 3. Few more visual examples of 3D objects with directional vectors after the grouping stage. From Left to Right: bottle, lamp, airplane, and chair.

3.3 Weights-Averaging of Checkpoints

Several methods save the model during training at every increment in the overall or mean accuracy and only use the best checkpoint for inference. We observe the best results of the PointNeXt model on the ScanObjectNN dataset around 190 to 210 epochs in a training cycle of 250. We noticed several epochs past the 210th epoch, achieved comparable results to the best evaluation at the 210th epoch. Following the idea of weights averaging from model soups [14], we take the best checkpoints from the training session, average the weights and use the weight-averaged checkpoint for inference. Weight averaging benefited in stabilizing the weights and greatly improved the test accuracy.

4   Experiments

We evaluate the impact of additional local neighborhood features i.e., radius normalized-distance and directional vectors, in the PointNeXt model using the benchmark synthetic dataset ModelNet40 [15]. ModelNet40 dataset has 12,311 shapes spread across 40 classes. We use the dataset's official split of 9843 objects for training and 2468 for testing. Additionally, we evaluate the PointNeXt model using two robust real-world datasets, i.e., the ScanObjectNN [16] and 3DGrocery100 (currently available to the authors, and the dataset will be made public before the conference through a project page).

ScanObjectNN contains 14,510 objects obtained by perturbing the original 2902 objects. ScanObjectNN is derived from 700 unique scenes from two popular scene meshes datasets; SceneNN [17] with 100 objects and ScanNet [18] with 1513 objects. Although the ScanObjectNN dataset has six variants, we use the hardest variant represented by PB_T50_RS (following PointNeXT and other recent models). PB_T50_RS variant constitutes perturbed objects with 50% translation and involves rotation and scaling.

3DGrocery100 dataset at a very high level contains three categories, i.e., Fruits, Vegetables, and Packages with 34, 28, and 38 classes, 37,587, 27,707, and 22,604 point clouds, respectively. The Fruits subset named Apple10 contains ten apple classes and 24 non-apple fruit classes. We consider each of the three categories, i.e., Fruits (non-apples), Vegetables, and Packages, along with the Full dataset as subset datasets. We train the PointNeXT model with and without our proposed additional local neighborhood features using the five subset datasets of 3DGrocery100. We evaluate the impact of the local neighborhood features in improving the classification results on these subsets of the grocery dataset on the PointNeXt model. Finally, we discuss and analyze our experimental results on Apple10, Fruits, Vegetables, Packages, and Full subsets.

5   Experimental Results

In this section, we show the benefits of using the radius r-normalized distance and directional vectors as additional neighborhood features in the PointNeXt model when trained and tested on ScanObjectNN, ModelNet40, and 3DGrocery100 datasets. Our experimental results encourage using these already computed features in future models.

Table 1 shows the impact of radius normalized distance and directional vector concatenation to local neighborhood features in the PointNeXt model trained on the ScanObjectNN dataset. In addition, we show inference results of weights averaging the best two checkpoints. Using the PointNeXt [20] settings, we train PointNet++ without and with the additional local neighborhood features and report 3D classification accuracy.

Table 1. Additive study of sequentially adding our strategies for classification on ScanObjectNN.

Improvements	Overall Accuracy (%)	Δ	Mean Average Accuracy (%)	Δ
PointNeXt [20]	87.7 ± 0.4	-	85.8 ± 0.6	–
+ r-normalized distance	87.9 ± 0.4	+0.2	86.4 ± 0.3	+0.6
+ directional vectors	88.2 ± 0.2	+0.3	86.9 ± 0.2	+0.5
+ best two average	88.4 ± 0.2	+0.2	87.1 ± 0.3	+0.2
Overall Best	88.6	0.5 ↑	87.4	1.0 ↑
PointNet++	83.5	-	81.7	–
PointNet++ (+ours)	86.0	+2.5↑	84.1	+2.4↑
Pix4Point [22]	86.2	–	83.9	–
Pix4Point [22](+ours)	86.4	+0.2↑	84.0	+0.1↑

Table 2. 3D Classification accuracy of PointNeXt model with our approach on ModelNet40. OA = Overall Accuracy, mAcc = Mean Average Accuracy

Improvements	OA (%)	Δ	mAcc (%)	Δ
PointNet++	90.2	-	86.6	–
PointNet++(+ours)	91.1	**+0.9**	89.1	**+2.5**
PointNeXt [20]	93.3	–	91.0	–
PointNeXt + (ours)	93.5	**+0.2**	91.0	**+0.0**

Table 2 shows a comparable overall improvement in the classification accuracy of the PointNeXt model trained on the ModelNet40 dataset. Most recent models are saturated with ModelNet40 results (in the range of 93–94), and an improvement of 0.2% is good. Table 3 shows the impact of radius normalized distance and directional vector concatenation to neighborhood features in the PointNeXt model trained on all five subsets of the 3DGroceyr100 dataset (without colors). Improvements in classification accuracy on synthetic and real-world datasets show the broad applicability of the additional features and the novelty of our work.

Table 3. 3D point cloud classification on PointNeXt model on all subsets of 3DGrocery100.

Models↓/Subsets→	Apple10	Fruits	Vegetables	Packages	Full
#Classes	10	24	28	38	100
#Point Clouds	12905	24682	27707	22604	87898
Train	9706	18406	20720	17214	66032
Test	3199	6276	6987	5390	21866
PointNeXt [20]	21.6	40.6	48.4	81.4	47.7
PointNeXt+(ours)	22.6	45.4	51.8	83.0	50.5
↑	**+1.0**	**+4.8**	**+3.4**	**+1.6**	**+2.8**

6   Ablation Study

6.1   Distance vs. r-normalized Distance

From Eq. 1, we know the distance of the neighborhood points to the anchor points is normalized using the radius r used for querying the points. While existing methods only compute the distance to check whether a point is within the ball for querying, here we provide experimental results to show that the normalized distance as additional neighborhood feature gives better mean accuracy, as shown in Table 4.

6.2   Weight Averaging (Same Training Session)

During a training session, we saved the checkpoints at the top 15 best results. A systematic study of averaging the weights of these checkpoints during inference, is shown

Table 4. Comparison of distance vs. radius r-normalized distance as an additional neighborhood feature. Model Used: PointNeXt [20] and Dataset Used: ScanObjectNN's hardest variant.

Local Features	OA (%)	mAcc (%)
distance	88.3	86.5
r-normalized distance	88.3	86.7

in Table 5. From our study, we gathered that averaging the weights of top-2 checkpoints further improved classification accuracy. Unlike ensemble of models the averaging of weights and saving the averaged model does not consume extra storage space.

Table 5. Ablation study on weight averaging of checkpoints for classification on ScanObjectNN.

Improvements	Overall Accuracy (%)	Δ	Mean Average Accuracy (%)	Δ
PointNeXt	88.1	–	86.4	–
Weight averaging (of checkpoints) from same training session				
+ top-2	88.4	**+0.3**	87.1	**+0.7**
+ top-3	88.3	+0.2	87.1	+0.7
+ top-5	88.2	+0.1	86.3	-0.1
+ top-10	88.0	0.1	86.4	+0.0
+ top-15	87.9	-0.2	86.3	-0.1

Table 6. Cost of additional neighborhood features in terms of the number of parameters, GFLOPs, and Throughput. Tested with 1024 input points and a batch size of 128.

	PointNeXt [20]	PointNeXt + (ours)	Δ
Params (M)	1.3671	1.3690	+0.0019
GFLOPs	1.64	1.65	+0.01
Throughput (ms/sample)	0.042	0.045	+0.003

6.3 Cost of Additional Neighborhood Features

The cost of adding local neighborhood features is negligible based on three measures; the number of parameters (in millions), GFLOPs, and throughput (millisecond/sample). Despite the four additional features - radius r-normalized distance (1) and directional vectors (3) - already computed in the set abstraction layers, there is an increase in all three measures, as shown in Table 6. The throughput in milliseconds (ms) per sample increases by a negligible 0.003 (i.e., 3 microseconds/sample) due to the division operation (to normalize the distance) and additional network computations from the extra features. The overall computational cost is still below the PointNet++ model (1.69 GFLOPs as shown in table 6 of TR-Net [21]). Improvements in the accuracy and a minimal increase in the number of parameters, GFLOPs, and computational costs are encouraging.

7 Conclusion

In this paper, we demonstrate the use of local neighborhood features as additional local features for point cloud classification. We also introduce an inference strategy to benefit the overall results. We present the benefits of neighborhood features and inference strategy on the state-of-the-art PointNeXt model using three datasets, ModelNet40, ScanObjectNN, and 3DGrocery100. The neighborhood features significantly improved the classification accuracy, particularly in the case of real-world datasets. Improvements of **0.5%**, **1%**, **4.8%**, **3.4%**, **1.6%**, **2.8%**, and **0.2%** in the overall accuracy on the real-world ScanObjectNN (hardest variant), 3DGrocery100 dataset's subsets Apple10, Fruits, Vegetables, Packages, Full, and synthetic dataset ModelNet40, respectively on PointNeXt model are significant. Our ablation study provides a detailed analysis to help understand the trade-offs in using additional local neighborhood features. Our results and ablation study strongly encourage the use of these features in future models.

References

1. Qi, C.R., Su, H., Mo, K., Guibas, L.J.: Pointnet: deep learning on point sets for 3d classification and segmentation. In: Proceedings of the IEEE Conference on Computer Vision and Pattern Recognition, pp. 652–660 (2017)
2. Qi, C.R., Yi, L., Su, H., Guibas, L.J.: Pointnet++: deep hierarchical feature learning on point sets in a metric space. In: Advances in Neural Information Processing Systems, pp. 5099–5108 (2017)
3. Liu, Y., Fan, B., Xiang, S., Pan, C.: Relation-shape convolutional neural network for point cloud analysis. In: Proceedings of the IEEE/CVF Conference on Computer Vision and Pattern Recognition, pp. 8895–8904 (2019)
4. Hua, B.S., Tran, M.K., Yeung, S.K.: Pointwise convolutional neural networks. In: Proceedings of the IEEE Conference on Computer Vision and Pattern Recognition, pp. 984–993 (2018)
5. Shen, Y., Feng, C., Yang, Y., Tian, D.: Mining point cloud local structures by kernel correlation and graph pooling. In Proceedings of the IEEE Conference on Computer Vision and Pattern Recognition, pp. 4548–4557 (2018)
6. Bronstein, M.M., Bruna, J., LeCun, Y., Szlam, A., Vandergheynst, P.: Geometric deep learning: going beyond Euclidean data. IEEE Signal Process. Mag. **34**(4), 18–42 (2017)
7. Le, T., Duan, Y.: Pointgrid: A deep network for 3d shape understanding. In: Proceedings of the IEEE Conference on Computer Vision and Pattern Recognition, pp. 9204–9214) (2018)
8. Duan, Y., Zheng, Y., Lu, J., Zhou, J., Tian, Q.: Structural relational reasoning of point clouds. In: Proceedings of the IEEE/CVF Conference on Computer Vision and Pattern Recognition, pp. 949–958 (2019)
9. Liu, Y., Fan, B., Meng, G., Lu, J., Xiang, S., Pan, C.: Densepoint: learning densely contextual representation for efficient point cloud processing. In: Proceedings of the IEEE/CVF International Conference on Computer Vision, pp. 5239–5248 (2019)
10. Xu, M., Zhou, Z., Qiao, Y.: Geometry sharing network for 3d point cloud classification and segmentation. In: Proceedings of the AAAI Conference on Artificial Intelligence, vol. 34, no. 07, pp. 12500–12507, April 2020
11. Sheshappanavar, S.V., Kambhamettu, C.:. A novel local geometry capture in pointnet++ for 3d classification. In: Proceedings of the IEEE/CVF Conference on Computer Vision and Pattern Recognition Workshops, pp. 262–263 (2020)

12. Sheshappanavar, S.V., Kambhamettu, C.: Dynamic local geometry capture in 3d point cloud classification. In: 2021 IEEE 4th International Conference on Multimedia Information Processing and Retrieval (MIPR), pp. 158–164. IEEE, September 2021

13. Lan, S., Yu, R., Yu, G., Davis, L.S.: Modeling local geometric structure of 3d point clouds using geo-CNN. In: Proceedings of the IEEE/CVF Conference on Computer Vision and Pattern Recognition, pp. 998–1008 (2019)

14. Wortsman, M., et al.: Model soups: averaging weights of multiple fine-tuned models improves accuracy without increasing inference time. In: International Conference on Machine Learning, pp. 23965–23998. PMLR, June 2022

15. Wu, Z., et al.: 3D shape nets: a deep representation for volumetric shapes. In: Proceedings of the IEEE Conference on Computer Vision and Pattern Recognition, pp. 1912–1920 (2015)

16. Uy, M.A., Pham, Q.H., Hua, B.S., Nguyen, T., Yeung, S.K.: Revisiting point cloud classification: A new benchmark dataset and classification model on real-world data. In: Proceedings of the IEEE/CVF International Conference on Computer Vision, pp. 1588–1597 (2019)

17. Hua, B.S., Pham, Q.H., Nguyen, D.T., Tran, M.K., Yu, L.F., Yeung, S.K.: Scenenn: a scene meshes dataset with annotations. In: 2016 Fourth International Conference on 3D Vision (3DV), pp. 92–101. IEEE, October 2016

18. Dai, A., Chang, A. X., Savva, M., Halber, M., Funkhouser, T., Nießner, M.: Scannet: richly-annotated 3d reconstructions of indoor scenes. In: Proceedings of the IEEE Conference on Computer Vision and Pattern Recognition, pp. 5828–5839 (2017)

19. Wang, Y., Sun, Y., Liu, Z., Sarma, S.E., Bronstein, M.M., Solomon, J.M.: Dynamic graph CNN for learning on point clouds. ACM Trans. Graphics (tog) 38(5), 1–12 (2019)

20. Qian, G., et al.: PointNeXt: revisiting PointNet++ with improved training and scaling strategies. arXiv preprint arXiv:2206.04670 (2022)

21. Liu, L., Chen, E., Ding, Y.: TR-Net: a transformer-based neural network for point cloud processing. Machines 10(7), 517 (2022)

22. Qian, G., Zhang, X., Hamdi, A., Ghanem, B.: Pix4point: image pretrained transformers for 3d point cloud understanding. arXiv preprint arXiv:2208.12259 (2022)

3D Point Cloud Registration
for GNSS-denied Aerial Localization
over Forests

Daniel Sabel[1,2(✉)], Torbjörn Westin[1], and Atsuto Maki[2]

[1] Spacemetric AB, Sollentuna, Sweden
[2] KTH Royal Institute of Technology, Stockholm, Sweden
dosabel@kth.se

Abstract. This paper presents a vision-based localization approach for Unmanned Aerial Vehicles (UAVs) flying at low altitude over forested areas. We address the task as a point cloud registration problem using local 3D features with the intention to exploit the shape and relative arrangement of the trees. We propose a 3D descriptor called SHOT-N which is an adaptation of the state-of-the-art SHOT 3D descriptor. SHOT-N leverages constraints in the extrinsic parameters of a gimballed, nadir-looking camera. Extensive experiments were performed with semi-simulated point cloud data based on real aerial images over four forested areas. SHOT-N is shown to outperform two state-of-the-art 3D descriptors in terms of the rate of successful registrations. The results suggest a high potential of the approach for aerial localization over forested areas.

Keywords: GNSS-denied · aerial navigation · point cloud registration · visual navigation · natural environments

1 Introduction

Unpiloted Unmanned Aerial Vehicles (UAVs) commonly navigate based on inertial navigation and Global Satellite Navigation Systems (GNSSs), such as GPS and Galileo. Inertial navigation allows continuous estimation of changes in a vehicle's velocity, position, and orientation, but suffers from unbounded drift errors. GNSS is used to constrain the drift errors in the inertial measurements and to provide geographic position. A weakness of GNSS is sensitivity to interference such as malicious jamming or spoofing [5,13,14], which can result in GNSS-denied situations. This may compromise flight safety, particularly in the case of autonomous or unpiloted UAVs.

In this work we consider the scenario of an unpiloted UAV navigating at low altitude over a forested area. This scenario is relevant to use cases such as reconnaissance and search and rescue missions. The UAV is assumed to be equipped with a GNSS-receiver, an inertial navigation system (INS) and a gimballed, nadir-looking monocular camera. Due to a jamming attack, the GNSS-receiver

© The Author(s), under exclusive license to Springer Nature Switzerland AG 2023
R. Gade et al. (Eds.): SCIA 2023, LNCS 13885, pp. 396–411, 2023.
https://doi.org/10.1007/978-3-031-31435-3_27

has been rendered unusable, forcing the UAV to rely on inertial navigation and vision-based methods for navigation.

Vision-based navigation generally exploits data from an onboard sensor, such as a camera, to compute the position and orientation, collectively called pose, of the UAV. The geographic position can be estimated by comparison of the sensor data with a georeferenced model of the environment, here called the "reference model". The reference model may consist of, e.g. satellite or aerial images, or of 3D models. A main challenge of this approach is the accurate registration of the sensor data with the reference model, given potentially large differences in, e.g. viewing perspective, color, illumination, and shadows.

Vision-based navigation is challenging in environments dominated by vegetation [15, 22]. Most of the research on vision-based navigation has been devoted to indoor or urban environments [4] that often contain a greater abundance of distinct features than do natural environments such as forests. Furthermore, forests may be affected by seasonal changes such as color variations and leaf-on/leaf-off condition, as well as the presence or absence of snow, which complicates the registration with the reference model.

An example of challenging differences between onboard sensor data and a reference model is shown in Fig. 1. The right image was captured from a UAV flying 90 m above the ground, while the reference image of same area (left) was captured by a high resolution mapping-grade camera from an altitude of 3 km. Differences due to viewing perspective, snow cover and shadow directions are evident.

Fig. 1. Left: Acquisition from 3 km altitude, on April 8, 2021. Right: Same area captured one year later from a UAV at 90 m altitude, on April 11, 2022.

In this work, we propose to exploit the 3D shape and relative arrangement of trees to overcome some of the challenges with matching images similar to those in Fig. 1. The novelty of our approach is primarily in addressing the pose estimation task as a point cloud registration problem using local 3D features. We thus rely only on geometric information for the registration process, aiming at a method insensitive to differences in color, illumination, and shadows. We also expect this approach to be less sensitive to perspective differences than single-view image matching approaches. We then propose SHOT-N, as an adaptation of the state-of-the-art 3D feature method SHOT [21], that exploits the a priori known observation geometry of the nadir-looking camera.

We evaluate the approach in controlled experiments with semi-simulated data, based on aerial images over four forested areas in Sweden.

Key Contributions

- Exploration of the challenging problem of aerial localization in non-urban areas. In particular, we exploit the forest structure and suggest that registration of raw point cloud data using local 3D features is a viable approach.
- The SHOT-N 3D feature descriptor, as an adaptation of SHOT, which exploits the a priori knowledge about the acquisition geometry of a nadir-pointing camera.

2   Related Work

3D point cloud registration is a fundamental problem in computer vision that has been extensively studied as a basis for, e.g. 3D reconstruction, pose estimation, and Simultaneous Localisation and Mapping (SLAM) [3]. The objective of the registration is to estimate the spatial transformation that aligns two separate point clouds. For an overview of the state of the art for point cloud registration, see [12,24].

Deep learning (DL) approaches represent current state of the art for many image-based problems. The processing of 3D point cloud data with DL is however not straightforward due to the inherent challenges of adapting DL methods to the unstructured nature of point cloud datasets [6,24]. PointNet [17] was a significant step to overcome these challenges and it can be argued that PointNet and its extensions represent the state of the art for 3D recognition tasks such as object classification and segmentation. PointNet takes as input unordered point cloud data and produces class labels for the entire input, or labels or segments for each point. The authors of [1] successfully adapted PointNet to point cloud registration with the use of a modified Lukas-Kanade framework and a recurrent neural network. The efficacy of DL methods for point cloud registration is evolving, while registration with partially overlapping data, to which our problem belongs, remains a challenging problem [24].

To the best of our knowledge, registration with 3D features has not yet been studied for the purpose of aerial navigation over non-urban areas. There are a few studies that instead of 3D features exploit height-patch or depth-patch matching for similar purposes, which we discuss below. In [8], a method for aerial localization insensitive to seasonal variations was sought. The authors assumed height information to be more robust to seasonal variations than approaches based on single-view matching, and matched local height-patch images derived from motion stereo against a georeferenced 3D model. Over a residential area, they reported a convergence zone for the a priori position error of approximately 30 m in diameter, outside of which their optimization method would fail. The authors noted that the convergence zone agreed well with the distance between houses in the height map. Another relevant study is that of [9], in which aircraft pose was estimated by matching height-patch images of the terrain below the vegetation against a Digital Elevation Model (DEM). They obtained the height patches by rasterizing LiDAR measurements. The approach in [6] identified so-called super-points for which point cloud data were rasterized to local depth-patch images, followed by max and mean filtering. A deep auto-encoder was used in the 2D-domain to generate local features for matching. The authors of [6] evaluated their method on challenging partial-view and cross-season point cloud data and demonstrated an improved rate of successful registration when compared with the general-purpose 3D feature FPFH [19].

An important distinction between our approach and the approaches in [6, 8, 9] is that we avoid rasterization of the point cloud data. Our motivation for this design choice is that the digitization and aggregation involved in rasterization may result in greater loss of structural information than the construction of 3D features directly from the raw point cloud data.

3   Method

A schematic of the processing steps involved in the camera pose estimation pipeline is shown in Fig. 2. Step 1 is the generation of the nominal 3D model of the environment observed by the UAV, which we call the local point cloud (PC). It can be computed with matching in a pair of overlapping images acquired by a monocular, nadir-pointing camera mounted on the UAV. In this study, we simulate local PCs with the use of aerial images captured from an airplane. The data generation is described in Sect. 4.2.

Steps 2 through 8 serve to georeference the local PC, i.e. to estimate the horizontal and vertical world coordinates of each point in the local PC by registering the local PC against a reference 3D model of the environment. Steps 2 through 8 are mainly based on general-purpose methods for point cloud registration. Step 9 is the computation of the camera pose which can be solved as a Perspective-n-Point problem, given the known correspondences between the points in the georeferenced local PC and the images used to generate those points in step 1. The evaluation in this work focuses on the efficacy of the point cloud registration pipeline, i.e. steps 2 through 8, to correctly georeference the local PC.

3.1   Registration Pipeline

The registration pipeline (steps 2 through 8) compares a local PC with reference PC data in order to estimate a rigid body transformation T_{geo} that serves to georeference each point in the local PC. We built our pipeline mainly using established methods for general point cloud processing, provided by the open-source software Point Cloud Library [10]. For simplicity, we assume local PCs and reference PCs to be square, so that their horizontal dimensions, here termed width and breadth, are equal.

The pipeline starts with the estimation of a surface normal (step 2) for each point in the local PC, approximated as the normal of a plane fitted to the points in the local neighborhood, defined as a sphere with a radius of R_{norm}. The plane

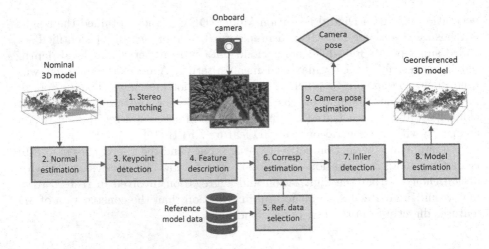

Fig. 2. Schematic of processing steps

fitting is solved as a least-squares problem through analysis of the eigenvalues and eigenvectors of the covariance matrix of the points in the local neighborhood [18].

To reduce computational cost, keypoint detection (step 3) is used to select a subset of the points in the local PC as keypoints, to which the subsequent feature processing is referred. While various keypoint detection methods exist, we employ a simplistic detection in order to simplify the interpretation of the results. Keypoints are selected in a regularly spaced horizontal grid with a sampling distance of s_{loc}, resulting in N_{loc} keypoints.

The feature description (step 4) produces, for each keypoint, a compact representation, called feature, of the shape in the local neighborhood around the keypoint. The set of features in the local PC is denoted $\phi_{loc}(i)$, $i \in \{1 \ldots N_{loc}\}$. We seek a feature description method (descriptor) that relies solely on shape and is robust to clutter, occlusion, and variations in point density, while still being sufficiently descriptive for effective feature correspondence estimation. In this work, we evaluate the efficacy of point cloud registration with the use of the descriptors FPFH [19] and SHOT [21,23], as well as our suggested variant of SHOT, called SHOT-N. FPFH was chosen as one of the most widely used descriptors. SHOT was chosen as it was designed specifically with the aim of being robust to noise and clutter, and point density variations. SHOT-N is described in Sect. 3.2. These descriptors all encode attributes of the geometric distribution of points in spherical neighborhoods, with radius R_ϕ. In accordance with our assumptions, they do not exploit any texture or color information. The set of features in the reference PC is denoted $\phi_{ref}(j)$, $j \in \{1 \ldots N_{ref}\}$.

In a real flight scenario, reference data covering the area in which UAV is expected to navigate are generated prior to the flight and stored onboard the UAV. For long distance flights, the comparison between the local PC and the entire reference dataset would be too computationally expensive for real-time processing. The reference data selection (step 5) mitigates this problem by restricting the search space, and thus the number of reference features N_{ref} involved. The width and breadth of the search space in the reference data (w_{search}) are computed as the sum of the width of the envelope of the local PC and the two-sided 3-sigma uncertainty in the horizontal position due to drift in the inertial navigation, according to $w_{search} = \sqrt{2}w_{loc} + 6\sigma_{pos}$, where w_{loc} is the width of the local PC. Here, we assume free heading rotation, hence the factor $\sqrt{2}$. In this study, we performed evaluations with three different drift scenarios (see Table 4).

Correspondence estimation (step 6) identifies tentative point-to-point correspondences between the local PC and the reference PC by comparing their features. As the local PC covers a smaller area than the reference PC, we divide the search space into N_{sub} overlapping subdivisions S_k, $k \in \{1 \dots N_{sub}\}$, each of which with a width and breadth equal to $\sqrt{2}w_{loc}$. For each subdivision S_k, each local feature $\phi_{loc}(i)$ is assigned the nearest neighbor reference feature $\phi_{ref}(j)$ in S_k. Nearest neighbors are found with a KD-tree search [2] and ℓ^2-norm distance in the feature space. The use of subdivisions in the correspondence estimation step allows for parallelization and thus reduction of the computation time. Finally, to reduce the number of false correspondences, only reciprocal correspondences are kept, i.e. correspondences for which $\phi_{ref}(j)$ does not have $\psi_{loc}(i)$ as its nearest neighbor are discarded.

The model to be estimated is a rigid body transformation with six degrees of freedom, consisting of 3-axis translation and 3-axis rotation that serves to geo-reference the local PC. A large fraction of the tentative correspondences from step 6 of the pipeline are expected to be erroneous, due to perturbations in the PC data and descriptor limitations. The inlier detection (step 6) uses conventional Random Sample Consensus (RANSAC) [7] with the aim of estimating the set of true correspondences. That is, for each subdivision S_k, the model is estimated [11] using three correspondences picked at random, and the model's score is recorded. The score is the number of correspondences that do not deviate from the model with more than a threshold Δ_{RANSAC}. The process is iterated either until a 99% probability of correct model is reached, or when a maximum number of iterations Γ_{RANSAC} has been performed. The correspondence inliers ϕ_{loc}^{in} and ϕ_{ref}^{in} are identified as the correspondences that contributed to the score for the highest-scoring model among all subdivisions.

In step 8, the final model T_{est}, expressed as a 4×4 matrix, is computed with the use of all correspondence inliers ϕ_{loc}^{in} and ϕ_{ref}^{in}.

3.2 SHOT-N

Both FPFH and SHOT estimate a 3-axis local reference frame (LRF), relative to which the points in the local neighborhood are analyzed to construct the feature signature. The importance of a repeatable LRF estimation in the presence of clutter and occlusions in the point cloud data was highlighted in [21] and scrutinized in [16]. We found, through experiments, that the z-axis ("up-direction") of the LRF in SHOT was significantly more repeatable that the corresponding horizontal axes. In our case the up-direction is known, as we assume a gimballed, nadir-looking camera. Thus, we only need to find a repeatable horizontal direction. We propose to find this horizontal direction by projecting the z-axis estimated with the SHOT LRF method to the plane given by the normal of the a priori known up-direction.

An outline of the proposed LRF estimation is given in Algorithm 1. Inspired by the LRF estimation in the SHOT descriptor, we exploit eigenvalue decomposition ($\mathbf{M} = \mathbf{VDV}^{-1}, \mathbf{V} = [\mathbf{m}_1\ \mathbf{m}_2\ \mathbf{m}_3]$) of the weighted covariance matrix $\mathbf{M}$ of the points in the local neighborhood S_p. As explained above, the up-direction $\mathbf{z}$ in our case is known a priori and so we only need to find a repeatable horizontal direction. We use $\mathbf{m}_1$ which corresponds to the smallest eigenvalue of $\mathbf{M}$, as we through experiments found it to provide higher repeatability than $\mathbf{m}_2$ and $\mathbf{m}_3$. The sign of $\mathbf{m}_1$ is disambiguated through conditioning on the dot product with $\mathbf{z}$. A horizontal direction $\mathbf{x}$ is then found by projecting the disambiguated eigenvector $\mathbf{m}_1'$ to the plane given by $\mathbf{z}$ as its normal. Finally, after normalization of $\mathbf{x}$, $\mathbf{y}$ is computed as $\mathbf{z} \times \mathbf{x}$.

With the exception of the LRF estimation method, our proposed descriptor called SHOT-N (short for SHOT-Nadir), is identical to SHOT.

It will be demonstrated that the benefit of SHOT-N over general-purpose descriptors appears to be that the LRF estimation makes the construction of the feature signature less sensitive to perturbations in the point cloud data. This is expected to increase the likelihood of successful feature correspondence estimation and thus also the chance of successful point cloud registration (Courtesy of Lantmäteriet, the Swedish Land Survey, https://www.lantmateriet.se/en/about-lantmateriet).

Algorithm 1. Local reference frame ($\mathbf{x}$, $\mathbf{y}$ and $\mathbf{z}$) estimation algorithm. Arrows indicate assignments.

$\mathbf{z} \leftarrow [0\ 0\ 1]^T$
$F = \{\text{set of keypoints}\}$
$S_p = \{\text{points in sphere of radius } R \text{ around keypoint } \mathbf{p}\}$
for all $p \in F$ **do**
    $\mathbf{M} \leftarrow \mathbf{0}$
    for all $\mathbf{q} \in S_p$ **do**
        $d \leftarrow \|\mathbf{p} - \mathbf{q}\|_2$
        $\mathbf{M} \leftarrow \mathbf{M} + (R + d)(\mathbf{p} - \mathbf{q})(\mathbf{p} - \mathbf{q})^T$
    end for
    Through eigenvalue decomposition, $\mathbf{M} = \mathbf{VDV}^{-1}$, where $\mathbf{V} = [\mathbf{m}_1\ \mathbf{m}_2\ \mathbf{m}_3]$.
    $\mathbf{m}_1' \leftarrow \text{sign}(\mathbf{z} \cdot \mathbf{m}_1)\mathbf{m}_1$
    $\mathbf{x} \leftarrow \mathbf{m}_1' - \mathbf{z}(\mathbf{z} \cdot \mathbf{m}_1')$
    $\mathbf{x} \leftarrow \mathbf{x}/\|\mathbf{x}\|_2$
    $\mathbf{y} \leftarrow \mathbf{z} \times \mathbf{x}$
end for

Table 1. Test areas and aerial images used for generating reference and local PCs.

Description	Reference data image	Local data image
Area A E 13°20′32″ N 58°03′46″ 0.5 km²		
Area B E 13°59′24″ N 55°46′38″ 0.5 km²		
Area C E 18°21′25″ N 59°47′27″ 0.6 km²		
Area D E 16°36′16″ N 60°57′32″ 0.6 km²		

4   Evaluation

The use case assessed in this work is a UAV flying over a mainly forested area at an altitude of 90 m above ground. The UAV is assumed to be equipped with an INS and a gimballed, nadir pointing calibrated camera. The camera has an angular field of view (AFOV) of 73.7°, which captures overlapping images. The UAVs GNSS receiver has been jammed, making the UAV dependent on vision-based methods for drift correction and positioning. With this use case in mind, we assessed the efficacy of the point cloud registration (steps 2-8) with the use of descriptors FPFH, SHOT and SHOT-N over four different test areas, with three different scenarios for the uncertainty in the horizontal position due to drift errors in the INS.

Keypoints in the local PC were selected with a sampling distance s_{loc} of 2 m. We set the feature radius R_ϕ to 10 m, with the intention of encoding the shape of individual trees as well as their relative arrangement. The surface normal radius R_{norm} was set to 3 m. Reference PC features were produced in the same way as the local PC features, with the exception of a finer keypoint sampling distance

Table 2. Specification of aerial images used for generation of point cloud data. Acquisition time is local time. Images were acquired with mapping-grade cameras onboard airplanes at altitudes ranging between 3200 m and 3900 m.

Test area	Purpose	Aerial images acquisition times	Resolution
A	Ref. PC mosaic	2015-08-14 12:28:31	25 cm
		2015-08-14 12:28:43	
	Local PC mosaic	2019-08-25 08:42:55	16 cm
		2019-08-25 08:42:59	
B	Ref. PC mosaic	2016-05-12 14:06:13	25 cm
		2016-05-12 14:06:23	
	Local PC mosaic	2020-06-03 14:33:37	15 cm
		2020-06-03 14:33:46	
C	Ref. PC mosaic	2019-07-20 09:08:35	15 cm
		2019-07-20 09:08:43	
	Local PC mosaic	2021-04-08 09:26:18	15 cm
		2021-04-08 09:26:26	
D	Ref. PC mosaic	2017-06-29 13:24:31	25 cm
		2017-06-29 13:24:41	
	Local PC mosaic	2019-06-06 08:47:43	16 cm
		2019-06-06 08:47:50	

of 1 m in the keypoint detection step. In the inlier detection step we used a Δ_{RANSAC} of 5 m. As correspondence estimation was performed in subsets of ϕ_{ref}, we can assume a higher inlier ratio than if correspondence estimation was performed with the full set of reference features. We therefore set Γ_{RANSAC} to 500, corresponding to a 99% probability of successful model estimation assuming an inlier ratio of 21%.

4.1   Test Areas

Overviews of the test areas, together with their centre coordinates and surface area, are presented in Table 1. The test sites are all dominated by forests, with different mixes of coniferous and deciduous tree species.

4.2   Datasets

Georeferenced point cloud mosaics covering the test areas were generated with motion stereo in pairs of overlapping aerial images with the use of a commercial software[1]. No subsequent morphological processing, such as outlier removal or noise reduction, was applied to the point cloud data. Specifications of the images used to generate the point clouds are provided in Table 2. The difference in acquisition dates between images used to generate reference PC and local

---

[1] Keystone software, http://spacemetric.com.

Table 3. Point cloud heights, presented as deviation from mean height. Height differences between reference and local PCs were computed over 10 m by 10 m cells and color coded to highlight significant differences.

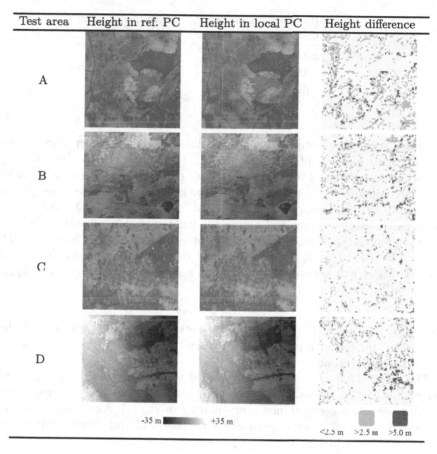

Test area	Height in ref. PC	Height in local PC	Height difference
A			
B			
C			
D			

-35 m ▬▬▬ +35 m   <2.5 m   >2.5 m   >5.0 m

PC mosaics varied between two and four years. Furthermore, the images were acquired at different time of the year and at different times of the day. An attempt to quantify the differences is provided in Table 3, showing point cloud heights and height differences. Note the significant height differences, which represent data perturbations that complicate point cloud registration.

We attribute the height differences between local PCs and reference PCs mainly to variations in the degree to which the motion stereo processing captured deciduous tree canopies.

We performed sets of experiments for three different scenarios for the standard deviation of the horizontal position (σ_{pos}), due to drift in the inertial measurements, as specified in Table 4. Point cloud registration experiments were performed at N evenly spaced horizontal coordinates ($p(i)$, $i \in \{1 \ldots N\}$) across each test area, with a sampling distance equal to w_{src}, so that experiments were

Table 4. Navigation drift scenarios and corresponding search space area. For all scenarios, an a priori scale error of 2% and a priori pitch and roll errors of 0.5° were assumed.

Drift scenario	Horizontal position uncertainty ($\pm 3\sigma_{pos}$)	Reference PC width and breadth
1	±21 m	137 m x 137 m
2	±43 m	180 m x 180 m
3	±64 m	222 m x 222 m

performed across the entire test areas without overlap between adjacent local PCs. For each location $p(i)$, a reference PC (P_{ref}) was created by cropping a part of the georeferenced reference PC mosaic (Table 3) centered at $p(i)$, with an extent according to Table 4. The local PC was generated as follows. First, a patch centered at $p(i)$ was cropped from the local PC mosaic (Table 3). We assumed a 50% image overlap along and across the flight trajectory, resulting in a w_{src} of 67 m, given the specified AFOV and flight altitude. This PC, here called P_{loc}^{gt}, was used as ground truth for measuring the registration accuracy. The nominal local PC (P_{loc}) was generated by applying a similarity transformation T_{gen} to P_{loc}^{gt} according to

$$P_{loc} = T_{gen}P_{loc}^{gt}, \tag{1}$$

where T_{gen} is composed of horizontal translation according to the magnitude of the position uncertainty in Table 4, pitch and roll rotations of 0.5°, a heading rotation of 45°, and an upscaling of 2%.

The point clouds P_{loc} and P_{ref} were fed to the registration pipeline (steps 2-8) in Fig. 2. The registration pipeline estimates a rigid body transformation T_{est} that serves to georeference every point j in $P_{loc}(j)$. An example of a pair of point clouds used in an experiment for scenario 3, and corresponding aerial images, are shown in Fig. 3.

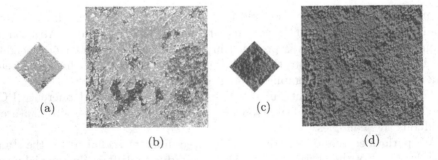

(a)   (c)

(b)   (d)

Fig. 3. Examples of PC data for drift scenario 3, and corresponding aerial images. The local PC (a) and the reference PC (b) are shown with colors representing height. The registration pipeline georeferenced the local PC with an *MTRE* of 1.7 m.

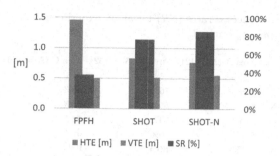

Fig. 4. Comparison of registration results for the evaluated feature descriptors, averaged over all test areas and drift scenarios.

4.3   Evaluation Metrics

We quantify the results with registration success rate (SR), horizontal translation error (HTE) and vertical translation error (VTE). A successful registration is defined as a Mean Target Registration Error ($MTRE$) [20] below 5 m. The $MTRE$ is the average Euclidian distance between all K points in the georeferenced local PC and the corresponding points in the local ground truth PC:

$$MTRE = \frac{1}{K} \sum_{j=1}^{K} \|T_{est}P_{loc}(j) - P_{loc}^{gt}(j)\|_2 \qquad (2)$$

We use $MTRE$ to define the success rate as it combines translation errors and orientation errors in a single metric. The SR is defined as the number of successful registrations divided by the number of registration experiments. The horizontal and vertical translation errors are computed according to Eqs. (3) and (4), where subscripts x, y and z indicate coordinates along a particular axis.

$$HTE = \frac{1}{K} \sum_{j=1}^{K} \sqrt{([T_{est}P_{loc}(j)]_x - [P_{loc}^{gt}(j)]_x)^2 + ([T_{est}P_{loc}(j)]_y - [P_{loc}^{gt}(j)]_y)^2}$$

$$(3)$$

$$VTE = \frac{1}{K} \sum_{j=1}^{K} \sqrt{([T_{est}P_{loc}(j)]_z - [P_{loc}^{gt}(j)]_z)^2} \qquad (4)$$

5   Results

The registration success rates (SR) for all experiments are shown in Table 5. Figure 4 compares the three descriptors, showing mean SR, and HTE and VTE for successful registrations. It can be noted that in the cases when registration failed, the HTE was typically at least an order of magnitude larger than for successful registrations.

Table 5. Registration success rates (SR) for all 36 experiments. N = number of point cloud registrations per experiment.

Drift scenario	Descriptor	Area A ($N = 121$)	Area B ($N = 132$)	Area C ($N = 156$)	Area D ($N = 156$)
1	FPFH	31%	26%	74%	24%
	SHOT	72%	76%	**98%**	61%
	SHOT-N	**83%**	**89%**	**98%**	**71%**
2	FPFH	18%	23%	72%	19%
	SHOT	72%	70%	97%	60%
	SHOT-N	**81%**	**86%**	**98%**	**71%**
3	FPFH	25%	25%	70%	24%
	SHOT	74%	71%	97%	62%
	SHOT-N	**83%**	**88%**	**99%**	**76%**

Overall, registration succeeded in 85.2% of the tests with the SHOT-N 3D feature descriptor, while the corresponding results for SHOT and FPFH were 76.2% and 37.1%, respectively (see Fig. 4). We attribute this to a high repeatability of the LRF in SHOT-N due to i) a known **z**-axis, and ii) a relatively stable estimation of a horizontal axis (see Algorithm 1).

The mean SR for drift scenario 3 with SHOT-N varied between 76% for area D and 99% for area C. These large variations are caused by differences between the local PC and reference PC data (see Table 3). As we did not observe any major disruptive changes such as large scale tree harvests in our datasets, we assess that the variations in SR between test areas primarily resulted from variations in how the motion stereo processing sampled the geometric structure. The degree to which deciduous tree canopies were sampled was significantly reduced during leaf-off conditions, as image texture was no longer dominated by the tree canopies, but rather by ground details such as tree shadows. Furthermore, the motion stereo processing occasionally failed to measure patches of the forest floor in small clearings that were in the shade due to a combination of forest structure and low solar angles. In the case of area C, we attribute the high SR to a combination of highly similar extrinsic camera parameters and matching shadow directions between the aerial images used for generation of the local and reference PCs.

The mean HTE-values for SHOT and SHOT-N were both approximately 0.8 m, whilst the corresponding value for FPFH was 1.5 m. The VTE averaged around 0.5 m for all descriptors. Mean HTE, VTE and SR were similar for the different drift scenarios (Table 5), suggesting a potential for handling larger search spaces.

6   Summary and Discussion

This paper presented a vision-based localization approach for Unmanned Aerial Vehicles flying at low altitude over forested areas. We addressed the task as a point cloud registration problem using local 3D features and proposed an adapted

3D feature descriptor called SHOT-N which exploited constraints in the extrinsic camera parameters. We performed extensive experiments with point cloud data from real aerial images over four forested areas. The point cloud data will be made available to the public.

SHOT-N outperformed two state-of-the-art 3D descriptors in terms of the rate of successful registrations, with an average success rate of 85%. Mean horizontal and vertical translation errors for successful registrations were 0.8 m and 0.5 m, respectively, which was similar to those obtained with state-of-the-art 3D descriptors.

The fact that the reference point clouds in our experiments were between two and four years older than the local point clouds demonstrates that, unless disruptive events like forest harvesting or fires occur, forest structures may in principle be sufficiently stable over such time periods to allow 3D feature-based localization.

The differences in registration success rates between the test areas highlights the importance of data quality. In comparison with terrain-referenced navigation that relies on ground topography, our method has the added advantage that it will work also in completely flat areas, as long as there are significant above-ground features such as trees. Assuming that the shape of the vegetation can be appropriately sampled with the use of motion stereo matching or LiDAR, the proposed approach should be insensitive to differences in color, illumination, and shadows. The results suggest a potential of the approach for low altitude aerial localization over forested areas.

7 Future Work

The computational cost of the registration pipeline must be reduced to achieve real-time performance. We found the computational cost to be dominated by the correspondence estimation step (up to 90%). Significant cost reduction could be achieved in this step, e.g. through parallel processing of the search space subsets and by reducing the number of features to match with the use of a more selective keypoint detector.

The approach presented in this work should be evaluated and adapted with point cloud data from low-altitude UAV images. The study of deep learning approaches for improved robustness to point cloud perturbations such as clutter and occlusions is a topic worth pursuing, e.g. considering the successes of PointNetLK [1] and its extensions.

Acknowledgements. This work was carried out in the project "Autonomous Navigation Support from Real-Time Visual Mapping", which was funded by Sweden's National Strategic Innovation Programme for Aeronautics (Innovair) through Sweden's Innovation Agency (Vinnova) (project nr. 2019-02746).

References

1. Aoki, Y., Goforth, H., Srivatsan, R.A., Lucey, S.: PointNetLK: robust & efficient point cloud registration using PointNet. In: Proceedings of the IEEE Computer Society Conference on Computer Vision and Pattern Recognition, pp. 7156–7165 (2019). https://doi.org/10.1109/CVPR.2019.00733
2. Bentley, J.L.: Multidimensional binary search trees used for associative searching. Commun. ACM **18**(9), 509–517 (1975). https://doi.org/10.1145/361002.361007
3. Campos, C., Elvira, R., Rodriguez, J.J., Montiel, J.M., Tardos, J.D.: ORB-SLAM3: An accurate open-source library for visual, visual-inertial, and multimap SLAM. IEEE Trans. Rob. **37**(6), 1874–1890 (2021). https://doi.org/10.1109/TRO.2021.3075644
4. Couturier, A., Akhloufi, M.A.: A review on absolute visual localization for UAV. Robot. Auton. Syst. **135**, 103666 (2021). https://doi.org/10.1016/j.robot.2020.103666. https://doi.org/10.1016/j.robot.2020.103666
5. Dovis, F.: GNSS interference threats and countermeasures. Artech House Publishers, Norwood, MA, United States (2015)
6. Elbaz, G., Avraham, T., Fischer, A.: 3D point cloud registration for localization using a deep neural network auto-encoder. In: Proceedings - 30th IEEE Conference on Computer Vision and Pattern Recognition, CVPR 2017, vol. 2017, pp. 2472–2481 (2017). https://doi.org/10.1109/CVPR.2017.265
7. Fischler, M.A., Bolles, R.C.: Random sample consensus: a paradigm for model fitting with applications to image analysis and automated cartography. Commun. ACM **24**(6), 381–395 (1981). https://doi.org/10.1145/358669.358692
8. Grelsson, B., Felsberg, M., Isaksson, F.: Efficient 7D aerial pose estimation. In: 2013 IEEE Workshop on Robot Vision, WORV 2013, pp. 88–95 (2013). https://doi.org/10.1109/WORV.2013.6521919
9. Hemann, G., Singh, S., Kaess, M.: Long-range GPS-denied Aerial Inertial Navigation with LIDAR Localization. In: IEEE/RSJ International Conference on Intelligent Robots and Systems (IROS) (May), pp. 1659–1666 (2016). https://doi.org/10.1109/IROS.2016.7759267
10. Holz, D., Ichim, A.E., Tombari, F., Rusu, R.B., Behnke, S.: Registration with the point cloud library: a modular framework for aligning in 3-D. IEEE Robot. Autom. Mag. **22**(4), 110–124 (2015). https://doi.org/10.1109/MRA.2015.2432331
11. Horn, B.K.P.: Closed-form solution of absolute orientation using unit quaternions. J. Opt. Soc. Am. **4**, 629–642 (1987)
12. Huang, X., Mei, G., Zhang, J., Abbas, R.: A comprehensive survey on point cloud registration, pp. 1–17 (2021). http://arxiv.org/abs/2103.02690
13. Jafarnia-Jahromi, A., Broumandan, A., Nielsen, J., Lachapelle, G.: GPS vulnerability to spoofing threats and a review of antispoofing techniques. Int. J. Navig. Observ. **2012**, 127072 (2012). https://doi.org/10.1155/2012/127072
14. Khan, S.Z., Mohsin, M., Iqbal, W.: On GPS spoofing of aerial platforms: a review of threats, challenges, methodologies, and future research directions. PeerJ Computer Science **7**, 1–35 (2021). https://doi.org/10.7717/PEERJ-CS.507
15. Patel, B., Barfoot, T.D., Schoellig, A.P.: Visual localization with google earth images for robust global pose estimation of UAVs. In: Proceedings - IEEE International Conference on Robotics and Automation, pp. 6491–6497 (2020). https://doi.org/10.1109/ICRA40945.2020.9196606
16. Petrelli, A., Di Stefano, L.: On the repeatability of the local reference frame for partial shape matching. Proceedings of the IEEE International Conference on Computer Vision, pp. 2244–2251 (2011). https://doi.org/10.1109/ICCV.2011.6126503

17. Qi, C.R., Su, H., Mo, K., Guibas, L.J.: PointNet: deep learning on point sets for 3D classification and segmentation. In: Proceedings - 30th IEEE Conference on Computer Vision and Pattern Recognition, CVPR 2017 2017-Janua, pp. 77–85 (2017). https://doi.org/10.1109/CVPR.2017.16

18. Rusu, R.B.: Semantic 3D object maps for everyday manipulation in human living environments. Doctoral thesis, Technische Universität München (2009). https://doi.org/10.1007/s13218-010-0059-6

19. Rusu, R.B., Blodow, N., Beetz, M.: Fast point feature histograms (FPFH) for 3D registration. In: IEEE International Conference on Robotics and Automation, pp. 3212–3217 (2009). https://doi.org/10.1109/robot.2009.5152473

20. Saiti, E., Theoharis, T.: An application independent review of multimodal 3D registration methods. Comput. Graph. (Pergamon) 91, 153–178 (2020). https://doi.org/10.1016/j.cag.2020.07.012

21. Salti, S., Tombari, F., Di Stefano, L.: SHOT: unique signatures of histograms for surface and texture description. Comput. Vis. Image Understanding 125, 251–264 (2014). https://doi.org/10.1016/j.cviu.2014.04.011

22. Sattler, T., et al.: Benchmarking 6DoF urban visual localization in changing conditions. In: IEEE Computer Society Conference on Computer Vision and Pattern Recognition (CVPR), pp. 8601–8610 (2018). http://arxiv.org/abs/1707.09092

23. Tombari, F., Salti, S., Di Stefano, L.: Unique signatures of histograms for local surface description. In: Daniilidis, K., Maragos, P., Paragios, N. (eds.) ECCV 2010. LNCS, vol. 6313, pp. 356–369. Springer, Heidelberg (2010). https://doi.org/10.1007/978-3-642-15558-1_26

24. Zhang, Z., Dai, Y., Sun, J.: Deep learning based point cloud registration: an overview. Virt. Real. Intell. Hardware 2(3), 222–246 (2020). https://doi.org/10.1016/j.vrih.2020.05.002

Cleaner Categories Improve Object Detection and Visual-Textual Grounding

Davide Rigoni[1,2]([✉]), Desmond Elliott[3], and Stella Frank[3,4]

[1] Department of Mathematics "Tullio Levi-Civita", University of Padua, Padua, Italy
[2] Bruno Kessler Foundation, Povo, Italy
davide.rigoni.2@phd.unipd.it
[3] Department of Computer Science, University of Copenhagen,
Copenhagen, Denmark
{de,stfr}@di.ku.dk
[4] Pioneer Center for AI, Copenhagen, Denmark

Abstract. Object detectors are core components of multimodal models, enabling them to locate the region of interest in images which are then used to solve many multimodal tasks. Among the many extant object detectors, the Bottom-Up Faster R-CNN [39] (BUA) object detector is the most commonly used by the multimodal language-and-vision community, usually as a black-box visual feature generator for solving downstream multimodal tasks. It is trained on the Visual Genome Dataset [25] to detect 1600 different objects. However, those object categories are defined using automatically processed image region descriptions from the Visual Genome dataset. The automatic process introduces some unexpected near-duplicate categories (e.g. "watch" and "wristwatch", "tree" and "trees", and "motorcycle" and "motorbike") that may result in a sub-optimal representational space and likely impair the ability of the model to classify objects correctly. In this paper, we manually merge near-duplicate labels to create a cleaner label set, which is used to retrain the object detector. We investigate the effect of using the cleaner label set in terms of: (i) performance on the original object detection task, (ii) the properties of the embedding space learned by the detector, and (iii) the utility of the features in a visual grounding task on the Flickr30K Entities dataset. We find that the BUA model trained with the cleaner categories learns a better-clustered embedding space than the model trained with the noisy categories. The new embedding space improves the object detection task and also presents better bounding boxes features representations which help to solve the visual grounding task.

Keywords: Object Detection · Visual Genome · Bottom-Up · Data Cleaning · Label Cleaning · Object Ontology

1 Introduction

Object detection is the task of locating and classifying the objects depicted in an image [32]. This is a core task in the field that is used whenever there is the need to localize and recognize objects in images, such as when an autonomous driving car needs to recognize road signs, people, and objects in the streets.

Beyond computer vision, object detectors are the cornerstone of multimodal vision and language (V&L) tasks, which require jointly reasoning over visual and linguistic input. Indeed, in order to reason about the objects in the image, it is first necessary to identify them. Examples of such tasks are the referring expression recognition and visual grounding [7,17,22,40,41,62], visual question answering [2,42,67], visual-textual-knowledge entity linking [11–13] and image-text retrieval [24,29,34,54,65]. In these V&L tasks, the object detector is used as a static black-box feature extractor. Therefore, it needs to be accurate and comprehensive in order to support the downstream multimodal tasks.

The Bottom-Up Faster R-CNN [1] (BUA) object detector is one of the most commonly-used black box object detectors in the field. Within the V&L literature, it is the defacto standard feature extractor used to represent the visual input [16]. BUA is pretrained on the Visual Genome dataset [25] to detect 1600 objects, e.g. "chair","horse", "woman", and also to predict their attributes, e.g. "wooden", "brown", "tall". Both the category and attribute set are derived from the freely annotated region descriptions in the Visual Genome dataset, rather than using pre-defined categories like in ImageNet [10] or COCO [31]. Anderson et al. did attempt to filter the categories and attributes to prevent near-duplicates, however, the resulting 1600 categories are still imperfect. There are synonymous categories ("wrist watch", "wristwatch"), categories representing single and plurals of the same concepts ("apple", "apples"), ambiguous, difficult to differentiate, categories ("trousers", "slacks", "chinos", "lift"), and categories that actually represent attributes such as "yellow" or "black". We argue that having to predict these noisy categories is likely to prevent the object detector from supporting downstream tasks well.

In this work, we propose a new set of categories that can be used to train the BUA object detector on the Visual Genome dataset. The new set is the result of a cleaning process performed manually by a native English speaker. Starting from the original 1600 noisy categories, the ambiguous categories were merged to build the final set of 878 clean categories. We then use these clean categories to re-train the BUA object detector. In addition to evaluating its object detection performance, we analyze the model's feature embedding space, and evaluate the benefits of using its features in a downstream referring expression comprehension grounding task. In our experiments, the BUA model trained with the cleaned categories detects objects better, and, examining its feature space representation, we find out that it learns a better-clustered embedding space than the model trained with the original noisy categories. The new embedding space produces better bounding boxes feature representations, which in turn can improve performance on a downstream visual-textual grounding task.

The contributions of this paper are summarized as follows:

1. Starting from the 1600 noisy categories developed by [1], we propose a cleaner set of 878 categories with less noise and fewer near-duplicates;
2. We show that a BUA detector trained on these cleaned categories improves object detection performance and produces a better visual embedding space compared to using the original noisy categories;

3. Finally, we show that using the new detector as a black-box feature extractor can improve performance on a downstream visual-textual grounding task.

2   Related Work

This paper relates to (a) work that adopts the Bottom-Up model [1] for the detection of objects depicted in images, especially for multimodal downstream tasks, and (b) work that addresses learning neural networks with noisy labels. We describe the Bottom-Up model itself in more detail in Sect. 3.

2.1   Bottom-Up for Object Detection

Many object detectors exist [9,39,48,49,55–58,63,66], that differ according to their ability to detect objects in the image, the computing power required for their use, and their ability to recognize a large set of different objects [1,38]. An object detector should be able to identify many different objects [61] and classify them correctly. The appeal of BUA features lies in part in the large number of object categories. Nevertheless, the increase in the number of objects to be recognized leads to a more challenging classification problem.

Starting with [1], in which the extracted object detector bounding boxes were used as input to a Visual Question Answering (VQA) model, much work on VQA adopted the BUA model as object detector [5,6,20,26,44,53,60,68]. BUA features have also been used for the Referring Expression Comprehension task [23,23,40,51,52,61]. In addition, many recent large pretrained Vison and Language models use BUA features as their visual representations [4,15,27,30, 33,46,47]. These models are used as the starting point for a wide variety of multimodal tasks, including image description, VQA, natural language visual reasoning, referring expression comprehension, etc [16,21,35].

All these works directly depend on the quality of the objects detected by the BUA model. Incorrect identification and/or classification of objects may have major repercussions in the resolution of downstream tasks, making it important to analyze in more detail the labels used to train the BUA model.

2.2   Noisy Label Sets

This work, aiming to improve data quality by improving label quality, is related to the branch of research area addressing noisy label effects during neural network training. However, most of this work addresses the problem of badly labeled data, i.e. noise at the instance level (see [45] for a recent survey).

We are interested in the problem of bad or noisy labels, rather than noisy data. [36] show that their framework for estimating noise in data labelling can also identify "ontological issues" with the labels themselves. Removing duplicate labels during training improves performance on ImageNet classification, in line with the object detection improvement we find in this paper. [3] identify and correct label issues in ImageNet for better, more robust model evaluation and comparison; removing "arbitrary" label distinctions ensures models are not rewarded for overfitting to spurious noise. [50] aim to discover a "basic level" label set, i.e. the labels corresponding to the human default or basic level categories, by merging labels that are often confused. They find that training an image classifier on these categories can improve downstream image captioning and VQA.

3   Recap: Bottom-Up Faster R-CNN

The Bottom-Up [1] model is based on the Faster R-CNN [39] object detector devised to recognize instances of objects belonging to a fixed set of pre-defined categories and localize them with bounding boxes. Faster R-CNN initially uses a vision backbone, such as ResNet [18] or a VGGNet [43], to extract image features from the image. Then Faster R-CNN applies a Region Proposal Network (RPN) over the input image, that predicts a set of class-agnostic bounding box proposals for each position in the image. The RPN aims to detect all the bounding boxes that contain an object, regardless of what the object is. Then, for each detected bounding box proposal, Faster R-CNN predicts a class-aware probability score and a refinement of the bounding box coordinates to better delimit the classified object. The Faster R-CNN multi-task loss function contains four components, defined over the classification and bounding box regression outputs for the Region Proposal Network and the final bounding boxes refinement.

The BUA object detector initializes its Faster R-CNN backbone weights from a ResNet-101 [19] model pre-trained on the ImageNet [10] dataset for solving the image classification task. The model is trained on the Visual Genome [25] dataset to predict 1600 different objects. Since the Visual Genome dataset also annotates a set of attributes for each bounding box in addition to the category it belongs to, the BUA model adds an additional trainable module for predicting attributes (in addition to object categories) associated with each object localized in the image. For this reason, the BUA model adds a multi-class loss component to the original Faster R-CNN losses to train the attribute predictor module.

The 1600 categories used to train the BUA model were set by [1]. The Visual Genome dataset annotations consist of image regions associated with region descriptions (natural language strings) and the attributes of the object depicted in it. [1] extract category labels from the region descriptions, but their procedure is underspecified (for example, it is unclear if they used a part-of-speech tagger to extract nouns and adjectives as labels for objects and attributes). They filtered the original set of 2500 object strings and 1000 attribute strings based on object detection performance, resulting in a set of 1600 categories and a set of 400 attributes. However, the remaining set of categories is still noisy. It contains plurals and singular of the same concepts, such as "dog" and "dogs", overlapping categories such as "animal", "cat", and "dog". Moreover, it contains near-duplicate categories such as "motorcycle" and "motorbike", unhelpful distinctions like "lady" and "woman", labels representing attributes such as "yellow" and abstract notions like "front". These noisy labels may result in a sub-optimal representational space and likely impair the ability of the model to classify objects correctly. Given that several labels equivalently express the same meaning, whenever the model needs to predict a category for an object appearing in the image, the model needs to split its predicted probabilities among all equivalent categories. This probability split occurs not only when two or more categories express the same meaning (e.g. "hamburger" and "burger") but also when the meanings expressed by the categories overlap substantially, such as the categories "pants", "trousers", and "slacks".

4   Cleaning the Visual Genome Category Set

In this paper, we propose a new set of categories to use for training the BUA object detector. This new label set is the outcome of a cleaning process applied to the 1600 original categories by the authors of this paper, which include native English speakers. This process aimed to combine ambiguous and low-frequency categories together. During the cleaning process, the categories were joined together according to the following principles:

1. **Plurals**: singular and plurals categories, such as "giraffe" and "giraffes". In most instances, these annotations represent the same concept and should be treated as the singular category. This led to 258 category merges.
2. **Tokenization**: categories with and without spaces, such as "wrist watch" and "wristwatch", should be treated as the same category. This resulted in 29 category merges.
3. **Synonyms**, such as "microwave" and "microwave oven", "hamburger" and "burger", express similar concepts with minor differences that are usually not important. Often, as in "microwave oven", these are compound phrases that can be identified automatically, though it is important to verify them manually (e.g. "surf" and "surf board" should not be merged).
4. **Over-specific** categories with substantial annotator disagreement where several words are used interchangeably, e.g. "pants", "trouser", "sweatpants", "jean", "jeans", and "slacks".

However, during the cleaning process, it was not always clear when to merge the categories since: (i) some categories are inherently ambiguous, such as "home"; (ii) some categories are abstract and don't have the meaning of a concrete object, such as "items", "front", "distance", "day"; (iii) some categories represent attributes rather than objects, such as "yellow" and "black".

For some ambiguous labels like "lot" or "lift", visual inspection of the labelled images showed that within VG, these labels were used mostly to refer to one concept: "lot" usually showed car parking and was merged with "parking lot", similarly "lift" was merged with "ski lift". In other cases, no single meaning predominated and these labels were left un-merged (e.g. "stand" was not merged with either "baseball stand" nor "tv stand"). The abstract and attribute categories were also left as they were. In this way, the adopted cleaning process defines a surjective function that maps the original labels set to cleaner labels set.

The cleaning process produces a new set of 878 categories from the original 1600 categories (Appendix 1). Figure 1 shows frequencies of objects appearing in the Visual Genome training split, where objects are either labeled according to the original label set (in blue) or the new cleaned label set (in orange). The new labels lead mostly to the removal of many low-frequency categories in the long tail, rather than creating new very frequent categories.

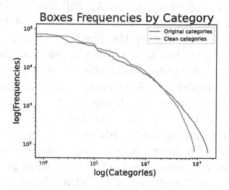

Fig. 1. LogLog plots of objects frequencies for each category. The frequencies are calculated on the training set annotations. The distribution of the original categories is in blue, and the new categories are in orange. The cleaning process did not generate high-frequency categories and at the same time removed many low-frequency categories. (Color figure online)

5   Experimental Setup

We train a BUA object detector matching the procedure of Anderson et al. [1], except that we use the new clean categories as object labels instead of the original noisy categories.

5.1   Datasets and Evaluation Metrics

Following [1], the training and test data for the models is the Visual Genome (VG) dataset [25]. It is a multipurpose dataset that contains annotations of images in the form of scene graphs that form fine-grained descriptions of the image contents. It supplies a set of bounding boxes appearing in the image, with labels such as objects and persons, together with their attributes, such as color and appearance, and the relations between them. The original VG labels were converted to object labels by [1], as described in Sect. 3. We note here that our BUA model is trained only using the VG training split, unlike some pre-trained models available, e.g. in the MILVLG repository, which use both training and validation splits for training.

To assess the object detectors' performance, we use the Mean Average Precision (AP) metric, which is the standard metric for measuring the accuracy of object detectors such as Faster R-CNN [39]. All evaluation results presented in this work are obtained on the VG test split. Average precision uses a Intersection over Union threshold of 0.5 to determine whether the predicted bounding box is sufficiently similar to the gold region. We distinguish between "macro" and "micro" (also known as "weighted") AP: MacroAP weights each category uniformly (macro-averaging class-wise precision) while MicroAP weights each category by the number of items in the category (equivalent to micro-averaging over all items, regardless of class). MacroAP will emphasize the effect of small categories, while MicroAP will be dominated by the most frequent categories.

Precision is indirectly affected by the number of categories in the label set: e.g. a random baseline over 100 categories will perform worse than a baseline over 10 categories. Since our objective in this paper is to compare models with different numbers of categories, this is an unavoidable confound. To mitigate against it, for the original model, which predicts labels in the original label set, we map its predictions to the clean label set. For example, if the model predicts "motorcycle" in the original label set, this prediction gets mapped to the same category ID as the model's "motorbike" predictions, because these two labels have been collapsed in the clean label set. This results in mapped predictions with the same number of categories as the clean label set predictions, which means that comparison between label sets is fairer. However, this procedure also removes all errors due to confusing the two labels that have been merged in clean (e.g. if the original gold label for the "motorcycle" prediction was "motorbike", this incorrect prediction is now counted as correct), which makes it a very strict evaluation.

5.2   Random Baseline

We also compare against a BUA detector trained with a randomly merged category set. The randomly merged set was created by randomly selecting pair of categories in the original set to combine until we reached the same number of categories adopted in the clean set (i.e. 878). This procedure leads to a distribution of category sizes that is very similar to the clean label set, see Appendix 1. However, the randomly merged categories will include semantically very distinct objects, e.g. bananas and motorcycles are in the same category. This allows us to separate the effect of having cleaner categories from the effect of simply having fewer categories.

5.3   Implementation Details

For the development of this work, we used the code available in the MILVLG[1] repository, which is a Pytorch implementation of the original Caffe[2] model. In

---

[1] https://github.com/MILVLG/bottom-up-attention.pytorch.
[2] https://github.com/peteanderson80/bottom-up-attention.

Table 1. BUA object detection results on the Visual Genome dataset. The model trained on the clean categories, "BUA Clean", achieves better object detection performance than the model trained on the original categories. "BUA Original→Clean-878" and "BUA Original→Random-878" are results from models trained on the original categories whose predictions are mapped to clean and random label set respectively, to match label set size (878 labels in both cases).

Model	Implementation	Visual Genome (%)	
		MacroAP50↑	MicroAP50↑
BUA Original	Caffe	9.37	15.14
BUA Original	PyTorch	9.10	15.93
BUA Original→Clean-878	PyTorch	10.72	17.34
BUA Clean	PyTorch	11.01	17.60
BUA Original→Random-878	PyTorch	9.49	15.79
BUA Random	PyTorch	9.46	15.61

particular, the MILVLG code allows to train, evaluate, and extract bounding boxes from images using both the Detectron2 framework[3] as well as the original Caffe model weights. When not explicitly indicated, we use BUA implemented with Detectron2. Between 10 and 100 bounding boxes are extracted for each image in input. We use the default MILVG hyper-parameters, apart from setting the batch size to 8, and training only on the training data split. We did not re-tune the model hyper-parameters when training on the new label set and used the same default hyper-parameters from the model trained on the original 1600 categories. The object detectors are trained for 180K iterations. All experiments were performed in a distributed parallel system using a V100 32GB GPU.[4]

6   Experiments

Our experiments compare BUA models trained on the new smaller label set with the original BUA model using the original label set. We compare these two models in terms of performance on the original object detection task, the properties of the embedding space learned by the detector, and the utility of the features in a visual grounding task on the Flickr30K Entities dataset. We expect the removal of label ambiguity in the new label set to lead to better performance on object detection and visual grounding.

6.1   Object Detection

We test object detection on the Visual Genome test set: see Table 1. The model trained on the new labels, BUA Clean, outperforms the BUA Original model by nearly two points on macro and micro AP.

---

[3] https://github.com/facebookresearch/detectron2.
[4] https://github.com/drigoni/bottom-up-attention.pytorch.

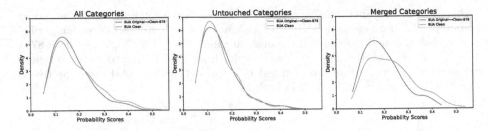

Fig. 2. KDE plots for the probability values of the argmax category predicted by the model. The plots on the left consider all the categories, the plots in the center consider just the categories that we did not merge during the cleanup process (i.e. "Untouched"), and the last plots on the right consider only the merged categories. Overall, the cleaned categories lead to higher confidence values than the original categories.

To check how much of this improvement is due to simply having a smaller label set, we also compare both against the random (i.e. BUA Random) baseline (where categories were iteratively merged to the same number of labels as the clean set) and against the same original predictions, but with predicted labels mapped to the clean set (e.g., predictions for "egg" and "eggs" are mapped to the same label, as in the clean set). The BUA Random results are slightly worse than the BUA Original model, indicating that fewer labels on their own are not enough to micro or macro AP. Mapping the original predictions to the new labels improves both metrics, indicating that many of the mistakes in the BUA Original model are due to confusion between labels that are merged in the clean set. However, performance does not reach the level of BUA Clean model, demonstrating that using better labels at training time is important. Since we see this improvement in both micro and macro AP, the new labels do not only improve frequent categories (reflected in MicroAP) or infrequent categories (MacroAP).

Figure 2 shows how noise in the category set affects the prediction confidence of the model. By "prediction confidence", we mean the probability assigned to the argmax category predicted by the model when it detects an object. These maximum probability detections play an important role in determining which detections to use in downstream tasks.[5] We find that the BUA detector trained on the cleaned categories produces more high confidence predictions than a detector trained on the original noisy categories. Closer inspection shows that this difference is due to higher confidence when predicting objects in the new merged clean categories. This confirms our hypothesis that the original categories result in probability mass being split across multiple synonymous labels, and this issue is resolved by the new cleaned categories. We do not see the same behavior with random categories (Appendix 2).

These results support the hypothesis that noise and repetition in the original label set make it difficult to learn good distinguishing features between cate-

---

[5] In V&L pretraining, it is common to use the (10-100) most confident regions [16] detected in each image.

gories. They also imply that it is necessary to retrain the object detector on cleaner labels to fully improve its detection capabilities on downstream tasks.

Our experiments also show differences in the performance of the BUA Original model as implemented in Caffe and Pytorch, despite the fact that Pytorch is meant to be a reimplementation of the Caffe version. We will see similar behaviour in the visual grounding experiment later on, where the difference between the two implementations is more substantial.

6.2   Feature Space Analysis

In this section, we attempt to characterise the differences in feature space, given features from a model trained with the clean label (i.e. Clean) set vs the original model (i.e. Original). The features are from the ResNet-101's `pool5_flat` layer; these are the most common representation used for downstream tasks (e.g. visual grounding). For each image in the VG validation set, the features corresponding to the bounding box proposals are extracted. We test two confidence thresholds: with th=0.05, the models return approximately 280,000 bounding box feature vectors, whereas with th=0.2, we only evaluate approximately 100,000 features. (Different models return slightly different but comparable numbers of proposals.)

In order to be useful for downstream tasks, we expect that bounding boxes that contain similar objects should have similar features and the same predicted categories. We test this using nearest neighbors and cluster analyses.

Nearest Neighbors. The local structure of the feature space can be examined using a nearest neighbors analysis: for each point in the embedding space (i.e. bounding box features), we calculate the proportion of K (with $K = 1, 5$, and 10) nearest neighbors that share the same category. This analysis is not affected by the different number of labels in the several sets and therefore it allows us to fairly compare models' embedding spaces. We expect the embedding space of the model trained with cleaner categories to be clustered better than the other embedding spaces. In other words, we expect that each point has more neighbors that share the same category when using cleaned labels.

Table 2 reports the results of this analysis, considering features extracted with different threshold values (i.e. 0.05 and 0.2) and considering either all features or only features from different images ("Filtered Neighbors"). This step removes features that might be from highly overlapping regions of the same image.

Overall, as expected, the bounding boxes extracted by the model trained on the cleaned label set have higher proportions of nearest neighbors that share the same category. This difference is substantial and consistent across different values of K, thresholds. Table 3 shows that the improvement is due to better neighborhoods of features with merged labels, and only in some case better features of unmerged, original labels.

The random features (i.e. Random) present results very similar to those obtained with the Original features, but with a small improvement. Surprisingly, this improvement is most evident for features of categories that are the

Table 2. Proportion of K-nearest neighbors that share the same predicted category. Results were obtained with the models trained on the original, the random, and the clean categories. Overall, at each value of K, the embedding space of the model trained on clean categories is better clustered than those of models trained on the original and random labels.

K	Th.	All Neighbors (%)			Filtered Neighbors (%)		
		Original	Random	Clean	Original	Random	Clean
1	0.05	12.15±12.25	12.36±11.15	17.30±14.79	37.32±15.07	37.83±12.32	42.34±15.82
5	0.05	24.33±13.38	24.91±12.01	29.74±15.10	34.16±13.78	34.68±12.24	39.09±15.09
10	0.05	27.76±13.23	28.37±11.87	32.96±14.85	32.91±13.71	33.48±12.19	37.84±15.11
1	0.2	51.02±22.74	51.88±20.91	55.36±20.03	69.22±18.99	70.03±16.76	71.96±17.54
5	0.2	60.40±19.75	61.47±17.84	63.92±19.00	65.12±19.68	66.12±17.54	68.29±18.58
10	0.2	60.55±20.18	61.71±18.20	64.16±19.31	62.95±20.43	64.05±18.34	66.32±19.39

same between Original and Random (Appendix 3), rather than the categories that were merged in Random, suggesting that there is an advantage to training on fewer labels overall.

Surprisingly, when features from the same image are ignored (Filtered Neighbors), the percentage of neighbors who share the same category increases dramatically. This indicates that BUA features tend to place visually similar regions (from the same image) close together, regardless of their semantic content (their predicted object label).

In conclusion, the analysis on the neighbors verified our main claim: when the BUA object detector is trained with the original noisy labels, it results in a sub-optimal representational space that can be improved simply by retraining the model on cleaner labels set.

Distances We examine the global structure of the feature space by looking at the distances between items with the same label (intra-category) and the distances between the category centroids (inter-category). If the feature space is organised by categories, then intra-category distances should be small, while inter-category distances should be larger.

Table 4 reports the inter and intra-categyr distances for features from the models trained with the original, clean, and random labels. Intra-category distance is the average Euclidean distance between features with the same predicted label, while inter-category distance is the average Euclidean distance between the centroids of each category (all averages are macro-averages over categories). We see that the Clean labels lead to categories that are clustered more closely together, evident in a lower average intra-category distance, compared to both the Original and Random labels. Counter to our hypothesis, inter-category distance is lower when using Clean labels, especially compared to the Original labels, and also slightly lower than Random labels. This indicates that the global feature space is also more compact overall. Surprisingly, across all feature spaces

Table 3. Proportion of K-nearest neighbors that share the same predicted category, comparing models trained using the original versus the clean categories. (See Table 6 for a comparison with random categories.) "Th." indicates the threshold values adopted for bounding box extraction. "Merged" refers to original categories that are merged into one new clean category. "Untouched" refers to those categories not merged with others during the cleaning process, and "All" refers to all the categories. Overall, the clean features are better clustered than the original features.

Th.	K	Categories	All Neighbors (%)		Filtered Neighbors (%)	
			Original	Clean	Original	Clean
0.05	1	All	12.15±12.25	17.30±14.79	37.32±15.07	42.34±15.82
0.05	1	Untouched	9.19±9.47	8.56±8.84	32.20±16.21	32.86±15.18
0.05	1	Merged	12.71±12.62	19.03±15.12	38.28±14.65	44.22±15.26
0.05	5	All	24.33±13.38	29.74±15.10	34.16±13.78	39.09±15.09
0.05	5	Untouched	19.71±12.27	20.35±11.77	28.62±24.39	29.48±13.68
0.05	5	Merged	25.19±13.40	31.60±14.99	35.19±13.41	40.99±14.63
0.05	10	All	27.76±13.23	32.96±14.85	32.91±13.71	37.84±15.11
0.05	10	Untouched	22.55±12.64	23.33±12.26	26.97±14.15	27.95±13.58
0.05	10	Merged	28.73±13.12	34.87±14.57	34.01±13.34	39.80±14.62
0.2	1	All	51.02±22.74	55.36±22.03	69.22±18.99	71.96±17.54
0.2	1	Untouched	43.34±21.95	41.37±21.45	62.26±23.11	60.92±22.20
0.2	1	Merged	52.14±22.64	57.29±21.40	70.23±18.09	73.48±16.22
0.2	5	All	60.40±19.75	63.92±19.00	65.12±19.68	68.29±18.58
0.2	5	Untouched	51.88±21.68	50.58±20.88	56.33±23.38	55.51±22.08
0.2	5	Merged	61.64±19.14	65.75±17.97	66.40±18.74	70.05±17.32
0.2	10	All	60.55±20.18	64.16±19.31	62.95±20.43	66.32±19.39
0.2	10	Untouched	50.83±22.63	49.92±21.42	52.89±23.80	52.12±22.38
0.2	10	Merged	61.97±19.39	66.12±18.15	64.42±19.46	68.28±18.09

(Original, Clean, and Random) the intra-category distances are higher than the inter-category distances, suggesting that features from different categories are highly intermingled.

In order to control for label set and category size, we map the original features to the clean (i.e. "Orig.→Clean-878") or random (i.e. "Orig.→Random-878") set of categories, ensuring the same number of points in each label category, as well as the same number of labels. This results in a higher intra-category average distance, compared to the original categories, which indicates that features from merged labels are not mapped to nearby parts of the space. Notably, the clean mapping leads to only very slightly lower, intra-category distances compared to the random mapping.

Table 4. Intra-category (average pairwise of points with the same label) and inter-category (average distance between categegory/label centroid) Euclidean distances in different feature spaces. Results were obtained with the models trained on original (i.e Orig.), clean, and random label sets. The model trained on cleaner labels presents lower distances in both the intra-categories and the inter-categories analysis.

Analysis	Orig.	Orig.→Clean-878	Clean	Orig.→Random-878	Random
Intra-Category	49.69	52.10	45.37	52.96	47.77
	±8.64	±8.10	±6.98	±8.63	±7.87
Inter-Category	47.97	NA	39.76	NA	40.19
	±5.31		±4.94		±5.87

Overall, our analysis of the local neighborhoods shows a positive effect of the clean label set, with more neighbors with the same label. However, the analysis of the global feature space suggests that the BUA features are not well separated according to object semantics, regardless of the label set used.

6.3 Visual Grounding Results

In this section, we investigate the utility of the features extracted with the BUA model in a visual grounding task, namely Referring Expression Comprehension, on the Flickr30K Entities dataset. Our expectation is that features extracted with the models trained on the new categories will be more coherent and useful than those extracted with the model trained on the original set of categories, leading to better performance on this downstream task.

As our visual grounding model, we use the Bilinear Attention Network [23] (BAN) model, which, even if no longer state of the art, obtains relatively good results on the Flickr30k Entities dataset. The advantage of using the BAN model is that it is a simple model that uses a straightforward fusion component to merge the text and visual information, and that requires only the Flickr30k Entities dataset for training (other models that achieve higher scores are pre-trained on much larger data sets and have more complex architecture [14, 22, 28, 59, 64]). BAN implements a simple architecture that uses only the 2048-dimensional bounding box features extracted from the object detector as the visual input features; it does not use the label predicted from the features. On the text side, the model initializes each word with its GloVe [37] embedding and uses a GRU [8] to generate a representation for the sentence. The visual and textual representations are then fused together through a bilinear attention networks. The simple fusion component allows us to see the effect of different visual feature spaces more clearly. We use the code provided by the authors[6], and no hyper-parameters were changed from the original model. The experiments were performed using an A5000 24GB GPU.

---

[6] https://github.com/jnhwkim/ban-vqa

Table 5. Visual Grounding results obtained with the Bilinear Attention Networks (BAN) [23] model on the Flickr30k Entities dataset. "R@K" refers to the Recall metric with the top K predictions, while "UB" refers to the upper bound results that can be achieved with the bounding boxes extracted with the indicated threshold. The features extracted with the model trained on the clean labels set consistently perform better than the original features.

| Features | Threshold | Test Set (%) | | | | N. Bounding Boxes | | |
		R@1↑	R@5↑	R@10↑	UB↑	Min	Max	Test
[23]	0.2	69.80	84.22	86.35	87.45	10	91	30034
Original	0.2	73.32	84.21	85.67	86.53	2	89	20916
Clean	0.2	73.41	85.08	86.52	87.31	2	93	21923
Original	0.1	74.72	86.06	88.71	90.70	5	100	36792
Clean	0.1	75.43	86.76	89.56	91.22	7	100	36719
Original	0.05	75.41	85.46	88.86	92.38	12	100	59256
Clean	0.05	75.75	85.88	89.52	92.67	11	100	56731

Table 5 reports the results obtained in the visual grounding task by the BAN model trained using the features extracted by both the models trained on the original (i.e. Original) and new cleaner (i.e. Clean) label sets. Whenever BAN is trained using the Clean features, the performance of the model increases compared to the BAN model trained on the Original features. The improvement is small but consistent across bounding box thresholds and recall levels.

We also see that the BUA PyTorch implementation of the BAN model always achieves better performance than the Caffe implementation, even with fewer bounding boxes. This result implies that the implementation code used to train the object detector strongly impacts the results of the visual grounding task, although, in the object detection task, there is only a small improvement[7].

In conclusion, the results obtained with the BAN model on the visual grounding task suggest that the BUA model trained using a cleaner set of labels presents not only a well-clustered embedding space but also a more useful features representations able to improve downstream tasks.

7   Conclusion and Future Work

This paper introduced a new set of 878 category labels to retrain the BUA model, which refines the originally noisy 1600 categories by merging labels that are synonymous or have highly related meanings. We investigated the effect of using the cleaner label set in terms of performance on the original object detection task,

---

[7] The extracted features used in the BAN paper are not made available by the authors. However, some "reproducibility" features (slightly different) were made available by third users (https://github.com/jnhwkim/ban-vqa/issues/44) who successfully reproduced the main paper results.

showing that the model trained on the new set of labels improves its object detection capabilities. We also analyzed the embedding space in the object detector trained on the cleaned categories and showed that it is better clustered than the embedding space derived from the original categories. Finally, we evaluated the utility of the new model as black-box feature extractor for a downstream visual-textual grounding task with the Bilinear Attention Network model. The results show that features from the new object detector can consistently improve the BAN model across commonly used object detection thresholds.

Future work involves studying the effect of using the improved label set on large pretrained language-and-vision models, such as VILBERT [33] and LXMERT [47]. Since these models use the bounding box category labels predicted by the object detector in their loss function, in addition to using the features as their visual input, removing label noise should benefit these models.

In this work, we merged the noisy categories using a skilled human annotator, which may have introduced some unwanted human bias or error into the cleaning process. Nevertheless, our approach highlights the advantage of using improved label sets, both for core object detection and downstream multimodal task performance. Future work could generate alternative cleaned categories by merging similar ones, e.g. using a framework similar to Confidence Learning [36].

Acknowledgements. This work was supported in part by the Pioneer Centre for AI, DNRF grant number P1. Davide Rigoni was supported by a STSM Grant from Multi-task, Multilingual, Multi-modal Language Generation COST Action CA18231. We acknowledge EuroHPC Joint Undertaking for awarding us access to Vega at IZUM, Slovenia.

Appendix 1: Frequencies by Categories

We introduced both the set of clean and random categories deriving from the original ones. The original label set is defined by 1600 categories, while both the new clean and the random sets are defined by 878 categories. Figure 3 shows frequencies of objects appearing in the Visual Genome training split, where objects are either labeled according to the original label set (in blue), the new cleaned label set (in orange), or the random label set (in brown). The new label sets lead mostly to the removal of many low-frequency categories in the long tail, rather than creating new very frequent categories. Surprisingly, the random procedure that generated the random label set also removed the long tail of low-frequencies categories.

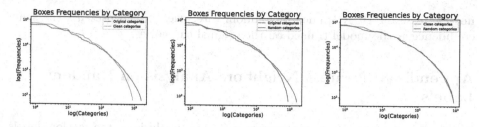

Fig. 3. LogLog plots of objects frequencies for each category. The frequencies are calculated on the training set annotations. The distribution of the original categories is in blue, the new categories are in orange, and the random categories are in brown. The cleaning process did not generate high-frequency categories and at the same time removed many low-frequency categories for both cleaner and random label sets. (Color figure online)

Appendix 2: Prediction Confidence

In Fig. 4 it is reported the KDE plots for the probability values of the argmax category predicted by the original, clean, and random label sets.

We find that the BUA detector trained on the cleaned categories produces more high confidence predictions than a detector trained on the original noisy categories. Closer inspection shows that this difference is due to higher confidence when predicting objects in the new merged clean categories. However, this is

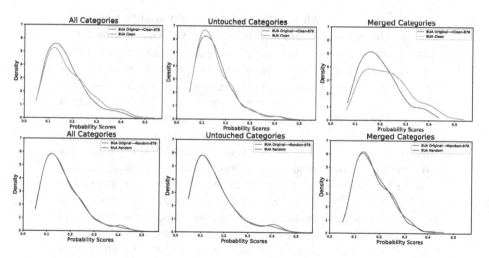

Fig. 4. KDE plots for the probability values of the argmax category predicted by the model. The plots on the left consider all the categories, the plots in the center consider just the categories that we did not merge during the cleanup process (i.e. "Untouched"), and the last plots on the right consider only the merged categories. Overall, the cleaned categories lead to higher confidence values than the original categories, while there is no difference between original and random categories.

not the case for BUA trained on random categories, which presents the same confidence as the model trained on the original categories.

Appendix 3: Nearest Neighbors Analysis on Random Labels

In this section, we perform the nearest neighbors analysis on the random labels focusing on the "Merged", "Untouched", and "All" categories. Table 6 reports the results of this analysis, considering features extracted with different threshold values (i.e. 0.05 and 0.2) and considering either all features or only features from different images ("Filtered Neighbors"). This step removes features that might be from highly overlapping regions of the same image.

The random features present results very similar to those obtained with the original features, but with a small improvement. In other words, there is an advantage to training on fewer labels overall. However, the improvement given by clean labels is much greater than that obtained with the random labels, strengthening the importance of training BUA with clean categories.

Table 6. Proportion of K-nearest neighbors that share the same predicted category, comparing models trained using the original versus random categories (cf. Table 3). The random features present small improvements over the original features, suggesting that there is a small advantage in training with fewer labels; however clean labels help more.

Th.	K	Categories	All Neighbors (%)		Filtered Neighbors (%)	
			Original	Random	Original	Random
0.05	1	All	12.15±12.25	12.36±11.15	37.32±15.07	37.83±12.32
0.05	1	Untouched	10.06±11.91	10.32±12.13	35.81±13.91	36.33±14.03
0.05	1	Merged	13.16±12.29	11.35±10.50	38.05±15.55	38.56±12.90
0.05	5	All	24.33±24.38	24.91±12.01	34.16±13.78	34.68±12.24
0.05	5	Untouched	22.66±12.60	23.12±12.61	33.09±12.88	33.54±12.78
0.05	5	Merged	25.13±13.66	25.77±11.61	34.68±14.16	35.23±11.93
0.05	10	All	27.76±13.23	28.37±11.87	32.91±13.71	33.48±12.19
0.05	10	Untouched	26.40±12.34	26.98±12.04	31.89±12.78	32.42±12.71
0.05	10	Merged	28.42±13.60	29.04±11.55	33.39±14.12	33.99±11.89
0.2	1	All	51.02±22.74	51.88±20.91	69.22±18.99	70.03±16.76
0.2	1	Untouched	45.05±21.50	46.30±21.68	65.93±17.39	66.70±17.10
0.2	1	Merged	53.84±22.73	54.70±19.93	70.98±19.37	71.72±16.33
0.2	5	All	60.40±19.75	61.47±17.84	65.12±19.68	66.12±17.54
0.2	5	Untouched	56.60±18.22	57.61±17.99	61.87±18.10	62.75±17.67
0.2	5	Merged	62.33±20.20	63.42±17.45	66.79±20.22	67.82±17.23
0.2	10	All	60.55±20.18	61.71±18.20	62.95±20.43	64.05±18.34
0.2	10	Untouched	57.05±18.44	58.14±18.24	59.76±18.69	60.69±18.39
0.2	10	Merged	62.31±20.78	63.51±17.91	64.56±21.06	65.75±18.07

Appendix 4: Clean Labels

The cleaning process produces a new set of 878 categories from the original 1600 categories, which we report below.

```
1:yolk,525:egg,324:eggs
2:goal
3:bathroom,1574:restroom
4:macaroni
6:toothpick
10:parrot
11:tail fin,1468:fin
13:calculator
15:toilet,85:toilet seat,302:toilet tank,385:toilet bowl,444:toilet lid
16:batter,5:umpire,14:catcher,474:baseball player,1210:baseball players,794:players
    ,92:player,78:tennis player,377:soccer player,207:pitcher
1254:referee
17:stop sign,17:stopsign,1437:sign post,941:traffic sign,589:street sign,817:signs
    ,129:sign,245:stop
1474:bus stop
18:cone,576:cones,560:traffic cone,658:safety cone
19:microwave,19:microwave oven
20:skateboard ramp
21:tea
23:products
25:kettle,67:tea kettle
26:kitchen
27:refrigerator,27:fridge
28:ostrich
29:bathtub,196:bath tub,306:tub
1168:blind,30:blinds
31:court,39:tennis court
314:urinals,32:urinal
34:bed,893:beds,947:bedding,660:bedspread,1343:bed frame
35:flamingo
36:giraffe,38:giraffes,471:giraffe head
37:helmet
1229:laptops,41:laptop,1124:laptop computer
42:tea pot,562:teapot
43:horse,187:horses,1319:pony
44:television,44:tv
1351:short,45:shorts
46:manhole,1014:manhole cover
47:dishwasher,148:washer
49:sail
125:parasail,1569:parachute
51:man,1511:young man,683:men,774:guy,1441:male
52:shirt,1404:tshirt,1404:t shirt,1404:t-shirt,1226:dress shirt,1099:tee shirt,1157:
    sweatshirt,653:undershirt,233:tank top,133:jersey,1288:blouse
686:cars,53:car,955:passenger car,1334:sedan
1479:police car
54:cat,185:cats,477:kitten,1117:kitty
55:garage door
56:bus,380:buses
57:radiator,1006:heater
58:tights
60:racket,60:racquet
251:home plate
1362:home
895:base
61:plate,956:plates,1378:paper plate,540:saucer,587:dishes,788:dish
65:ocean,1214:sea
63:beach
327:sand
1587:shoreline,816:shore
64:trolley
66:headboard,66:head board
68:wetsuit,217:wet suit
69:tennis racket,69:tennis racquet
70:sink,692:sinks,1123:bathroom sink,1424:basin
815:trains,71:train,1448:passenger train,899:train front,626:train car,1182:train
    cars,490:carriage,637:locomotive,1275:caboose,1318:railroad
73:sky,1217:weather
1273:skies
75:train station,272:train platform,319:platform,387:station
76:stereo
77:bats,301:bat,657:baseball bat
79:toilet brush
80:lighter
83:hair dryer
142:elephants,84:elephant
86:zebra,88:zebras
87:skateboard,87:skate board,1224:skateboards
89:floor lamp,1426:table lamp,1083:lamps,225:lamp,161:chandelier,905:light fixture
91:woman,749:women,858:lady,996:she,1486:ladies,1245:mother,1539:bride
93:tower
685:bicycles,94:bicycle,506:bikes,100:bike
95:magazines,1096:magazine
```

```
96:christmas tree
495:umbrellas ,97:umbrella ,1523:parasol
151:cows,98:cow,428:bull ,793:cattle ,583:ox,1202:calf
280:herd
99:pants,1492:pant ,781:trouser ,1111:sweatpants ,973:jean ,48:jeans ,651:snow pants ,503:
    ski pants ,1344:slacks
102:living room
103:latch
104:bedroom
1204:grapes ,105:grape
106:castle
107:table ,1301:tables ,875:end table ,200:coffee table
108:swan
109:blender
110:orange ,408:oranges
219:teddy bears ,111:teddy bear ,1293:teddy ,270:stuffed animals ,767:stuffed animal ,647:
    stuffed bear
113:meter ,1481:meters ,211:parking meter
115:runway
262:ski boots ,117:ski boot
118:dog,338:dogs ,1532:puppy
119:clock ,1393:clocks ,7:alarm clock ,1274:clock hand,509:clock face
1023:hour hand
120:hair ,505:mane,1187:bangs
121:avocado
123:skirt
124:frisbee
126:desk
128:mouse,486:computer mouse
134:reigns ,574:bridle ,24:halter ,1388:harness
1321:hot dogs ,1321:hotdogs ,135:hot dog ,135:hotdog,1384:sausage
136:surfboard ,136:surf board ,351:surfboards
163:glasses ,138:glass
1493:wine glasses ,614:wine glass
625:sunglasses ,990:eye glasses ,800:eyeglasses
1327:shades ,620:shade
1139:snow board ,139:snowboard
140:girl ,754:girls ,953:little girl
141:plane ,532:planes ,489:airplanes ,132:airplane ,536:aircraft ,803:jets ,545:jet
143:oven ,679:oven door ,198:stove
1233:range
146:area rug ,335:rug ,467:carpet
344:bears ,147:bear ,131:polar bear ,283:cub
149:date
150:bow tie ,578:necktie ,655:neck tie ,268:tie
152:fire extinguisher
153:bamboo
154:wallet
156:truck ,839:trucks
158:boat ,234:boats ,59:sailboat ,59:sail boat ,421:ship ,719:yacht ,988:canoe ,1143:kayak
159:tablet
160:ceiling
162:sheep ,164:ram ,231:lamb
705:kites ,165:kite
166:salad ,868:lettuce ,1398:greens
167:pillow ,332:pillows ,842:pillow case ,675:throw pillow
168:fire hydrant ,168:hydrant
169:mug,232:cup ,850:coffee cup
170:tarmac ,1495:asphalt ,831:pavement
171:computer ,1032:computers ,1053:cpu
172:swimsuit ,1174:swim trunks ,388:bikini ,1008:bathing suit
173:tomato ,665:tomatoes ,426:tomato slice
174:tire ,1456:tires
175:cauliflower
177:snow
178:building ,670:buildings ,581:skyscraper ,1193:second floor
1581:sandwiches ,179:sandwich ,1052:sandwhich
180:weather vane ,753:vane
181:bird ,1000:birds
182:jacket ,381:coat ,1521:ski jacket ,566:suit jacket ,836:blazer
183:chair ,699:chairs ,552:office chair ,390:lounge chair ,157:beach chair ,504:seat ,1022:
    seats ,242:stool ,1015:stools ,1325:recliner
184:water ,1429:ocean water
186:soccer ball ,1235:balls ,568:ball ,481:tennis ball ,674:baseball
189:barn
190:engine ,619:engines ,567:train engine ,1093:jet engine
191:cake ,12:birthday cake ,273:cupcake ,764:frosting
192:head
193:head band ,368:headband
780:skiers ,194:skier ,1009:skiier
195:town
197:bowl ,1027:bowls
199:tongue
1241:floors ,201:floor ,1556:tile floor ,1310:flooring
519:uniforms ,202:uniform
203:ottoman ,424:sofa ,137:couch ,228:armchair
204:broccoli
205:olive ,1148:olives
206:mound,459:pitcher 's mound
1530:jug
```

```
208:food ,703:meal
209:paintings ,346:painting
210:traffic light ,1347:traffic lights
212:bananas ,531:banana ,554:banana peel ,464:banana bunch ,266:banana slice
958:peel
213:mountain ,457:mountains ,1304:mountain top ,984:mountain range ,1487:peak ,1375:
     mountainside
1161:landscape
214:cage
218:radish
221:suitcase ,221:suit case ,429:suitcases ,297:luggage
507:drawer ,222:drawers
1069:grasses ,223:grass ,488:lawn ,963:turf
101:field ,1286:grass field ,418:pasture
667:soccer field ,114:baseball field ,763:infield ,729:outfield ,22:dugout
289:apples ,224:apple
226:goggles ,1246:ski goggles
510:boys ,227:boy
229:ramp
269:burners ,230:burner
235:hat ,798:cowboy hat ,487:cap ,721:baseball cap ,1153:beanie ,1149:ball cap
922:brim
239:visor
236:soup
238:necklace
240:coffee
241:bottle ,379:bottles ,1554:beer bottle ,931:wine bottle ,476:water bottle
1267:surfers ,244:surfer
1203:back pack ,246:backpack
1498:pack
247:shin guard ,876:shin guards
248:wii remote ,432:remotes ,805:remote ,348:remote control ,723:controller ,812:game
     controller ,1208:controls ,1589:control ,1303:wii
1101:walls ,249:wall ,62:rock wall ,1220:stone wall ,1279:brick wall
250:pizza slice ,127:pizza ,914:pizzas
1466:slices ,1005:slice
252:van ,1281:minivan ,669:suv ,704:station wagon
253:packet
1402:earring ,254:earrings
255:wristband ,569:wrist band
797:track ,256:tracks
257:mitt ,1256:baseball mitt ,1454:catcher's mitt ,1049:baseball glove
258:dome
259:snowboarder
260:faucet ,1328:tap
261:toiletries
263:room
806:snowsuit ,265:snow suit
591:benches ,267:bench ,1191:park bench
271:zoo
717:curtains ,274:curtain ,872:drape ,188:drapes
275:ear ,524:ears
276:tissue box ,1198:tissues ,1519:tissue
277:bread ,384:bun
792:toast
329:scissor ,278:scissors
412:vases ,279:vase
281:smoke
284:tail ,443:tails
285:cutting board
286:wave ,713:waves ,1311:surf
288:windshield
290:mirror ,1363:side mirror
291:license plate ,1541:license
382:trees ,292:tree ,1185:pine trees ,688:pine tree ,1436:tree line
1562:tree branch ,1356:tree branches ,933:tree trunk
1575:twig ,1271:twigs
999:branches ,1067:branch
833:wheels ,293:wheel ,791:front wheel ,666:back wheel
294:ski pole ,890:ski poles
295:clock tower
296:freezer
299:mousepad ,1257:mouse pad
300:road ,584:roadway ,122:highway ,1056:dirt road ,309:street ,353:lane ,1137:intersection
304:neck
305:cliff
307:sprinkles
308:dresser ,303:vanity
310:wing ,1232:wings ,145:tail wing
311:suit
761:outfit
312:veggie ,861:veggies
460:palm tree ,313:palm trees
1040:doors ,315:door ,1490:glass door
316:propeller
317:keys ,840:key
411:skatepark ,318:skate park
320:pot ,1551:pots
321:towel ,363:towels ,1195:hand towel
322:computer monitor ,220:monitors ,50:monitor ,597:computer screen ,116:screen
```

454:window,979:windows,1422:side window,537:front window,282:skylight,1586:panes
455:bridge
1432:overpass
456:corn
458:beer
609:ski,462:skis,1337:skiis
465:tennis shoe,1173:tennis shoes,748:sneakers,904:sneaker,771:shoes,243:shoe,520:cleat,1340:cleats,706:boots,873:boot
468:eye,547:eyes
469:urn
470:beak
473:mattress
475:wine
478:archway,1549:arches,636:arch
929:candles,479:candle
480:croissant
482:dress
483:column,1496:columns
1238:utensil,484:utensils,765:forks,264:fork,1179:butter knife,757:knife,557:spoon,517:chopsticks,542:silverware,1316:knives
622:cellphone,485:cell phone,463:phone,813:smartphone,582:telephone,514:iphone
498:ipod
492:cabinets,611:cabinet,558:cabinet door,819:cupboards,1330:cupboard
493:lemons,678:lemon
494:grill
496:meat,1380:beef
497:wagon
499:bookshelf,863:book shelf,848:shelf,887:bookcase,1163:shelves
501:roof
502:hay
508:game
555:baseball game
74:match,638:tennis match,974:tennis
511:rider
512:fire escape
1535:pans,516:pan,1295:skillet
588:hills,518:hill,1132:hill side,1175:hillside,513:slope,1025:ski slope
521:costume
522:cabin
523:police officer,431:policeman,855:officer,826:police
1268:arrows,528:arrow\scriptsize
529:toothbrush
533:garden,768:yard
534:forest,409:woods,1228:wood
535:brocolli
538:dashboard
1222:statues,539:statue,682:monument,1332:sculpture
571:fruits,543:fruit
544:drain
546:speaker,1058:speakers
549:lid
550:soap
601:rock,551:rocks,1087:stone,967:stones,845:boulder,1457:boulders
553:door knob,976:doorknob,698:knob,607:knobs
556:asparagus
559:pineapple
561:nightstand,561:night stand
563:taxi,1265:taxi cab,901:cab
564:chimney
565:lake
865:pickles,570:pickle
572:pad,1369:pads,33:knee pads,994:knee pad,747:kneepad
575:breast
880:head light,577:headlight,590:headlights
579:skater,298:skateboarder
580:toilet paper
1160:socks,585:sock
586:paddle,1464:oar
593:card
807:bushes,595:bush,1336:shrubs,1305:shrub,287:hedges,215:hedge
596:rice
1183:spoke,598:spokes
599:flowers,663:flower,689:bouquet
600:bucket
603:pear,1491:pears
604:sauce,608:mustard,786:ketchup,1566:condiments
605:store,404:shop,1131:storefront
866:stand
610:stands,985:bleachers
612:dirt,466:ground,1272:soil,1476:pebbles,1477:mud
613:goats,712:goat
617:pancakes
673:kid,618:kids,1063:children
621:feeder
624:blanket,446:comforter,1200:quilt
627:magnet,641:magnets
629:sweater,407:hoodie,645:vest
630:signal
632:log
633:vent,1043:air vent

```
825:masts ,1046:mast
828:biscuit
1074:toys ,829:toy
1346:doll
832:outside
834:driver
835:numbers ,992:number
838:cabbage
841:saddle
843:goose ,383:geese
844:label
846:pajamas
847:wrist
849:cross
854:air
856:pepperoni
857:cheese
859:kickstand
936:countertop ,860:counter top
862:baseball uniform
867:netting ,1570:mesh
112:net ,883:tennis net
870:lime
884:animal ,871:animals
874:railing ,1475:railings
237:fence ,1412:wire fence ,1390:fence post ,1283:fencing
1563:tusks ,879:tusk
881:walkway ,885:boardwalk
882:cockpit
891:parking lot ,852:lot
573:dispenser ,896:soap dispenser
897:banner
898:life vest ,727:life jacket
1180:words ,900:word ,1597:text
903:exhaust pipe
1248:power line ,906:power lines
908:scene
909:buttons ,1089:button
910:roman numerals ,960:roman numeral ,769:numeral ,756:numerals
911:muzzle
912:sticker ,1170:stickers ,1387:decal
913:bacon
917:stairs ,877:steps ,1484:staircase ,1423:stairway
918:triangle
921:beans ,1135:bean
924:letters ,1472:letter ,1122:lettering
926:menu
983:fingers ,927:finger ,733:thumb
930:picnic table
932:pencil
934:nail
935:mantle
176:fireplace
937:view
938:line ,1155:lines ,1560:baseline
1467:arms ,942:arm
944:stabilizer
945:dock ,1138:pier
946:doorway
950:canal
951:crane
952:grate
954:rims ,1066:rim
957:background
1349:strings ,961:string
920:rope
1297:cable
1165:cord ,1528:cords
962:tines
964:armrest
966:leash
1147:stop light ,968:stoplight
970:front
948:end
971:scarf
972:band
975:pile ,1192:stack
977:foot ,916:feet
980:restaurant
981:booth
987:pastry ,741:pastries
989:sun ,1002:sunset
993:fish
995:fur
998:rod
1001:printer
1003:median
1007:prongs
1010:rack
1012:blade ,1592:blades
```

```
1213: hotel
1215: cover
1216: tarp
1218: notebook
1221: closet
1223: bank
1225: butter
1227: knee
1230: cuff
1231: hubcap
1234: structure
1236: tunnel
1237: globe
1239: dumpster
1240: cd , 928: dvds , 1510: disc
1242: wrapper
1243: folder
1244: pocket
1249: wake
1204: rose , 1250: roses
1252: reflection
1253: air conditioner
1255: barricade
1258: garbage can , 1360: trash bin , 889: trash bag , 878: trashcan , 526: trash can
1259: buckle
1260: footprints
1262: muffin
1263: bracket
1264: plug
1269: control panel , 1506: panel
1270: ring
1276: playground
1277: mango
1278: stump
1280: screw
1312: cloth
1285: clothes , 1342: clothing
1287: plumbing
1289: patch
1290: scaffolding
1291: hamburger , 1150: burger
1296: cycle
1299: bark
1300: decoration
1302: palm
1306: hoof
1307: celery
1308: beads
1309: plaque
1540: spray
1543: passengers , 1313: passenger
1314: spot , 1599: spots
1315: plastic
1317: case
1320: muffler
1322: stripe , 1392: stripes
1323: scale
1324: block , 1503: blocks
1326: body
1329: tools
1331: wallpaper
1333: surface
1335: distance
1338: lift , 1164: ski lift
1339: bottom
1341: roll
1348: symbol
1350: fixtures
1352: paint
1353: candle holder
1354: guard rail
1355: cyclist
1357: ripples
1358: gear
1359: waist
1364: brush
1370: ham
1372: reflector
1373: figure
1376: black
1409: brick , 1377: bricks
1379: stick
1381: patio
82: gazebo
1383: back
1386: farm
1389: monkey
1391: door frame
1428: pony tail , 1394: ponytail
1395: toppings
```

```
1396:strap
1399:chin
1400:lunch
1403:area
1405:cream
1408:lanyard
1410:hallway
1411:cucumber
1413:fern
1414:tangerine
1418:wheelchair
1419:chips
1420:driveway
1421:tattoo
1425:machine
1427:radio
1430:inside
1431:cargo
1433:mat
1435:flower pot
1438:tube
1439:dial
1440:splash
1442:lantern
1443:lipstick
1445:tongs
1446:ski suit
1449:bandana
1450:antelope
1458:mannequin
1459:plain
1460:layer
1463:piece
1465:bike rack
1470:hood
1473:dot
1478:claws
1480:crown
1483:entrance
1485:shrimp
1488:vines ,1577:ivy
1489:computer keyboard ,886:keypad ,72:keyboard
1494:stall
1497:sleeve
1499:cheek
1501:land
1502:day
1504:courtyard
1505:pedal
1507:seeds
1508:balcony
1509:yellow
1513:crumbs
1514:spinach
1515:emblem
1516:object ,1548:objects
1518:cardboard
1524:terminal
1525:surfing
1526:streetlight ,1526:street light ,1368:street lamp
1527:alley
1531:antenna
1534:diamond
1536:fountain
1537:foreground
1538:syrup
1546:shack ,1571:hut
1547:trough
1550:streamer
1552:border
1557:page
1558:pin
1559:items
1564:donkey
1568:envelope
1572:butterfly
1576:pilot
1578:furniture
1579:clay
1582:lion
1583:shingles
1590:lock
1591:microphone
1593:towel rack ,1561:hanger
1594:coaster
1595:star
1600:buoy
```

References

1. Anderson, P., et al.: Bottom-up and top-down attention for image captioning and visual question answering. In: Proceedings of the IEEE Conference on Computer Vision and Pattern Recognition, pp. 6077–6086 (2018)
2. Antol, S., et al.: VQA: visual question answering. In: ICCV, pp. 2425–2433 (2015)
3. Beyer, L., Hénaff, O.J., Kolesnikov, A., Zhai, X., van den Oord, A.: Are we done with imagenet? https://doi.org/10.48550/ARXIV.2006.07159. https://arxiv.org/abs/2006.07159
4. Bugliarello, E., Cotterell, R., Okazaki, N., Elliott, D.: Multimodal pretraining unmasked: a meta-analysis and a unified framework of vision-and-language BERTs. arXiv preprint arXiv:2011.15124 (2020)
5. Cadene, R., Ben-Younes, H., Cord, M., Thome, N.: MUREL: multimodal relational reasoning for visual question answering. In: Proceedings of the IEEE/CVF Conference on Computer Vision and Pattern Recognition, pp. 1989–1998 (2019)
6. Cadene, R., Dancette, C., Cord, M., Parikh, D., et al.: RUBi: reducing unimodal biases for visual question answering. In: Advances in Neural Information Processing Systems 32 (2019)
7. Chen, K., Gao, J., Nevatia, R.: Knowledge aided consistency for weakly supervised phrase grounding. In: Proceedings of the IEEE Conference on Computer Vision and Pattern Recognition, pp. 4042–4050 (2018)
8. Cho, K., et al.: Learning phrase representations using RNN encoder-decoder for statistical machine translation. arXiv preprint arXiv:1406.1078 (2014)
9. Dai, X., et al.: Dynamic head: unifying object detection heads with attentions. In: Proceedings of the IEEE/CVF Conference on Computer Vision and Pattern Recognition, pp. 7373–7382 (2021)
10. Deng, J., Dong, W., Socher, R., Li, L.J., Li, K., Fei-Fei, L.: ImageNet: a large-scale hierarchical image database. In: 2009 IEEE Conference on Computer Vision and Pattern Recognition, pp. 248–255. IEEE (2009)
11. Dost, S., Serafini, L., Rospocher, M., Ballan, L., Sperduti, A.: Jointly linking visual and textual entity mentions with background knowledge. In: Métais, E., Meziane, F., Horacek, H., Cimiano, P. (eds.) NLDB 2020. LNCS, vol. 12089, pp. 264–276. Springer, Cham (2020). https://doi.org/10.1007/978-3-030-51310-8_24
12. Dost, S., Serafini, L., Rospocher, M., Ballan, L., Sperduti, A.: On visual-textual-knowledge entity linking. In: ICSC, pp. 190–193. IEEE (2020)
13. Dost, S., Serafini, L., Rospocher, M., Ballan, L., Sperduti, A.: VTKEL: a resource for visual-textual-knowledge entity linking. In: ACM, pp. 2021–2028 (2020)
14. Dou, Z.Y., et al.: Coarse-to-fine vision-language pre-training with fusion in the backbone. arXiv preprint arXiv:2206.07643 (2022)
15. Frank, S., Bugliarello, E., Elliott, D.: Vision-and-language or vision-for-language. On Cross-Modal Influence in Multimodal Transformers. (2021). https://doi.org/10.18653/v1/2021.emnlp-main.775 (2021)
16. Gan, Z., Li, L., Li, C., Wang, L., Liu, Z., Gao, J., et al.: Vision-language pre-training: basics, recent advances, and future trends. Found. Trends® Comput. Graph. Vis. 14(3–4), 163–352 (2022)
17. Gupta, T., Vahdat, A., Chechik, G., Yang, X., Kautz, J., Hoiem, D.: Contrastive learning for weakly supervised phrase grounding. In: Vedaldi, A., Bischof, H., Brox, T., Frahm, J.-M. (eds.) ECCV 2020. LNCS, vol. 12348, pp. 752–768. Springer, Cham (2020). https://doi.org/10.1007/978-3-030-58580-8_44

18. He, K., Zhang, X., Ren, S., Sun, J.: Deep residual learning for image recognition. In: 2016 IEEE Conference on Computer Vision and Pattern Recognition (CVPR). IEEE (2016). https://doi.org/10.1109/cvpr.2016.90

19. He, K., Zhang, X., Ren, S., Sun, J.: Deep residual learning for image recognition. In: 2016 IEEE Conference on Computer Vision and Pattern Recognition, CVPR 2016, Las Vegas, NV, USA, 27–30 June 2016, pp. 770–778. IEEE Computer Society (2016). https://doi.org/10.1109/CVPR.2016.90

20. Jing, C., Jia, Y., Wu, Y., Liu, X., Wu, Q.: Maintaining reasoning consistency in compositional visual question answering. In: Proceedings of the IEEE/CVF Conference on Computer Vision and Pattern Recognition, pp. 5099–5108 (2022)

21. Kafle, K., Shrestha, R., Kanan, C.: Challenges and prospects in vision and language research. Front. Artif. Intell. **2**, 28 (2019)

22. Kamath, A., Singh, M., LeCun, Y., Synnaeve, G., Misra, I., Carion, N.: MDETR - modulated detection for end-to-end multi-modal understanding. In: 2021 IEEE/CVF International Conference on Computer Vision, ICCV 2021, Montreal, QC, Canada, 10–17 October 2021, pp. 1760–1770. IEEE (2021). https://doi.org/10.1109/ICCV48922.2021.00180

23. Kim, J.H., Jun, J., Zhang, B.T.: Bilinear attention networks. In: Advances in Neural Information Processing Systems 31 (2018)

24. Kiros, R., Salakhutdinov, R., Zemel, R.S.: Unifying visual-semantic embeddings with multimodal neural language models. arXiv preprint arXiv:1411.2539 (2014)

25. Krishna, R., et al.: Visual genome: connecting language and vision using crowd-sourced dense image annotations. Int. J. Comput. Vision **123**(1), 32–73 (2017)

26. Li, L., Gan, Z., Cheng, Y., Liu, J.: Relation-aware graph attention network for visual question answering. In: Proceedings of the IEEE/CVF International Conference on Computer Vision, pp. 10313–10322 (2019)

27. Li, L.H., Yatskar, M., Yin, D., Hsieh, C.J., Chang, K.W.: VisualBERT: a simple and performant baseline for vision and language. arXiv preprint arXiv:1908.03557 (2019)

28. Li, L.H., et al.: Grounded language-image pre-training. In: Proceedings of the IEEE/CVF Conference on Computer Vision and Pattern Recognition, pp. 10965–10975 (2022)

29. Li, W.H., Yang, S., Wang, Y., Song, D., Li, X.Y.: Multi-level similarity learning for image-text retrieval. Inf. Process. Manage. **58**(1), 102432 (2021)

30. Li, X., et al.: OSCAR: object-semantics aligned pre-training for vision-language tasks. In: Vedaldi, A., Bischof, H., Brox, T., Frahm, J.-M. (eds.) ECCV 2020. LNCS, vol. 12375, pp. 121–137. Springer, Cham (2020). https://doi.org/10.1007/978-3-030-58577-8_8

31. Lin, T.-Y., et al.: Microsoft COCO: common objects in context. In: Fleet, D., Pajdla, T., Schiele, B., Tuytelaars, T. (eds.) ECCV 2014. LNCS, vol. 8693, pp. 740–755. Springer, Cham (2014). https://doi.org/10.1007/978-3-319-10602-1_48

32. Liu, L., et al.: Deep learning for generic object detection: a survey. Int. J. Comput. Vision **128**(2), 261–318 (2020)

33. Lu, J., Batra, D., Parikh, D., Lee, S.: ViLBERT: pretraining task-agnostic visiolinguistic representations for vision-and-language tasks. In: Wallach, H.M., Larochelle, H., Beygelzimer, A., d'Alché-Buc, F., Fox, E.B., Garnett, R. (eds.) Advances in Neural Information Processing Systems 32: Annual Conference on Neural Information Processing Systems 2019, NeurIPS 2019 (December), pp. 8–14, 2019. Vancouver, BC, Canada, pp. 13–23 (2019). https://proceedings.neurips.cc/paper/2019/hash/c74d97b01eae257e44aa9d5bade97baf-Abstract.html

34. Mao, J., Xu, W., Yang, Y., Wang, J., Yuille, A.L.: Deep captioning with multimodal recurrent neural networks (m-RNN). In: Bengio, Y., LeCun, Y. (eds.) ICLR (2015)
35. Mogadala, A., Kalimuthu, M., Klakow, D.: Trends in integration of vision and language research: a survey of tasks, datasets, and methods. J, Artif. Intell. Res. **71**, 1183–1317 (2021)
36. Northcutt, C., Jiang, L., Chuang, I.: Confident learning: estimating uncertainty in dataset labels. J. Artif. Intell. Res. **70**, 1373–1411 (2021)
37. Pennington, J., Socher, R., Manning, C.D.: Glove: Global vectors for word representation. In: Proceedings of the 2014 Conference on Empirical Methods in Natural Language Processing (EMNLP), pp. 1532–1543 (2014)
38. Redmon, J., Farhadi, A.: YOLO9000: better, faster, stronger. In: Proceedings of the IEEE Conference on Computer Vision and Pattern Recognition, pp. 7263–7271 (2017)
39. Ren, S., He, K., Girshick, R., Sun, J.: Faster R-CNN: towards real-time object detection with region proposal networks. In: Advances in Neural Information Processing Systems 28 (2015)
40. Rigoni, D., Serafini, L., Sperduti, A.: A better loss for visual-textual grounding. In: Hong, J., Bures, M., Park, J.W., Cerný, T. (eds.) SAC 2022: The 37th ACM/SIGAPP Symposium on Applied Computing, Virtual Event, 25–29 April 2022, pp. 49–57. ACM (2022). https://doi.org/10.1145/3477314.3507047
41. Rohrbach, A., Rohrbach, M., Hu, R., Darrell, T., Schiele, B.: Grounding of textual phrases in images by reconstruction. In: Leibe, B., Matas, J., Sebe, N., Welling, M. (eds.) ECCV 2016. LNCS, vol. 9905, pp. 817–834. Springer, Cham (2016). https://doi.org/10.1007/978-3-319-46448-0_49
42. Shih, K.J., Singh, S., Hoiem, D.: Where to look: Focus regions for visual question answering. In: CVPR, pp. 4613–4621 (2016)
43. Simonyan, K., Zisserman, A.: Very deep convolutional networks for large-scale image recognition. arXiv preprint arXiv:1409.1556 (2014)
44. Singh, A., et al.: Towards VQA models that can read. In: Proceedings of the IEEE/CVF Conference on Computer Vision and Pattern Recognition, pp. 8317–8326 (2019)
45. Song, H., Kim, M., Park, D., Shin, Y., Lee, J.G.: Learning from noisy labels with deep neural networks: A survey. IEEE Transactions on Neural Networks and Learning Systems, pp. 1–19 (2022). https://doi.org/10.1109/TNNLS.2022.3152527
46. Su, W., et al.: VL-BERT: pre-training of generic visual-linguistic representations. arXiv preprint arXiv:1908.08530 (2019)
47. Tan, H., Bansal, M.: LXMERT: learning cross-modality encoder representations from transformers. In: Proceedings of the 2019 Conference on Empirical Methods in Natural Language Processing and the 9th International Joint Conference on Natural Language Processing (EMNLP-IJCNLP), pp. 5100–5111 (2019)
48. Tian, Z., Shen, C., Chen, H., He, T.: FCOS: fully convolutional one-stage object detection. In: Proceedings of the IEEE/CVF International Conference on Computer Vision, pp. 9627–9636 (2019)
49. Wang, C.Y., Bochkovskiy, A., Liao, H.Y.M.: YOLOv7: trainable bag-of-freebies sets new state-of-the-art for real-time object detectors. arXiv preprint arXiv:2207.02696 (2022)
50. Wang, H., Wang, H., Xu, K.: Categorizing concepts with basic level for vision-to-language. In: Proceedings of the IEEE Conference on Computer Vision and Pattern Recognition (CVPR) (2018)

51. Wang, L., Huang, J., Li, Y., Xu, K., Yang, Z., Yu, D.: Improving weakly supervised visual grounding by contrastive knowledge distillation. In: Proceedings of the IEEE/CVF Conference on Computer Vision and Pattern Recognition, pp. 14090–14100 (2021)
52. Wang, Q., Tan, H., Shen, S., Mahoney, M.W., Yao, Z.: MAF: multimodal alignment framework for weakly-supervised phrase grounding. arXiv preprint arXiv:2010.05379 (2020)
53. Wang, R., Qian, Y., Feng, F., Wang, X., Jiang, H.: Co-VQA: answering by interactive sub question sequence. In: Findings of the Association for Computational Linguistics: ACL 2022, pp. 2396–2408 (2022)
54. Wang, S., Wang, R., Yao, Z., Shan, S., Chen, X.: Cross-modal scene graph matching for relationship-aware image-text retrieval. In: Proceedings of the IEEE/CVF Winter Conference on Applications of Computer Vision (WACV) (2020)
55. Wang, X., Zhang, S., Yu, Z., Feng, L., Zhang, W.: Scale-equalizing pyramid convolution for object detection. In: Proceedings of the IEEE/CVF Conference on Computer Vision and Pattern Recognition, pp. 13359–13368 (2020)
56. Xu, M., et al.: End-to-end semi-supervised object detection with soft teacher. In: Proceedings of the IEEE/CVF International Conference on Computer Vision, pp. 3060–3069 (2021)
57. Yang, J., Li, C., Gao, J.: Focal modulation networks. arXiv preprint arXiv:2203.11926 (2022)
58. Yang, Z., Liu, S., Hu, H., Wang, L., Lin, S.: RepPoints: point set representation for object detection. In: Proceedings of the IEEE/CVF International Conference on Computer Vision, pp. 9657–9666 (2019)
59. Yao, Y., et al.: PEVL: position-enhanced pre-training and prompt tuning for vision-language models. arXiv preprint arXiv:2205.11169 (2022)
60. Yu, Z., Yu, J., Cui, Y., Tao, D., Tian, Q.: Deep modular co-attention networks for visual question answering. In: Proceedings of the IEEE/CVF Conference on Computer Vision and Pattern Recognition, pp. 6281–6290 (2019)
61. Yu, Z., Yu, J., Xiang, C., Zhao, Z., Tian, Q., Tao, D.: Rethinking diversified and discriminative proposal generation for visual grounding. In: Proceedings of the 27th International Joint Conference on Artificial Intelligence, pp. 1114–1120 (2018)
62. Zhang, H., Niu, Y., Chang, S.F.: Grounding referring expressions in images by variational context. In: Proceedings of the IEEE Conference on Computer Vision and Pattern Recognition, pp. 4158–4166 (2018)
63. Zhang, H., et al.: DINO: DETR with improved denoising anchor boxes for end-to-end object detection. arXiv preprint arXiv:2203.03605 (2022)
64. Zhang, H., et al.: Glipv2: unifying localization and vision-language understanding. arXiv preprint arXiv:2206.05836 (2022)
65. Zhang, Q., Lei, Z., Zhang, Z., Li, S.Z.: Context-aware attention network for image-text retrieval. In: Proceedings of the IEEE/CVF Conference on Computer Vision and Pattern Recognition (CVPR) (2020)
66. Zhang, S., Chi, C., Yao, Y., Lei, Z., Li, S.Z.: Bridging the gap between anchor-based and anchor-free detection via adaptive training sample selection. In: Proceedings of the IEEE/CVF Conference on Computer Vision and Pattern Recognition, pp. 9759–9768 (2020)
67. Zhou, B., Tian, Y., Sukhbaatar, S., Szlam, A., Fergus, R.: Simple baseline for visual question answering. arXiv preprint arXiv:1512.02167 (2015)
68. Zhou, L., Palangi, H., Zhang, L., Hu, H., Corso, J., Gao, J.: Unified vision-language pre-training for image captioning and VQA. In: Proceedings of the AAAI Conference on Artificial Intelligence, vol. 34, pp. 13041–13049 (2020)

Correction to: Image Analysis

Rikke Gade⦿, Michael Felsberg⦿, and Joni-Kristian Kämäräinen⦿

Correction to:
R. Gade et al. (Eds.): SCIA 2023, LNCS 13885,
https://doi.org/10.1007/978-3-031-31435-3

In the originally published version of the book cover and front matter the conference edition number was incorrect. The conference edition number has been corrected as 22.

The updated original version of the book can be found at
https://doi.org/10.1007/978-3-031-31435-3

Author Index

Printed in the United States
by Baker & Taylor Publisher Services